AI in Chemical Engineering

Industry 4.0 is revolutionizing chemical manufacturing. Today's chemical companies are swiftly embracing the digital era, recognizing the significant benefits of interconnected products, production equipment, and personnel. As technology advances and production volumes grow, there is an increasing need for new computational tools and innovative solutions to address everyday challenges. *AI in Chemical Engineering: Unlocking the Power Within Data* introduces readers to the essential concepts of machine learning and their application in the chemical and process industries, aiming to enhance efficiency, adaptability, and profitability. This work delves into the transformation of traditional plant operations into integrated and intelligent systems, providing readers with a foundation for developing and understanding the tools necessary for data collection and analysis, thereby gaining valuable insights and practical applications.

- Introduces the principles and applications of unsupervised learning and discusses the role of machine learning in extracting information from plant data and transforming it into knowledge.
- Conveys the concepts, principles, and applications of supervised learning, setting the stage for developing advanced monitoring systems, complex predictive models, and advanced computer vision applications.
- Explores implementation of reinforced learning ideas for chemical process control and optimization, investigating various model structures and discussing their practical implementation in both simulation and experimental units.
- Incorporates sample code examples in Python to illustrate key concepts.
- Includes real-life case studies in the context of chemical engineering and covers a wide variety of chemical engineering applications from oil and gas to bioengineering and electrochemistry.
- Clearly defines types of problems in chemical engineering subject to AI solutions and relates them to subfields of AI.

This practical text, designed for advanced chemical engineering students and industry practitioners, introduces concepts and theories in a logical and sequential manner. It serves as an essential resource, helping readers understand both current and emerging developments in this important and evolving field.

AI in Chemical Engineering
Unlocking the Power Within Data

By
José A. Romagnoli, Luis Briceño-Mena,
and Vidhyadhar Manee

CRC Press
Taylor & Francis Group
Boca Raton London New York

CRC Press is an imprint of the
Taylor & Francis Group, an **informa** business

Designed cover image: Shutterstock

First edition published 2025
by CRC Press
2385 NW Executive Center Drive, Suite 320, Boca Raton FL 33431

and by CRC Press
4 Park Square, Milton Park, Abingdon, Oxon, OX14 4RN

CRC Press is an imprint of Taylor & Francis Group, LLC

Library of Congress Cataloging-in-Publication Data
Names: Romagnoli, José A. (José Alberto), author. | Briceno-Mena, Luis, author.
Title: AI in chemical engineering : unlocking the power within data / by José A. Romagnoli and Luis Briceno-Mena.
Other titles: Artificial intelligence in chemical engineering
Description: First edition. | Boca Raton : CRC Press, 2025. | Includes bibliographical references and index. |
Identifiers: LCCN 2024023523 (print) | LCCN 2024023524 (ebook) | ISBN 9781032597003 (hardback) | ISBN 9781032597034 (paperback) | ISBN 9781003455905 (ebook)
Subjects: LCSH: Chemical engineering--Data processing. | Artificial intelligence--Engineering applications. | Chemical processes--Data processing. | Artificial intelligence--Industrial applications.
Classification: LCC TP184 .R66 2025 (print) | LCC TP184 (ebook) | DDC 660.0285/63--dc23/eng/20241009
LC record available at https://lccn.loc.gov/2024023523
LC ebook record available at https://lccn.loc.gov/2024023524

ISBN: 978-1-032-59700-3 (hbk)
ISBN: 978-1-032-59703-4 (pbk)
ISBN: 978-1-003-45590-5 (ebk)

DOI: 10.1201/9781003455905

Typeset in Times
by KnowledgeWorks Global Ltd.

To my wife Liliana, my children José and Juliana, my grandchildren Josefina, Cruz, Miguel, and Sebastian and the memory of my parents.

José A. Romagnoli

To my wife Carla, and my parents Marlene and José.

Luis Briceño-Mena

To my family.

Vidhyadhar Manee

With our everlasting love and gratitude.

"From research labs to industrial floors,
AI and chemicals, an unexplored shore,
Unlocking new discoveries, with leaps and bounds,
Revolutionizing, with innovation's sounds."

A poem produced by ChatGPT using GPT-3 to illustrate the creative capabilities of large language models. Prompt: "write a poem about artificial intelligence and chemical engineering".

Contents

PART II Supervised Learning

Foreword

Professor José A. Romagnoli is a renowned pioneer in AI research within the chemical engineering community. His passion and dedication to teaching process systems engineering principles have spanned several decades. The textbook, *Introduction to Process Control*, co-authored with Professor Ahmet Palazoglu in 2005, has been an invaluable asset for students and industrial practitioners seeking to understand the application of process control concepts in industrial processes.

Since 2005, the availability of industrial data, advancements in AI and machine learning algorithms, and the affordability of computing resources have propelled the Industry 4.0 evolution in the chemical industry. In response to these developments, Professor José A. Romagnoli, Dr. Luis Briceño-Mena, and Dr. Vidhyadhar Manee have authored the timely textbook, *AI in Chemical Engineering: Unlocking the Power Within Data*, to articulate how AI and data will accelerate innovation and enhance productivity in chemical engineering.

The textbook is structured into four parts. Chapters 1 and 2 lay out data as the foundation of AI. Part I: Unsupervised Learning (Chapters 3–5) introduces dimensionality reduction and clustering techniques, and case studies. Part II: Supervised Learning (Chapters 6–10) showcases predictive models, deep learning, hybrid modeling, and case studies. As illustrated in Parts I and II, these unsupervised and supervised techniques with curated data have been a "gold standard" in the chemical industry. Part III (Chapters 11–13) explains how recent developments in AI, namely, reinforcement learning and generative machine learning, are inspiring research and industrial applications in chemical engineering.

I had the honor of collaborating with José on two AI projects to advance deep learning methodologies for image analysis and implement reinforcement learning for a batch process. José's contribution has been instrumental in pushing the boundaries of AI applications in chemical engineering and I am grateful that José, Luis, and Vidhyadhar have written this textbook with plenty of hands-on examples, relevant industrial case studies, and open-sourced codes. This textbook is a comprehensive guide that explains the Why, What, and How of AI will unlock Power within data. It is poised to impact the chemical engineering community for decades to come.

By Leo H. Chiang
Senior R&D Fellow, AI and data analytics
The Dow Chemical Company

About the Authors

José A. Romagnoli

José A. Romagnoli is currently the endowed Gordon & Mary Cain Chair Professor and M. Gautreaux/Ethyl Chair Professor at the Department of Chemical Engineering at Louisiana State University. Previously he was the ICI Chair in Process Systems Engineering at the University of Sydney where he was also the Director of the Laboratory for Process Systems Engineering. He received his bachelor's degree (Chemical Engineering) from the National University of the South (Argentina) in 1974 and PhD (Chemical Engineering) from the University of Minnesota in 1980. He is author of more than 300 international publications and author of the books *Data Processing and Reconciliation for Chemical Process Operations* published by Academic Press International and *Introduction to Process Control* published by CRC Press. He is a member of the Australian Academy of Technological Sciences and Engineering, Fellow of the Institute of Engineers Australia, and consultant for several international companies. He was awarded the Centenary Medal by the Prime Minister of Australia for his contributions to Chemical Engineering. His research interests cover all aspects of Process Systems Engineering, especially: machine learning, modeling of complex systems, advanced model-based control, and intelligent process monitoring and supervision.

Luis Briceño-Mena

Luis Briceño-Mena is part of Dow's Research and Development team in the Machine Learning Optimization and Statistics group, where he works on hybrid modeling to tackle challenges in process and product development. Luis got his undergraduate degree in Chemical Engineering in 2015 from the University of Costa Rica, with an undergraduate thesis on reactor engineering for advanced oxidation processes for wastewater treatment. He worked as Lecturer for the School of Chemical Engineering at the University of Costa Rica from 2016 to 2019, teaching statistics, thermodynamics, and reaction kinetics. In 2023, Luis received his PhD degree in Chemical Engineering from Louisiana State University. His dissertation work studied hybrid modeling for the understanding and optimization of electrochemical systems.

Vidhyadhar Manee

Vidhyadhar Manee is a Senior Scientist in Process Research at Boehringer Ingelheim Pharmaceuticals, Inc. He is currently working on applying artificial intelligence (AI) techniques to speed up the synthesis of active pharmaceutical ingredients. Prior to joining Boehringer, he worked in Dr. José A. Romagnoli's lab in the Department of Chemical Engineering at Louisiana State University (LSU). At LSU, he received his PhD for developing AI-based process monitoring and control tools for crystallization. Outside of work, he likes to read non-fiction and practice meditation and cognitive behavioral therapy.

Preface

The Chemical Process Industry plays a crucial and ever-growing role in the world's economy. In the United States alone, the volume of total chemicals produced per year is expected to double over the next decade. As technology evolves and production volumes rise, challenges like resilience, sustainability, and of course scale grow larger as well. This landscape demands for effective, fast, and robust solutions that can only be achieved with the help of computational methods. Thus, it is essential for both academia and industry to provide the foundation for practical tools capable of meeting these emerging needs. Industry 4.0 is transforming chemical manufacturing, and today's companies are rapidly adapting to the digital world, recognizing the power of connectivity among products, production equipment, and personnel.

Smart Manufacturing and Industry 4.0 initiatives aim to achieve adaptability and resource efficiency. Their technological foundation is composed of cyber-physical systems that combine communications, information technology (IT), data, and physical elements. These systems utilize core technologies, including sensors and sensor networks, internet communication infrastructure, intelligent real-time process and event management, and data provisioning. Machine learning techniques are utilized in smart manufacturing to complement physical science to improve system performance and decision-making. Advanced techniques are necessary to convert data collected at different stages of the manufacturing process into insightful knowledge for smart manufacturing. Data-driven methods derived from machine learning are a core component of this approach.

The impact of changing industrial needs and advances in computer technology on process operations has been profound and continues to shape the way we operate. Industrial goals of improved productivity, efficiency, and product quality have generated a demand for more effective operational strategies in the production line, while developments in digital computers and communications have revolutionized process operations and allowed for the implementation of more advanced tools.

This book is intended primarily for chemical engineering students at the advanced undergraduate or graduate level, as well as industrial practitioners, to familiarize them with the key concepts of machine learning and their implementation. It also aims to engage them in the appreciation of the evolution of traditional plant operation into an integrated and smart operational environment. Accordingly, the book is divided as follows.

Chapter 2 builds a foundation by discussing concepts about data along with data preprocessing methods that are critical to ensure meaningful and reliable results.

Part I (Chapters 3 through 5) introduces the principles and applications of unsupervised learning, discussing the role of machine learning in extracting information from plant data and transforming it into knowledge. It also presents key concepts, definitions, and general terminology.

Part II (Chapters 6 through 10) covers the concepts, principles, and applications of supervised learning and deep learning, setting the stage for developing advanced monitoring systems, complex predictive models, and advanced computer vision applications.

Part III (Chapters 11 and 12) is devoted to introducing and exploring the concepts and implementation of reinforcement learning ideas for chemical process control and optimization. It investigates a number of models structures and discusses their practical implementation in both simulation and experimental units.

Finally, Chapter 13 introduces the ideas of generative modeling and their practical implementation. As an emerging approach in chemical engineering applications, we aim to provide the basic intuitions of these methods, grounded in the concepts introduced in the previous chapters.

This book represents the collection of many years of experience in research, development, and technology transfer within the Process Systems Engineering group at LSU (PSE@LSU) in the area of machine learning. It illustrates the work of several present and former students involved in industrial sponsored research and development projects in the exciting field of artificial intelligence (AI) applications to chemical processes.

We extend our heartfelt gratitude to the many individuals who played a pivotal role in shaping and bringing this book to fruition. Foremost among them, we offer our sincerest thanks to our current and former graduate students, whose unwavering support proved invaluable throughout the various stages of crafting the content for this book.

In particular, we owe a profound debt of gratitude to Dr. G. Robertson, Dr. D. Thomas, Dr. W. Zhu, Dr. Y. Ma, Dr. M. Mandis, M.Sc. Estelle Seghers, Z. Webb, M. Nnadili, K. Territo, and T. Olayiwola for their pioneering applications of machine learning in their research projects, which are extensively explored in numerous chapters.

Additionally, we wish to acknowledge the contributions of Dr. J. Chebeir and Dr. S. Salas in developing some of the software components utilized in our case studies. Furthermore, we express our deep appreciation to Prof. Roberto Baratti and Prof. Stefania Tronci for their invaluable contributions to our work on crystallization and the development of the virtual twin distillation column. We are also grateful for their continuous friendship throughout the years.

Lastly, we extend our gratitude to the dedicated faculty and staff at Louisiana State University, whose support and conducive environment played a vital role in the successful completion of this book.

Furthermore, we wish to express our sincere appreciation to our invaluable industrial partners for their unwavering support and invaluable feedback, both of which have played a pivotal role in shaping the content of this book. In particular, we extend our heartfelt thanks to SupeZet for their generous financial backing of our research endeavors in the field of machine learning applied to pyrolysis reactors. We would like to convey our deepest gratitude to Mr. Zhang, President of SupeZet, for his unwavering belief in our capacity to see these ideas through to fruition.

In addition, we would like to recognize Professor Wei Sun, our esteemed academic collaborator in this project, for her significant contributions to several of the subjects covered within the pages of this book. We are also grateful for her enduring friendship throughout the years.

We must also acknowledge the consistent support of ExxonMobil Chemicals at Baton Rouge in this field, spanning numerous years, which has paved the way for countless students to successfully graduate. It is worth noting that some of the ideas discussed in this book have been successfully implemented and even brought to market through our collaboration with these esteemed industrial partners.

Finally, we would like to express our gratitude to our families for their continued support throughout the various stages of preparing the manuscript, and for their understanding of the many days and nights spent in this pursuit.

José A. Romagnoli,
Baton Rouge, Louisiana

Luis Briceño-Mena,
Houston, Texas

Vidhyadhar Manee,
Ridgefield, Connecticut

1 Smart Manufacturing and Machine Learning

1.1 SMART MANUFACTURING

Since the middle of the 18th century, when the industrial revolution began, the manufacturing industry has significantly impacted people's way of life [1]. Mass production with assembly lines and electricity has made the manufacturing of complex consumer products possible. The chemical industry has been a long-time beneficiary of the great industrial revolutions, from the basic steam power to programmed automation, which has enabled significant improvements in productivity.

As information and communication technology (ICT) has developed, automation has brought numerous advances, including highly delicate assembly lines with robotics [1]. However, as chemical process operations become more complicated, there are increasing demands for designing more advanced sensors, anomaly detection platforms, and control systems to further improve the process safety assessment and economic gain. With the advancements of cutting-edge ICTs, countries with advanced manufacturing industries are recognizing their potential and identifying several major areas of interest, including IoT (Internet of Things), CPS (cyber-physical system), cloud computing, and more [1] (see Figure 1.1).

In 2023, the US government released a strategic report on the country's leadership in advanced manufacturing, outlining three goals for innovation in this sector. These goals are to develop and transition to new manufacturing technologies, educate the manufacturing workforce, and expand the domestic manufacturing supply chain. The report emphasizes the importance of cloud computing, data analytics, modeling,

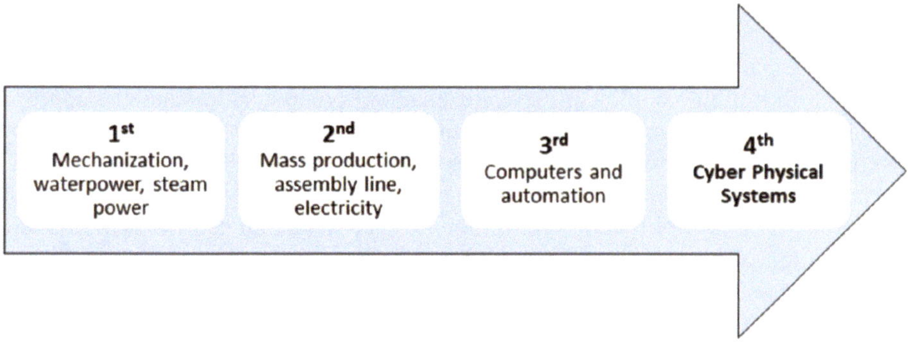

FIGURE 1.1 Industrial revolution from mechanization to cyber-physical system.

DOI: 10.1201/9781003455905-1

and Artificial Intelligence (AI) in facilitating the digitization of manufacturing processes and the implementation of the IoT.

In addition to the United States, other countries such as Germany, Japan, and China are also seeking innovative solutions through the application of digitization and AI. The German government, for example, has initiated a strategic initiative for digital transformation aimed at driving digital manufacturing forward by promoting interconnectivity between products, supply chains, and business models [2]. Similarly, a report by China's Academy of Sciences has identified eight key technologies with potentially innovative applications, including computer vision, natural language processing (NLP), cross-media analysis, intelligence adaptive learning, collective intelligence, auto-piloting systems, deep learning microchips, and brain-computer interface (BCI).

Therefore, the development of smart manufacturing has reached the level of national interest, kicking off an international race for the leadership in the next great advancement of the industrial revolution.

1.2 MACHINE LEARNING

Since the advent of computers, there have been numerous attempts to create machines that can learn and think like humans. This has led to the development of a plethora of algorithms that are effective in various tasks and with significant commercial applications [3].

Machine learning (ML), a branch of AI and computer science, aims to imitate the way humans learn by using data and algorithms to gradually improve its accuracy [4]. The primary goal of ML is to enable computers to improve automatically through experience by incorporating cross-disciplinary knowledge from computer science and statistics. With the development of new learning algorithms and the increasing accessibility of online data and computational power, ML is currently one of the fastest-growing fields [5].

A salient application of ML, known as data mining, exploits the vast amount of data that is collected daily through sensors and cameras. Analyzing this data can be challenging for humans, making it essential for computers to learn correlations, features, and patterns [5]. However, conventional ML techniques have limitations in processing raw data. To address this challenge, careful preprocessing and feature engineering are required for specific tasks [6].

Another subfield of ML, deep learning, employs multiple layers of data representation with multiple nonlinear modules, enabling computers to learn the appropriate feature engineering [6]. Deep learning allows computer intelligence to extract useful information from highly intricate datasets and nonlinear behaviors. As a result, modern society has witnessed the impact of ML and deep learning in various domains, ranging from web search filters to face detection and traffic monitoring systems [6]. Moreover, deep learning has also been utilized in industrial applications, such as crack detection in power plants using deep convolutional neural networks (ConvNets) from camera images [7]. Additionally, Zhu et al. [8] have utilized deep learning tools to analyze industrial pyrolysis reactor images to monitor the reaction process.

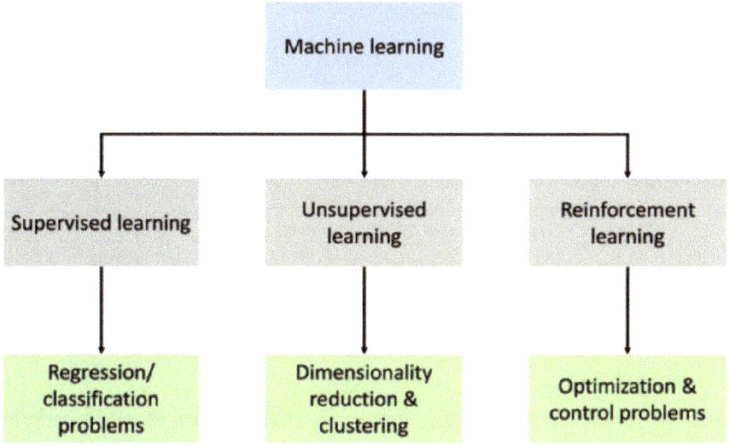

FIGURE 1.2 Machine learning hierarchy with supervised, unsupervised, and reinforcement learning.

Figure 1.2 depicts the overarching hierarchical structure of ML, highlighting its three main branches: supervised learning, unsupervised learning, and reinforcement learning (RL). Supervised learning typically involves classification or regression models, where it discerns correlations between input data and ground truth labels. On the other hand, unsupervised learning operates without the need for ground truth labels, focusing instead on grasping the distribution and patterns within the dataset autonomously. Lastly, RL functions as a behavioral learning model, receiving feedback from analysis to steer the user toward optimal outcomes.

1.3 PRINCIPLES OF MACHINE LEARNING AND ARTIFICIAL INTELLIGENCE

Currently, machines and devices are generating an enormous amount of data, ranging from cloud solutions to business management, resulting in the accumulation of exabytes annually. This data is projected to increase dramatically in the coming decade [9, 10]. As big data plays a pivotal role in the fourth industrial revolution [9], the chemical industry has been rapidly progressing toward advanced control, automation, and smart manufacturing utilizing data-driven methods. This new paradigm shift in manufacturing industries is known as "Industry 4.0" [9], which combines the digitization of factories, internet technologies, and data-driven processes, with the potential to significantly improve manufacturing efficiency and quality. The goal is to establish smart factories capable of tailoring specific products to customer requests, collecting and transmitting data in real-time [9].

Deep learning is currently one of the fastest-growing fields in data science, with numerous tools available to facilitate knowledge discovery and smart decision-making [6]. Advances in ML and deep learning have stimulated interest in data-driven optimization, with process engineers utilizing advanced techniques, including mathematical programming and ML, to optimize reactions in a data-driven approach [11–13].

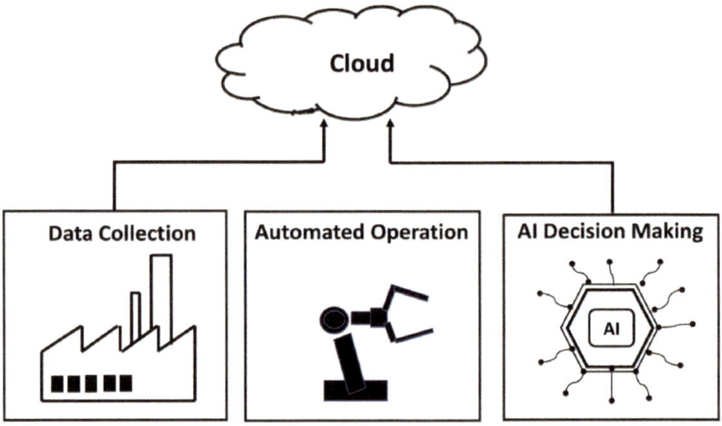

FIGURE 1.3 Schematic diagram of industrial application with data collection, data processing, and smart decision-making with AI models.

Figure 1.3 illustrates a schematic diagram of an industrial application with data collection and ML models. The data collected from the plant is communicated through the cloud and fed to AI models for smart decision-making, which guides automated operation. This iterative process allows real-time data to be used to update the AI model promptly to account for disturbances and process drifting.

With the revolutionary advancement of low-cost computational power, especially with the availability of parallel computing units, such as graphics processing units (GPU), which were originally designed for video gaming, ML has made significant progress in the past decade. Although algorithms have been invented and demonstrated huge potential in certain types of learning tasks [3], most of the advancements have come from computer science and statistical departments, with a lack of expertise in practical ML model applications in the manufacturing industry. Therefore, fundamental research with big data solutions and data-driven optimization is necessary for future industrial applications [9].

In March 2016, AlphaGo, a computer program, defeated the best human players in the game of Go, which is known for its significantly high complexity of combinatorial possibilities [14]. Since then, RL, one of the major branches of ML, has become famous for its superior decision-making abilities. RL is not only known for its high performance in relatively simple games like chess, but it has also demonstrated great potential in highly complicated environments, such as video games, robotics, and self-driving cars, in a self-learning manner [14, 15].

1.4 MACHINE LEARNING APPLICATIONS IN CHEMICAL MANUFACTURING

Chemical engineering is a significant component of modern society that uses principles from physics, chemistry, math, and biology to transform energy and material effectively. Currently, chemical engineering contributes approximately $5.7 trillion worth of gross domestic product (GDP) and supports 120 million jobs worldwide [16]. Chemical engineering involves aspects such as plant design, safety and hazard assessment, process design, modeling, and optimization. The chemical engineering industry covers a broad spectrum of work, ranging from petrochemical engineering and nano-materials engineering to genetic engineering. As such, the industry has been absorbing cutting-edge technology and knowledge base across disciplines, especially from chemistry, physics, math, and computer science.

The chemical industry is growing in the United States, but manufacturers face challenges and opportunities to remain competitive. Key challenges include globalization, supply chain resilience, rising energy costs, sustainability, and the need for workers who understand chemistry, engineering, and computing. These challenges are paired with new opportunities, particularly those enabled by advances in computing and cyberinfrastructure, that offer increased productivity, more energy efficiency, improved resiliency, and the ability to rapidly adapt to market needs.

Process operations in chemical industries are becoming more complicated, and plants can experience periods of abnormal behavior, even in the best-designed systems. Faults can result in a poor-quality product, equipment damage, or catastrophic failure of the process, which can cause loss of equipment and possible worker injury. The potential economic and human loss, the negative publicity generated by such events, as well as the commitment to achieve sustainability goals make the industry more sensitive to the issues of early detection of faults. However, the most immediate impact of a fault to a company is the financial loss caused by abnormal operations. During these periods of abnormal operations, operators must make rapid decisions to return the process to normal operation, often relying only on heuristics, experience, and instincts. Case studies show that operators do not always make the correct decision. According to Malkoff's study [17], the factors that affect operators' decisions in addressing and assessing abnormal behaviors include not only safety training and stress but also the "Christmas tree effect", where a major fault event can cause more alarms than human operators can effectively process. These issues contribute to many tragic outcomes, such as Deepwater Horizon and Piper Alpha. It has been reported that the total cost attributable to preventable losses in petrochemical companies in the United States is well over $20 billion per year. The cost is much higher when similar situations in other industries, such as pharmaceuticals, specialty chemicals, or power, are included.

Consequently, process systems engineers face the challenge of designing and implementing intelligent systems that can assist operators in making supervisory decisions. Developing intelligent systems that can provide automated operator assistance for complex plants is a great incentive, and process industries view this as the next major challenge in control systems research and application.

In the current chemical industry, the distributed control system (DCS) handles normal operations and adjusts to disturbances in the system. However, the DCS is unable to control many complex abnormal situations and faulty conditions. In such cases, it is essential for the operator to assess the situation quickly, determine the appropriate cause(s), and take necessary corrective actions. Failure to act can lead to significant economic, safety, and environmental impacts.

Thus, it is of paramount importance to develop reliable process monitoring and fault diagnosis methods at the top of the current system to ensure process safety. Electrical sensors have facilitated data collection from the plant, leading to a data revolution in chemical engineering. The accessibility of process data has driven the development of data-based methods that use a large amount of historical data to present and learn prior knowledge [18–21].

Statistical feature extraction methods, notably principal component analysis (PCA) and partial least squares (PLS), have played a pivotal role in data-driven approaches [22]. Over the years, advanced techniques have emerged to tackle diverse process systems. One such innovation is the recursive PCA (RPCA) method, which imbues self-adaptation into the original PCA model. Additionally, the nonlinear PCA (NLPCA) method enhances the efficacy of nonlinear processes through the incorporation of nonlinear kernels [23].

While traditional statistics remain crucial for data analysis, their adequacy falters when confronted with the complexities of large datasets. In contrast, contemporary data mining techniques, exemplified by neighbor graphs dimensionality reduction methods, are computational algorithms meticulously crafted to unveil meaningful patterns and relationships within extensive datasets. These innovative methods prove indispensable in identifying intricate connections that traditional approaches may struggle to discern [24, 25].

Model-based approaches have been widely employed to tackle an array of decision-making, real-time prediction, and optimization of challenges arising in process operations. However, traditional methods rooted in models encounter issues related to process uncertainty, parameter estimation errors, and substantial effort [26]. Furthermore, these approaches often heavily depend on first principle models and meticulous process identification on the actual system. This reliance poses challenges for real-time applications due to computational costs and adaptability concerns.

In contrast, emerging ML techniques, such as deep learning and RL, offer a more efficient approach to executing optimization and control tasks in real-time. These novel techniques operate swiftly, requiring only a forward run of the trained network. This stands in stark contrast to the intricate processes associated with traditional model-based methodologies [27–30].

In this book, we will delve into the realm of data preprocessing, exploring a variety of tools while also examining the key branches of ML as visually represented in Figure 1.2. Our primary focus will revolve around the three main facets of ML and their practical applications in chemical processes, as illustrated in Figure 1.4. We'll commence our exploration by immersing ourselves in the domain of unsupervised learning and its role in extracting valuable insights from plant data. Special attention will be given to the assessment of plant operational status. As we transition to supervised learning, our focus will pivot toward fault detection, predictive model creation, and the application of ML techniques to innovate new sensors. Additionally, we'll briefly explore reinforced learning concepts, which are employed for controlling complex systems and optimizing operations. Lastly, we'll delve into generative ML, unlocking entirely new avenues for research and discovery.

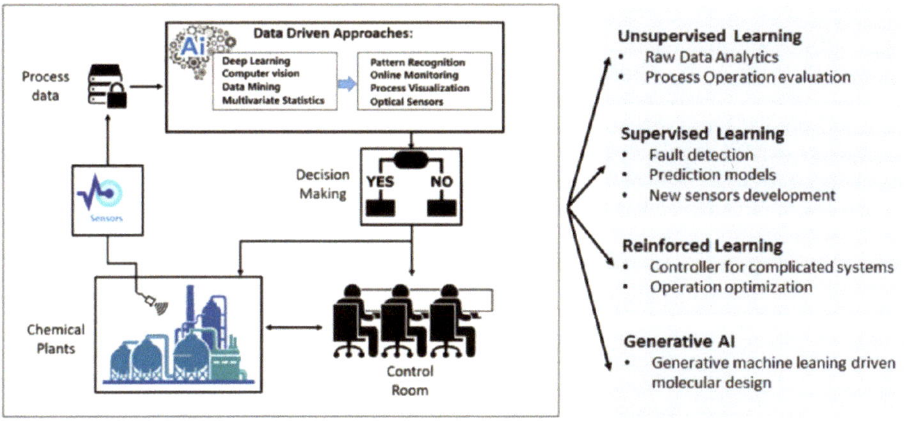

FIGURE 1.4 Machine learning (ML) principles applied to chemical engineering.

1.5 CONCLUSIONS

The evolution from mechanization to CPSs underscores the journey of manufacturing toward enhanced efficiency, safety, and innovation. With the integration of advanced technologies like IoT, cloud computing, and AI, organizations are embracing the potential of smart manufacturing to drive economic growth and global competitiveness.

Moreover, the chapter elucidates the principles and applications of ML in chemical manufacturing, emphasizing its pivotal role in process optimization, fault detection, and decision-making. By harnessing the power of data-driven methods, ML facilitates real-time monitoring, predictive analytics, and intelligent automation, thereby improving operational efficiency and product quality. The synergy between traditional statistical approaches and modern ML techniques offers a comprehensive framework for addressing the complexities of industrial processes.

Moving forward, the convergence of smart manufacturing and ML presents boundless opportunities for innovation and sustainability in the chemical industry. Through collaborative efforts in research, education, and technology adoption, stakeholders can navigate challenges, capitalize on opportunities, and pave the way for a smarter, more resilient manufacturing ecosystem. As industries embrace digital transformation and data-driven methodologies, the journey toward Industry 4.0 accelerates, heralding a new era of intelligent manufacturing and economic prosperity.

REFERENCES

1. Kang H. S., Lee J. Y., Choi S., Kim H., Park J. H., Son J. Y., Kim B. H., Do Noh S. (2016), Smart manufacturing: Past research, present findings, and future directions. International Journal of Precision Engineering and Manufacturing-Green Technology, 3(1), pp. 111–128.
2. Rojko A. (2017), Industry 4.0 concept: Background and overview. International Journal of Interactive Mobile Technologies (iJIM), 11(5), pp. 77–90.

3. Mitchell T. M. (1997), Machine Learning, McGraw-Hill Science/Engineering/Math. https://dl.acm.org/doi/10.5555/541177
4. Kotsiantis S. B., Zaharakis I., Pintelas P. (2007), Supervised machine learning: A review of classification techniques. Emerging Artificial Intelligence Applications in Computer Engineering, 160, pp. 3–24.
5. Jordan M. I., Mitchell T. M. (2015), Machine learning: Trends, perspectives, and prospects. Science, 349(6245), pp. 255–260.
6. LeCun Y., Bengio Y., Hinton G. (2015), Deep learning. Nature, 521(7553), p. 436.
7. Schmugge S. J., Rice L., Lindberg J., Grizziy R., Joffey C., Shin M. C. (2017), Crack segmentation by leveraging multiple frames of varying illumination.2017 IEEE Winter Conference on Applications of Computer Vision (WACV), March 24–31, Santa Rosa, CA, pp. 1045–1053.
8. Zhu W., Ma Y., Benton M., Romagnoli J. A., Zhan Y. (2019), Deep learning for pyrolysis reactor monitoring: From thermal imaging toward smart monitoring system. AIChE Journal, 65(2), pp. 582–591.
9. Yin S., Kaynak O. (2015), Big data for modern industry: Challenges and trends [point of view]. Proceedings of the IEEE, 103(2), pp. 143–146.
10. Lasi H., Fettke P., Kemper H.-G., Feld T., Hoffmann M. (2014), Industry 4.0. Business & Information Systems Engineering, 6(4), pp. 239–242.
11. Ning C., You F. (2019), Optimization under uncertainty in the era of big data and deep learning: When machine learning meets mathematical programming. Computers and Chemical Engineering, 125, pp. 434–448.
12. Ning C., You F. (2017), Data-driven adaptive nested robust optimization: General modeling framework and efficient computational algorithm for decision making under uncertainty. AIChE Journal, 63(9), pp. 3790–3817.
13. Bertsimas D., Gupta V., Kallus N. (2018), Data-driven robust optimization. Mathematical Programming, 167(2), pp. 235–292.
14. Shin J., Badgwell T. A., Liu K. H., Lee J. H. (2019), Reinforcement learning–overview of recent progress and implications for process control. Computers and Chemical Engineering, 127, pp. 282–294.
15. Mnih V., Kavukcuoglu K, Silver D., Rusu A. A., Veness J., Bellemare M. G., Graves A., Riedmiller M., Fidjeland A. K., Ostrovski G., Petersen S., Beattie C., Sadik A., Antonoglou I., King H., Kumaran D., Wierstra D., Legg S., Hassabis D. (2015), Human-level control through deep reinforcement learning. Nature, 518(7540), p. 529.
16. Oxford Economics (2019), The Global Chemical Industry: Catalyzing Growth and Addressing Our World's Sustainability Challenges. Report for International Council of Chemical Associations.
17. Malkoff D. B. (1987), A framework for real-time fault detection and diagnosis using temporal data. Artificial Intelligence in Engineering, 2(2), pp. 97–111.
18. Himmelblau D. M. (2008), Accounts of experiences in the application of artificial neural networks in chemical engineering. Industrial & Engineering Chemistry Research, 47(16), pp. 5782–5796.
19. Venkatasubramanian V., Rengaswamy R., Yin K., Kavuri S. N. (2003), A review of process fault detection and diagnosis: Part I: Quantitative model-based methods. Computers and Chemical Engineering, 27, pp. 293–311.
20. Venkatasubramanian V., Rengaswamy R., Kavuri S. N. (2003), A review of process fault detection and diagnosis: Part II: Qualitative models and search strategies. Computers and Chemical Engineering, 27, pp. 313–326.
21. Venkatasubramanian V., Rengaswamy R., Kavuri S. N., Yin K. (2003), A review of process fault detection and diagnosis: Part III: Process history-based methods. Computers & Chemical Engineering, 27, pp. 327–346.

22. Qin S. J. (2012), Survey on data-driven industrial process monitoring and diagnosis. Annual Reviews in Control, 36(2), pp. 220–234.
23. Chiang L. H., Russell E. L., Braatz R. D. (2000), Fault Detection and Diagnosis in Industrial Systems. Springer Science & Business Media.
24. Kohonen T., Honkela T. (2007), Kohonen network. Scholarpedia, 2(1), p. 1568.
25. Robertson G., Thomas M., Romagnoli J. A. (2015), Topological preservation techniques for nonlinear process monitoring. Computers and Chemical Engineering, 76, pp. 1–16.
26. Sahinidis N. V. (2004), Optimization under uncertainty: State-of-the-art and opportunities. Computers and Chemical Engineering, 28(6–7), pp. 971–983.
27. Zhou Z., Li X., Zare R. N. (2017), Optimizing chemical reactions with deep reinforcement learning. ACS central Science, 3(12), pp. 1337–1344.
28. Ma Y., Zhu W., Benton M. G., Romagnoli J. A. (2019), Continuous control of a polymerization system with deep reinforcement learning. Journal of Process Control, 75, pp. 40–47.
29. Badgwell T. A., Lee J. H., Liu K. H. (2018), Reinforcement learning–overview of recent progress and implications for process control. Computer Aided Chemical Engineering, 44, pp. 71–85.
30. Li H., Collins C. R., Ribelli T. G., Matyjaszewski K., Gordon G. J., Kowalewski T., Yaron D. (2018), Tuning the molecular weight distribution from atom transfer radical polymerization using deep reinforcement learning. Molecular Systems Design Engineering, 3(10), p. 1039.

2 Data and Data Pretreatment

2.1 INTRODUCTION

Data mining and machine learning (ML) provide methods for analyzing large amounts of multivariate data accumulated in modern chemical process databases. However, in many cases the data is not in a form suitable for the desired application and must be preprocessed before it can be fed to the algorithms. Typical activities in data preprocessing include the removal of outliers, the identification and approximation of missing data, and normalization so that the engineering used in measurements does not exert an influence over the results. Additionally, errors from sensor instruments and sampling techniques must also be addressed as they can give rise to outliers and missing data. Furthermore, these activities must be automated and incorporated in the analysis and modeling pipeline for an efficient implementation.

Beyond the basic need of cleaning up data, in more advanced case studies, preprocessing can address other crucial issues with the data. In batch process monitoring for example, methods developed for nearly steady-state processes cannot be applied due to time-dependent variations in the data and the way it occurs in discrete batches in a time-varying trajectory. Likewise, in applying ML to process monitoring and fault diagnosis studies, challenges arise from imbalance in training datasets. In fault detection applications, for example, often a large amount of normal, steady-state data is available with a relatively small amount of fault data. Datasets with such characteristics require specialized algorithms for their proper use.

This chapter reviews basic concepts about data and data preprocessing, including methods for outlier removal, missing data approximation, undersampling and oversampling, and scaling.

2.2 WHAT IS DATA?

Given that data is fundamental for the implementation of most of the methodologies presented in this book, a formal definition for it and as well as some adjacent concepts is granted.

Data is a collection of data objects described by a number of attributes [1]. Attributes, also known as features or just variables, are properties or characteristics of an object that may vary from one object to another as well as over time. Each element in the collection is also called record or data point. Thus, data can be defined as the set $\mathcal{D} = \{a_{1,1}, \ldots, a_{n,m}\}$ that contains n records, each with m attributes. The term dataset is often used in reference to a subset of the data. Also, in the context of modeling, where the distinction between inputs and outputs is important, some attributes

DOI: 10.1201/9781003455905-2

may be designated as inputs while others will be designated as outputs. This distinction of course, responds to the causal relationships between the attributes. However, the nature of these correlations may not be known to the modeler, so the selection will depend on the problem at hand, the domain knowledge of the modeler, and the purpose of the model.

Data objects can take many forms as they come from many sources. Examples of data objects are audio recordings, text excerpts, pictures, and, perhaps more typical in engineering applications, measurements. To analyze the attributes of these data objects, we assign symbols or numbers to them. The assignment of symbols or numbers to attributes is governed by rules called *scales*. Consider, for example, a sample of water boiling in a beaker over a hotplate. The attribute *temperature*, which is a property of the water sample, can be assigned values using the Kelvin scale. This places the minimum possible value to zero, and the most likely value around 373 (boiling point of pure water at atmospheric pressure).

In the temperature example, if we were to choose a different scale, say, Celsius, the minimum possible value is no longer zero, and the most likely value would be around 100 instead of 373. Furthermore, we can consider the mass of the water sample as, say, 0.2 kilograms. Now, there are two different ways of analyzing the attribute *temperature*, one with only positive values and one that allows negative values, and each of these ways will result in the values taken by the attribute *temperature* being much bigger that those taken by the attribute *mass*. This can result in the attribute temperature overpowering the attribute mass during the analysis, which is undesired. A simple problem such as handling the scales of the different attributes illustrates how we often need to transform the raw data before it can be used effectively. These transformations are discussed later in this chapter.

2.3 PROCESS DATA

Instruments in the plant collect data that in turn enable engineers to build a mental and phenomenological map of what is going on with the process. Because of this, instrumentation is often referred to as the engineers' *eyes*, and is the cornerstone of the data analysis and ML effort. However, process data must often be treated and transformed before we can proceed with a proper analysis. This section focuses on methods to treat plant data so we can make meaningful interpretations of the process.

Once process data has been collected, the first intuitive step is to present it in a graphical format. Typically, this is done through a time-series plot, where each process variable is observed sequentially in time. Such a plot is commonly referred to as a run chart. Figure 2.1 shows a time-series plot for 150 observations of the nitrogen oxides (NO_x) level in a wastewater treatment plant. The red line in the plot represents the mean of the data, also known as *centerline* in the control chart terminology [2]. The centerline is critical as it serves as a reference in identifying patterns and determining the magnitude of the process data.

In plant data, measurements inevitably exhibit variations arising from instrument calibration, sensitivity, and data transmission medium. The effect of these factors on the data reported by the instrument are typically lumped together and referred to a *measurement error.* Since measurement errors will always be present, it is useful

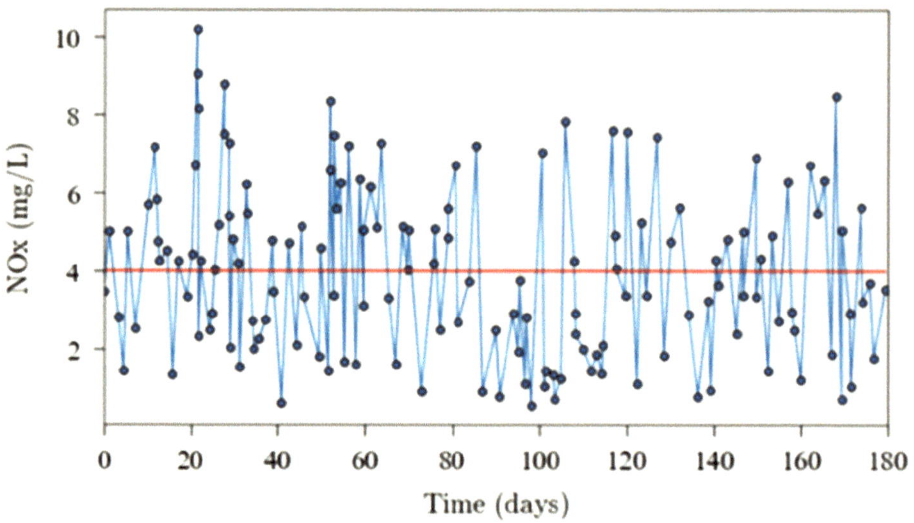

FIGURE 2.1 A time-series plot of data for NO_x level in a waste treatment plant. The red line represents the mean of the data.

to describe the quality of the data in terms of two important characteristics: *accuracy* and *precision*. Accuracy refers to the extent to which the average of repeated measurements on the same instrument differs from the true value. The difference between the average and the true value is known as the *systematic error* (or *bias*). Precision refers to the ability of a measurement system to replicate its own measurements. Precision is inversely related to the variations or dispersion of the replicated measurements.

Time-series plots and tendency metrics such as the central line in Figure 2.1 provide quick insights about the data but fail to fully capture its characteristics. Given a dataset, additional questions may arise beyond its central tendency, such as whether the observations are evenly spread within a particular range, whether the process outcome has equal chances of falling within any given interval, or where the data tend to group or concentrate if the observations are not evenly spread. These questions are related to the *distribution* of the data, which can be characterized through histograms and probability distribution function, as shown in Figure 2.2.

Figure 2.2a shows the dispersion of measurements, as well as their relative frequency of over the range of data. This type of chart, known as histogram, helps us understand the nature of the process under analysis. For example, Figure 2.2c shows the histogram for a *normally distributed* dataset with same central line, variability, and number of records. A normal distribution is a symmetric distribution in which the central values taken by the variable of interest have a high probability of occurring while the probability is lower for extreme values on either end of the range. When we compare the histogram for our plant with the one in Figure 2.2c, we can

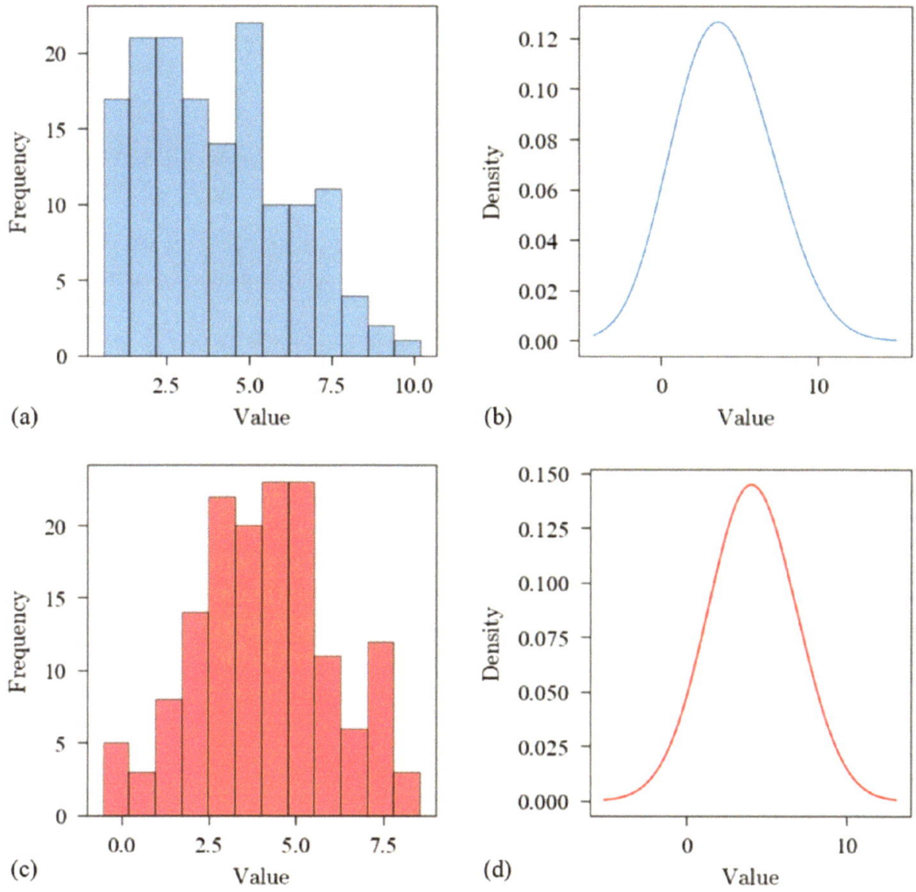

FIGURE 2.2 Data distribution: (a) histogram for the wastewater plant data, (b) probability distribution function for the wastewater plant data, (c) histogram for normal distribution, and (d) probability distribution function for a normally distributed dataset.

note that ours is skewed to the lower limit of the range, meaning that a lower measurement has a higher chance of occurring in our system. As we will see later in this book, data distribution plays a crucial role in the outcomes we can obtain in data analysis and ML applications, and the reader is encouraged to learn more about these topics from statistics references.

Now, if the number of observations was very large and the groups were made much narrower, the previous irregular step form of the histogram could be represented as a smoothed curve, which is known as the frequency curve. Sometimes, it is preferable to express the frequencies in terms of proportional frequencies instead of actual frequencies (probability distribution) or probability density function (PDF) curve. The PDF curve for the data in Figure 2.1 is given in Figure 2.2b. This example illustrates how a simple distribution diagram (histogram or PDF curve) provides insight about the data concentration or distribution around the centerline without considering the time dimension.

2.4 MODELING PROCESS DATA

Motivated by the previous discussions, data generated by a random process is represented by the following conceptual equation [3]:

$$\text{Observed date value} = \text{Mean (actual) level} + \text{Random deviation} \tag{2.1}$$

In other words, any observation or measurement comprises the contribution of the actual state of the process (mean level) and a random deviation component. The latter describes the random phenomena that might push what is actually observed above or below the mean level and, therefore, has a positive or a negative value.

It is often assumed that the average of a given set of random deviations is close to zero. Actually, one can prove that this average equals to zero exactly for a very large (infinite) number of observations. This is the evidence that will allow us to assume that the random deviations follow normal (Gaussian) distribution.

Often, the normal distribution is a good assumption to start with, but not something that should be taken for granted. If the data conform to the normal distribution, it can be easily characterized by two parameters, namely, the mean and the variance. This results in a very simple model. However, for most process signals, such a model may be overly simplistic.

Let us express the model for a random variable (process) y as follows:

$$y(i) = \bar{y} + \varepsilon(i) \quad i = 1, 2, \ldots \tag{2.2}$$

In this equation, ε denotes the random deviation at the sampling instant i and $\hat{y}$ is the underlying mean level. The random deviations $\varepsilon(i)$ is also commonly referred to as *random errors* or *disturbances*. To facilitate the development and interpretation of data analysis techniques that will follow, one has to make the following assumptions about random errors:

- The random errors are independent. Keeping in mind that the random errors occur sequentially over time, this assumption basically implies that the errors are generated from a random process.
- The random errors have an expected value (or mean) of zero, that is,

$$E(\varepsilon(i)) = 0 \tag{2.3}$$

where $E(\cdot)$ denotes the mean or expected value of a random variable.

- The random errors have a variance of σ_{ε}^2. This specification is often called the *assumption of constant dispersion*. The idea is that the vertical scatter of points about the systematic pattern tends to be the same everywhere.

- The random errors are normally distributed random variables. Using the notation introduced in Appendix D, this specification along with the other assumptions implies that

$$\varepsilon(i) \rightarrow N(0, \sigma_\varepsilon^2) \tag{2.4}$$

When referring to the traditional Gaussian distribution, the relationship between σ and the sample mean defines the distribution curve explicitly. Additionally, this curve definition establishes some sample population proportions. For instance, it is known that 68.26% of all the samples of a population will fall within $\pm 1\sigma$ (1-Sigma) of the mean, 95.46% within $\pm 2\sigma$ (2-Sigma), and 99.73% within $\pm 3\sigma$ (3-Sigma) of the mean. Shewhart demonstrated that plotting sample averages would produce a normal distribution, and this enables one to establish a strategy for quality monitoring.

Once statistical control is established, any data exceeding 3-Sigma (or any other predefined threshold) indicate a situation where the process is out of target (statistical control). Shewhart's observations led to the development of what is known today as univariate control charts.

For processes where multiple variables are measured and monitored (multivariate), using multiple Shewhart charts would be cumbersome and may lead to false alarms and undetected events. Consequently, the following chapters will discuss approaches to deal with multivariate data after introducing some issues with process data.

2.5 ISSUES WITH PLANT DATA

As we discussed previously, computer control systems are prevalent in process plants, making it easier to gather and store data from plant instrumentation. These data play a critical role in manufacturing initiatives; however, it's important to note that the quality of data collected is subject to variation. This is because the sampling and testing equipment, schedules, and techniques are exposed to various influences that can affect the data. As a result, the raw dataset often contains missing points, gross errors, and outliers that must be eliminated to obtain a usable dataset.

Table 2.1 provides a glimpse of a typical dataset, showcasing 10 variables that were recorded as part of a daily monitoring activity in a sewage treatment plant. This snapshot also highlights some of the issues that can arise with raw data. Upon examining Table 2.1, two types of errors become apparent. The first type is constant values that may or may not be erroneous, depending on one's understanding of the process. The second type is the presence of the term "nan" (not a number), indicating a lack of numerical data. This could be due to the data point being below the detection limit or being missing altogether. Since distinguishing between the two can be challenging, such a point is always considered a missing point.

Given that data is fundamental for the implementation of most of the methodologies presented in this book, in this chapter, our goal is to study a set of methodologies that can help engineers obtain a meaningful dataset from raw data, so that the conclusions based on the dataset are sound and reliable.

TABLE 2.1
Typical Plant Data

Date	V1	V2	V3	V4	V5	V6	V7	V8	V9	V10
950101	573	nan	13	433	168	7.4	248	560	33.3	27.3
950102	489	nan	4.3	372	245	2.1	182	560	86.5	27.3
950103	686	nan	nan	nan	nan	nan	nan	560	nan	16.1
950104	696	nan	10.1	544	140	7	379	560	53.9	16.1
950105	1194	nan	7.1	554	124	8.5	661	560	78	16.1
950106	968	nan	nan	709	153	nan	nan	560	nan	16.1
950107	1504	nan	14.7	902	nan	22.1	1357	560	61.4	16.1
950108	860	nan	8.7	427	148	7.5	367	560	49.1	16.1
950109	864	nan	7.2	886	440	6.2	766	560	123.1	16.1
950110	1236	nan	7.6	611	111	9.4	755	560	80.4	16.1
950111	984	nan	8.7	624	178	8.6	614	560	71.7	16.1
950112	778	nan	8.8	742	82	6.8	577	560	84.3	16.1
950113	519	nan	8.7	654	77	4.5	339	560	75.2	16.1
950114	663	nan	7.9	791	100	5.2	524	560	100.1	16.1
950115	680	nan	7.7	513	40	5.2	349	560	66.6	16.1
950116	719	nan	8.4	746	232	6	536	560	88.8	16.1
950117	937	nan	8.1	579	196	7.6	543	560	71.5	16.1
950118	1174	nan	9.7	1160	280	11.4	1362	560	119.6	16.1
950119	1398	nan	nan	nan	nan	nan	nan	560	nan	16.1
950120	968	nan	nan	nan	nan	nan	nan	560	nan	16.1
950121	3073	21	5	339	221	15.4	1042	560	67.8	16.1
950122	723	22	6.7	475	378	4.8	343	560	70.9	16.1
950123	636	18	7.6	946	193	4.8	602	560	124.5	16.1
950124	838	19	10.8	685	288	9.1	574	560	63.4	16.1
950125	955	nan	10.1	1001	173	9.6	956	560	99.1	16.1
950126	890	nan	12.3	548	333	10.9	488	560	44.6	16.1
950127	836	nan	14.4	543	231	12	454	560	37.7	16.1

Source: From [2].

2.6 DATA PRETREATMENT

To ensure accurate and effective process monitoring, data must be cleaned and pretreated. Three standard pretreatment operations are missing data approximation, outlier removal, and data sampling and normalization.

2.6.1 Missing Data Approximation

Missing data may arise in process databases from measurement instrument or software errors. They can often be removed without affecting data analysis, but in smaller datasets they can have a larger impact. In these cases, data must be filled in using interpolation techniques to facilitate further analysis. Depending on how much

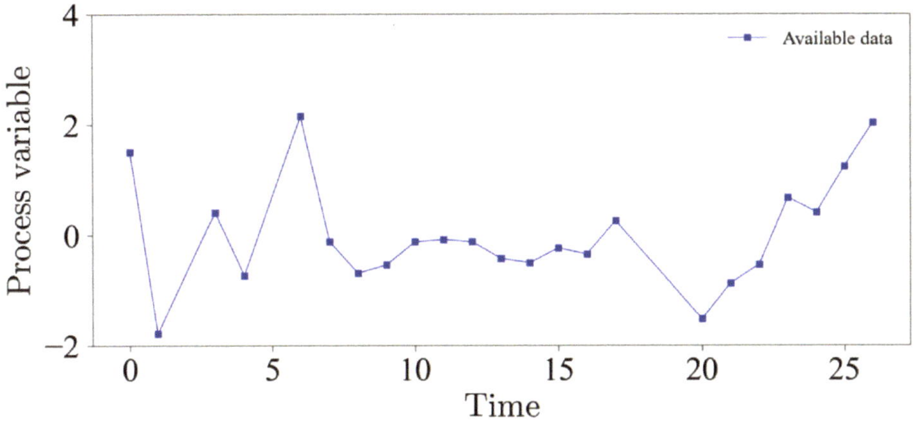

FIGURE 2.3 Process data for V4, a measured variable in a treatment plant.

knowledge there is about the domain, that is, the nature of the data being analyzed, regression methods can be used to fill in the missing data. It should be recognized however that there will be a trade-off between how well the regression method fits the existing data and how well it would represent the missing data (Figure 2.3).

The application of interpolation techniques is a matter of choice, and it will influence subsequent analyses. This point is illustrated in Figure 2.4, where the modified continuous data are compared to the true data that would have been obtained if the missing points were available. Figure 2.4 also illustrates the impact of assuming too much information. The first interpolation (the section between days 1 and 5) would be a misrepresentation of the system; however, its effect would be minimal on the overall data statistics. If the large portion of missing data, from days 17 to 20, were to be interpolated as shown, the mean of the data series would be lower than the actual one. This would lead to an erroneous view of the state of the system, which may be more serious than the calculated statistical confidence implies.

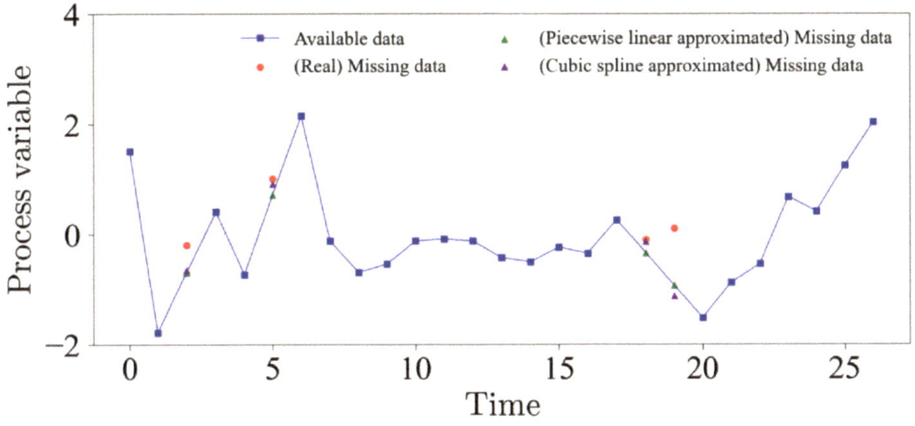

FIGURE 2.4 New data after interpolation using different methods.

2.6.2 Outlier Removal

An outlier is an anomaly that does not follow the statistical distribution of the bulk of the data. Outliers represent extreme values far outside the statistical distribution of practical measurements and can be heavily influential observations. Therefore, cleaning the data of outliers can dramatically affect the estimates, confidence region, and other tests performed on the data. Because of the effect outliers can have on the integrity of a dataset, their detection and removal are crucial before further analysis [3, 4]. On smaller datasets, outliers can be removed simply through visual inspection. For larger datasets however, this approach would be extremely inefficient and an algorithmic method should be used instead.

In general, an outlier removal method would include a statistical criterion, such as a number of standard deviations away from the mean, to identify an outlier, as well as a method to organize and/or navigate the data to search for anomalies. Chen and Romagnoli [4] used a mean minimum distance-based clustering method for outlier removal. Russell et al. [5] proposed a method for outlier removal using the T^2 statistic. Other techniques for outlier removal are reviewed in Rousseeuw Leroy [6].

2.6.2.1 Z-Score

Z-score is a simple method for outlier detection on each individual variable of the dataset. The Z-score, indicates how far the value of the data point/sample is from its mean for a specific variable. For example, a Z-score of 1 means the sample point is 1 standard deviation away from its mean. Typically, Z-score values greater than or less than +3 or −3, respectively, are considered outliers. The Z-score is expressed mathematically as:

$$z = \frac{x_i - \bar{x}}{\sigma} \tag{2.5}$$

where σ and $\bar{x}$ is the standard deviation and mean of the distribution of variable x, respectively, and x_i is the value of the variable x for the i-th sample.

2.6.2.2 *k*-Nearest Neighbors (*k*-NN) for Outlier Detection

Outliers can also be identified by their proximity to other data samples. The general notion is that samples that lie further away from the overall data distribution by some distance metric can be considered outliers. The *k*-nearest neighbors (*k*-NN) algorithm implements this idea by measuring the distances (typically Euclidean distance) from a point to its *k* closest neighbors. Then, the point is labeled as an outlier or not based on some criterion, such as the farthest distance to a neighbor being greater than a certain threshold. The elbow method can be applied to find a useful threshold. In the elbow method, all *k*-NN distances are sorted from smaller to greater and then the point at which these distances start growing exponentially is deemed as the threshold (see Figure 2.5.). The selection of the number of neighbors, *k*, has an impact on the result of the outlier removal operation as it will be illustrated in Example 2.1.

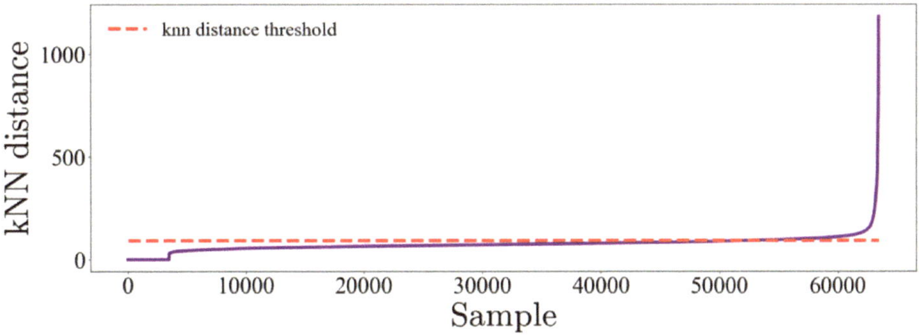

FIGURE 2.5 Elbow method to find the threshold for outlier identification using the k-NN's algorithm with $k = 10$.

Example 2.1: Effect of Outlier Removal

An industrial pyrolysis reactor's data corresponding to a whole cycle of operation will later be considered as an industrial case study to illustrate some of the tools described in this book. In this example, issues associated with outlier detection and elimination, as well as data sampling, will be discussed. The reactor operates in batches, with each operating cycle lasting approximately 90 days and requiring decoking afterward for further operations.

Figure 2.6a depicts the time evolution of hydrocarbon flowrates for two coils in the reactor (FC05034.PV1 and FC05032.PV1) during an operation cycle where the colors represent the time. The larger concentration of data at the top of the plot in Figure 2.6b corresponds to the main data distribution while the scattered points closer to the lower-left region can be identified as outliers.

Here, the k-NN algorithm is used to detect and eliminate outliers, with the number of k-neighbors strongly influencing the amount of data being classified as outliers. Figure 2.6b, illustrates the same data projected using a scatter plot, where the outliers (very low numbers in the flowrates) are clearly shown in the lower left corner of the graph.

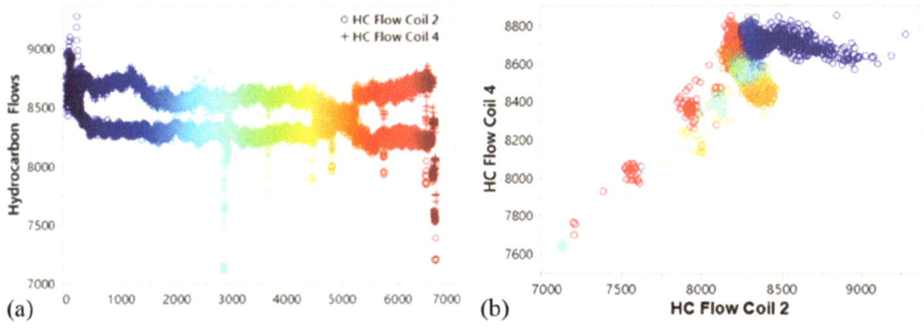

FIGURE 2.6 Data for Example 3.2: (a) raw data and (b) scatter plot of original data

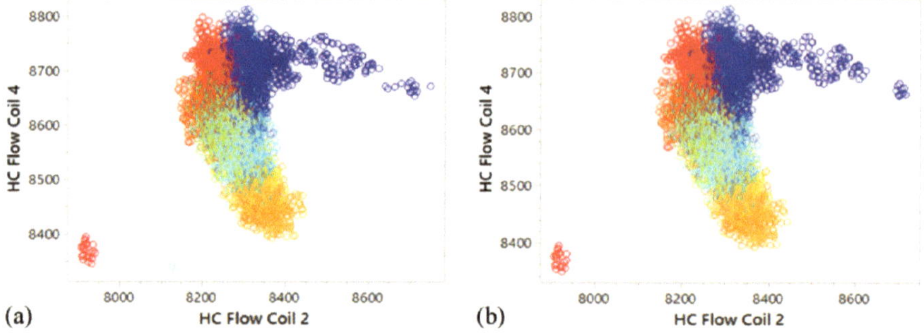

FIGURE 2.7 Effect of outlier removal: (a) $k = 5$ and (b) $k = 10$.

Figure 2.7a and b shows the results after implementing the k-NN algorithm for values of $k = 5$ and $k = 10$. For this example, increasing the number of neighbors beyond $k = 10$ does not significantly improve the data.

Eliminating the outliers (large tails) has significantly improved the visualization, and it will undoubtedly have a strong impact on the performance of the data mining tools to be discussed in the upcoming chapters. It is essential to pay attention to the scales when considering alternative outlier elimination methods. As the data cleaning increases with increasing k, the resolution of the graphics improves, enabling us to concentrate on the underlying process dynamics without being impeded by the presence of outliers.

2.6.3 Under-Sampling and Oversampling

2.6.3.1 Data Under-Sampling

ML learning applications require datasets to be not only large but also balanced. This means that the number of instances (i.e., data points) for each class (e.g., a particular plant state) must be even across classes [7, 8]. When this requirement is not met, two approaches can be taken: oversampling and under sampling. In oversampling, data augmentation methods are used to generate new data points, thus increasing the number of instances for a given class [9]. In under sampling, not all the available instances for a given class are used. This brings the number of instances of all classes to a similar level such that an overrepresented class does not exerts a heavier influence on the data analysis. When to use each of these two approaches will depend mainly on the characteristics of the underrepresented classes and it should be noted that both could be used on different regions of the dataset. The goal is to have enough data to train the model to capture the presence of even the most underrepresented classes. Once these classes have the adequate number of instances, which will depend largely on the data analysis method being used, then the rest of the dataset can be assessed for under sampling or oversampling.

When sampling is needed, the ability of the sampled data to accurately represent the original data is of paramount importance. With this consideration in mind, we can identify three sampling procedures that aim, in some way or another, to provide a useful representation of the original data, although with many fewer data points.

2.6.3.1.1 Random Sampling

The first approach is random sampling in which, after segmenting the complete dataset in intervals of equal size, a point is sampled at random from within each interval. The segmentation of the original dataset prevents regions of data from being overrepresented as each segment will have one representative point. Taking this idea further, the so-called sampling and average method takes the average of all points within each interval, thus including information from all points in each interval.

These two methods described above sample equally along the dataset, which is good enough if the data does not show rapid changes or all the classes have around the same amount of data points. In other applications however, a method capable of identifying regions in the dataset where less data are needed (i.e., monotone regions) and regions where more data are needed (i.e., rapidly variable regions) might be preferred. A method based on the Condensed Nearest Neighbor (c-NN) algorithm, which is an example of such approach, will be presented next [10].

2.6.3.1.2 Condensed Nearest Neighbor Sampling (c-NN)

In c-NN, data points that are similar are identified and classified. Using this information, a reduced dataset is produced in which the highly variable regions (more classes) are more densely sampled. The algorithm starts by taking a random point in the dataset and storing it in a subset S. Then, a second point is taken and classified using the NN rule with the stored values as reference set. If the second point is classified correctly, it is stored in a subset G; if not it, is stored in S. The procedure is repeated over the whole dataset. After the dataset is exhausted, the steps are repeated over the subset G. The algorithm terminates when G is exhausted or there are no more transfers to S. The new sampled set is S.

2.6.3.2 Data Oversampling

Problems in fault detection and diagnosis often have a problem with dataset imbalance. Normal data by definition will be more numerous than fault data, often by a sizable margin. In a dataset with 99% normal data, a default guess of normal would achieve 99% accuracy. In these cases, there is a higher cost of missing fault cases than missing normal cases. Dataset imbalance is an important issue with discriminant analysis and in training fault diagnosis algorithms where imbalance in the data can cause one fault class to dominate the others. To balance the representation of normal and faulty data in a data-based model for fault diagnosis, it is vital to make maximum use of the information in faulty data by oversampling fault data and under-sampling data from normal operations.

Ling and Li [8] performed a comprehensive study of the interaction of oversampling and under-sampling and how it affects classification performance using a dataset from product marketing. Different under-samplings of the majority class were tested and measured using lift analysis and the study determined that the best lift index is obtained when the classes are equally represented. Oversampling with replacement itself was found not to significantly improve minority class recognition. As the minority class is over-sampled by increasing amounts, the effect is to identify similar but more specific regions in the feature space as the decision region for the minority class. In the fault diagnosis case, for example, where fault regions of space must be distinguished from the normal operating region, sampling with replacement can actually make the minority decision region smaller and more specific as minority samples in the region are duplicated. SMOTE (Synthetic Minority Over-Sampling Technique) data, which is explained below, has larger decision regions, leads to larger decision regions with improved characterization of the shape and size of the fault regions.

SMOTE is an oversampling approach in which the minority class is over-sampled by creating "synthetic" examples rather than by oversampling with replacement. SMOTE creates new data along line segments between randomly chosen k-NN among the minority class of data. The user specifies k, the number of nearest neighbors to generate. For each minority data point, SMOTE finds its k-NN in the minority dataset, selects a number of them based on the percent oversampling specified (e.g., when $N = 200\%$, two neighbors are selected), and randomly generates a synthetic data point along the line between the data point and its nearest neighbor. This process generalizes and "fleshes out" the decision region of the minority class to a larger and less specific region [9].

To under-sample data from the majority class, simply randomly remove samples from the population until the size of the majority class become some specified percentage of the minority class. In SMOTE terminology, stating that "the majority class is under sampled at 200%" means that in the modified "SMOTEd" dataset, the number of data points from the minority class will be double the number of data points from the majority class. For example, if the minority class had 50 samples and the majority class had 200 samples and we under sample the majority at 200%, the majority class would end up having 25 samples. Equal representation between minority and majority classes occurs when the majority class is under sampled at 100% (i.e., the number of majority samples is 100% of the number of minority samples). To create a balanced dataset, generally both oversampling of the minority class and under-sampling of the majority class are required.

Example 2.2: Effect of Sampling

Let us consider the same data used in Example 2.1 with the flowrates of an industrial pyrolysis reactor to illustrate the effect of sampling. The effect of sampling real plant data is investigated using random sampling and c-NN with different values of k. Figure 2.8 illustrates the sampling effect on two of the process variables for

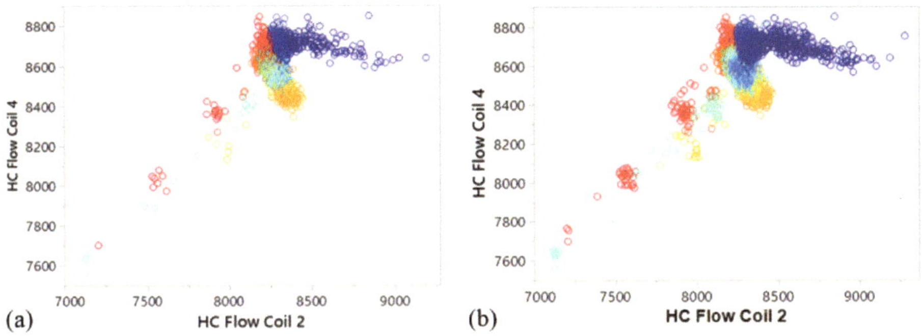

FIGURE 2.8 Effect of sampling on the process variables: (a) Random sampling and (b) c-NN sampling.

> the c-NN and random sampling approaches with k number of neighbors $k=5$ and with intervals = 3 for random sampling. For the c-NN approach the data size is reduced six-fold, from 7800 to 1400 data points; however, this reduction does not affect the quality of the dataset since the main events are still captured in any of the three conditions investigated. This is shown with the help of 2D scatter plots that clearly illustrate three main events captured by the data while the random sampling loses part of the information during the initial process transition. The number of neighbors slightly changes the size of the data but does not have further influence on the analytics. This agrees with the idea that in this approach the highly variable regions (more classes) are more densely sampled. Furthermore, since no steady state is actually reached (real plant conditions and the drifting nature of the pyrolysis reactor), the effect of c-NN is different from that observed with simulated data.

From the previous analysis, it can be seen that sampling strategy utilized to reduce the size of the dataset does affect the quality of the data in terms of capturing the main process dynamics. From the analysis, the c-NN approach, while reducing the size of the data, is also able to capture not just the main events (end of the cycle) but also other events, in more details, during the initial period.

2.6.4 Standardization and Normalization

After a dataset has been properly cleaned, including the removal of outliers and handling of missing data, and sampled as necessary, the next crucial step is scaling. In the context of process monitoring and other engineering applications that involve ML, scaling is essential to eliminate the effects of engineering units on the process statistics. For instance, consider a situation where a flow measurement changes from 0.1 kg/hr to 1 kg/hr, which is more significant than a temperature measurement changing from 400°C to 410°C. This difference in magnitude can cause the temperature measurement to have a greater influence on the data analysis, not necessarily because it is more important from a phenomenological point of view than the flow measurement, but simply because of its magnitude.

To compare scores between different types of variables, it's important to put them on the same scale during data preprocessing. There are two main scaling methods: standardization and normalization. Standardization assumes a known probability distribution, typically a Gaussian distribution, for the data. The Z-score method falls into this category, as it's based on the probability of a score occurring within the normal distribution of the data. Using the Z-score method allows for comparison of scores from different normal distributions of the data. On the other hand, normalization doesn't assume any distribution and includes methods such as over max, over min, over mean, and over min-max.

In both cases, you're transforming the values of numeric variables so that the transformed data points have specific helpful properties. The difference is that:

- in **normalization**, you are changing the *range* of your data, and
- in **standardization**, you are changing the *shape of the distribution* of your data.

2.6.4.1 Z-Score (Standardization)

Z-score, sometimes called standard score, is a measurement of how many standard deviations a point is away from the mean of its dataset, making it particularly good in handling outliers. The Z-score z for a point x is given by:

$$z = \frac{x - \mu}{\sigma} \tag{2.6}$$

where μ is the mean of all the data points and σ is the standard deviation.

The Z-score has numerous applications and can be used to perform a z-test, calculate prediction intervals, process control applications, comparison of scores on different scales, and more [11].

2.6.4.2 Normalization

Over range normalization is used to scale the data from 0 to 1 without assuming a known probability distribution. Over max or over min normalization is done by dividing the raw data with the maximum or minimum (minimum value must not be zero) values [11]. Over the mean normalization scales each variable on a 0 to 1 scale by dividing each value by the mean value of the variable. Over min-max normalization performs a linear transformation on the original data [12]. For a point x the scaled value x_{scaled} using a min-max scaler is given by:

$$x_{scaled} = \frac{x - x_{min}}{x_{max} - x_{min}} \tag{2.7}$$

where x_{min} and x_{max} are the minimum and maximum values of x in the dataset.

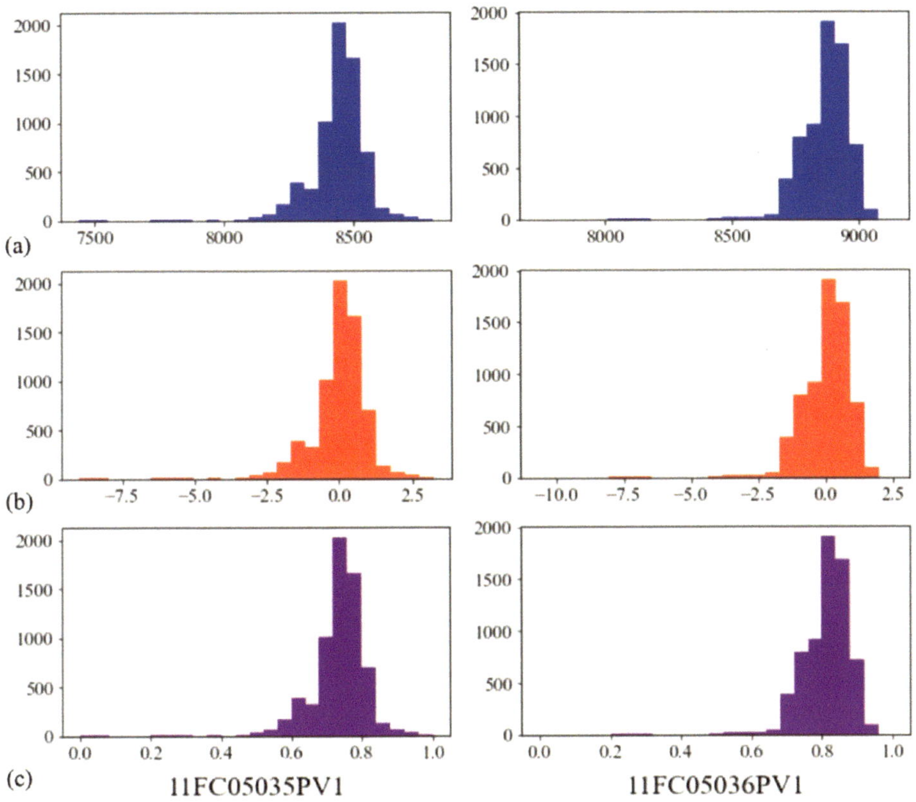

FIGURE 2.9 Data normalization/standardization: (a) original data distribution, (b) data distribution after standardization, and (c) data distribution after normalization.

Example 2.3: Effect of Normalization

To illustrate the effect of standardization and normalization, we will use the same data that was used in previous examples, corresponding to the two hydrocarbon flowrates. The effect of alternative approaches is shown in Figure 2.9 in terms of all reactor process variables. Also, given in Figure 2.10 is the projected data, after normalization, for the two specific flowrates analyzed previously. Scaling, either through standardization or normalization, can significantly impact data characterization as will be shown later.

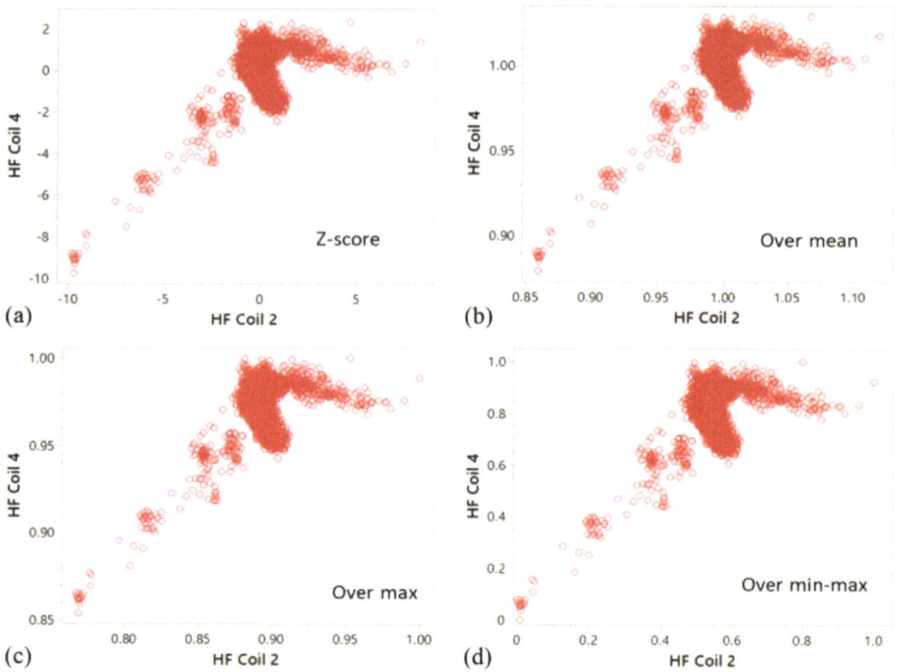

FIGURE 2.10 Scatter plot for hydrocarbon flowrates after standardization and normalization: (a) Z-score, (b) over mean, (c) over max, (d) over min-max.

2.7 CONCLUSIONS

Prior to implementing data analysis, it is essential to perform data preprocessing. In industrial applications where the dataset is excessively large or underrepresented faults occur, sampling can be employed to attain an appropriate dataset resolution. This approach provides a favorable trade-off between the dimension of the dataset, required resolution, and processing time.

Scaling, either through standardization or normalization, can significantly impact data characterization. As demonstrated in subsequent chapters using both simulated and real plant data, Z-scores deliver the best outcomes, as observed in SOM projections. The regions of SOM were better characterized by Z-scores (standardization) than over the range approaches (normalization), although similar regions were identified. Z-scores exhibit improved separation among regions.

Data cleaning and preprocessing are critical in any ML application. Additionally, the selection of methods for each task, such as outlier removal, missing data approximation, sampling, and normalization, can influence the data analysis outcome, which will be discussed in subsequent chapters.

REFERENCES

1. Tan P.-N., Steinbach M., Kumar V. (2016), Introduction to Data Mining. Pearson Education India.
2. Romagnoli J. A., Palazoglu A. (2020), Introduction to Process Control. 3 ed.: CRC Press.
3. Romagnoli J. A., Sanchez M. (1999), On Data Processing and Reconciliation for Chemical Processes Operation. Elsevier.
4. Chen J., Romagnoli J. A. (1998), A strategy for simultaneous dynamic data reconciliation and outlier detection. Computers & Chemical Engineering, 22(4), pp. 559–562.
5. Russell E. L., Chiang L. H., Braatz R. D. (2000), Data-driven methods for fault detection and diagnosis in chemical processes. Advances in Industrial Control. Springer-Verlag.
6. Rousseeuw P. J., Leroy A. M. (2003), Robust regression and outlier detection. Wiley Series in Probability and Statistics. John Wiley & Sons.
7. Mohammed R., Rawashdeh J., Abdullah M. (2020), Machine learning with oversampling and undersampling techniques: Overview study and experimental results. 11th International Conference on Information and Communication Systems (ICICS). IEEE, pp. 243–248.
8. Ling C. X., Li C. (1998), Data mining for direct marketing: Problems and solutions. Proceedings of the Fourth International Conference on Knowledge Discovery and Data Mining. AAAI Press, pp. 73–79.
9. Chawla N. V., Bowyer K. W., Hall L. O., Kegelmeyer W. P. (2002), SMOTE: Synthetic minority over-sampling technique. Journal of Artificial Inteligent Research, 16(1), pp. 321–357.
10. Hart P. (1968), The condensed nearest neighbor rule (corresp.). IEEE Transactions on Information Theory, 14(3), pp. 515–516.
11. Pedregosa F., Varoquaux G. V., Gramfort A., Michel V., Thirion B. (2011), Scikit-learn: Machine learning in python. Journal of Machine Learning Research, 12, pp. 2825–2830.
12. Han J., Pei J., Kamber M. (2011), Data Mining: Concepts and Techniques. Elsevier.

Part I

Unsupervised Learning for Knowledge Discovery

Most process monitoring methods are supervised, meaning they require organized data groups before training. However, unsupervised learning offers benefits for engineers seeking to identify different operating regions, including normal and faulty conditions, with limited knowledge of the process. Unsupervised learning is a machine learning (ML) area in which algorithms extract features and identifies commonalities in a dataset that has not been well labeled or classified. By leveraging unsupervised learning, engineers can uncover insights from unlabeled data and gain a more comprehensive understanding of the process.

DOI: 10.1201/9781003455905-3

3 Dimensionality Reduction

3.1 INTRODUCTION

The world generates an astonishing 1.2 zettabytes (1021) of data every day, comprising everything from social media posts, pictures, videos to scientific sensor measurements [1]. Despite this impressive volume, extracting valuable insights from large datasets remains a significant challenge.

Although traditional statistics are still essential for data analysis, they are not suitable for managing the intricacies of large datasets. In contrast, data mining techniques are computational algorithms specifically designed to detect meaningful patterns and relationships within vast datasets that would be challenging to identify using traditional methods. For example, Gardner and Bieker [2] conducted several case studies that demonstrated how data mining generated hypotheses successfully resolved complex manufacturing problems in semiconductor processes that traditional experimental and statistical approaches consistently overlooked.

Once raw measurements have been preprocessed, the challenge lies in analyzing the significant patterns within the resulting data. Historical data is often high dimensional, making it difficult to identify patterns within such a space. Moreover, conventional metrics like the Euclidean distance become less effective in high-dimensional spaces, which is commonly known as the "curse of dimensionality". Therefore, dimensionality reduction (DR) methods are often preferred for analyzing high-dimensional data. Rather than listing every DR method available, in this chapter, we will instead review several methods that have proven to be effective for process data based on our practical experience.

3.2 DATA MINING

Figure 3.1 outlines the data mining approach used. After the preprocessing step, DR techniques project the process data, removing redundant, correlated sensor measurements and combining them into lower-dimensional scores. DR may project data in two or three dimensions to enable visualization, or the technique may simply function to remove redundant information from process data. In some cases, DR may not be necessary if the data are of good quality.

Following projection by a DR technique, data clustering algorithms partition the data using any number of clustering techniques. Depending on the data and the parameters used to calculate the clustering, clusters found may or may not correspond to significant structures, therefore, cluster evaluation metrics are important to help the user judge the quality of the clusters extracted before more detailed analysis.

DOI: 10.1201/9781003455905-4

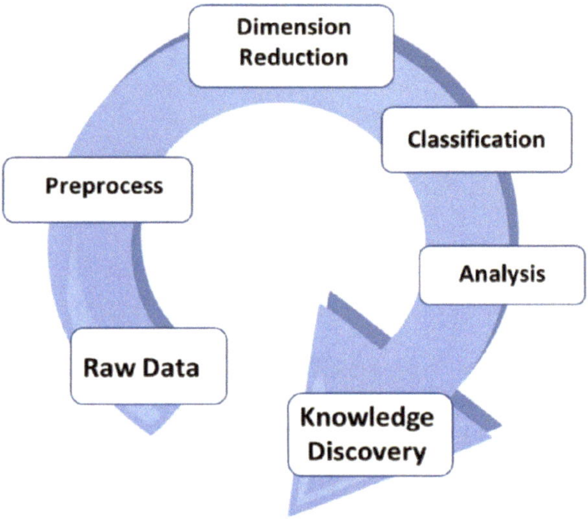

FIGURE 3.1 Typical flow of a data mining and knowledge discovery methodology.

In the cluster assignment step, the user analyzes the data within the clusters to establish connections to significant process events, such as faults or operating procedures. Once labeled with process events, the data can be employed for training and fitting in machine learning (ML) or other supervised fault detection or diagnosis algorithms. This process of extracting information from databases is referred to as knowledge discovery.

3.3 DIMENSIONALITY REDUCTION (DR) METHODS

DR is the process of reducing the number of variables or attributes that are taken into consideration. There are generally two main groups of DR methods: matrix factorization and neighbor graphs. In matrix factorization, the original high-dimensional space is decomposed into two lower-dimensional matrices, and then an optimization algorithm is implemented to minimize the reconstruction error. The reconstruction error is obtained from the difference between the actual values in the original space and those computed from the product of the two matrices [3]. On the other hand, in neighbor graphs, a mathematical graph is built in the high-dimensional space including all variables and then embedded in a lower-dimensional space using a topological manifold [4, 5].

3.3.1 Matrix Factorization Methods

3.3.1.1 Principal Component Analysis (PCA)

Principal component analysis (PCA) is a linear distance preservation technique which determines a set of orthogonal vectors that optimally capture the variability of the data. The orthogonal vectors are determined through an eigenvalue decomposition of the covariance matrix and arranged in order of the variance explained in the loading vector directions [6].

PCA reduces the data as follows. For a given training set of n observations and m sensor measurements stored in $X^{n\times m}$, its sample covariance matrix S can be calculated:

$$S = \frac{1}{n-1} X^T X = P \Lambda P^T \tag{3.1}$$

By finding the eigenvalues (Λ) of covariance matrix S, the projections of the observation in X into the lower-dimensional space are calculated in the score matrix:

$$T = XP \tag{3.2}$$

The number of principal components (PCs) is typically determined using the percent variance test. The percent variance test calculates the smallest number of loading vectors needed to explain a minimum percentage of the total variance. For examples, the number of PCs needed to model 90–95% of the variance of the original data.

PCA has many advantages over other DR methods. First, it does not require user-defined parameters, making its implementation straightforward. Second, by preserving the global structure of data in the high-dimensional space, we can retain the interesting variations in the data in the low-dimensional representation. This ensures that the points at the extremes of the projection also remain at the extremes in the high-dimensional space. However, PCA also has shortcomings. For example, it does not preserve neighborhoods or local structures. That is, if the data lives in clusters in the high-dimensional space, then the information will not necessarily be preserved when projected.

3.3.1.2 Independent Component Analysis

Independent component analysis (ICA) is used in multivariate signal separation for extracting hidden and statistically independent components (ICs) from the observed data and has been adapted for process monitoring tasks similar to PCA by Lee et al. [7]. Signal source separation recovers the independent signals after linear mixing. In other words, for mixed signals represented by x, independent signal sources s are linearly mixed by the matrix A by:

$$\boldsymbol{x} = \sum_{k=1}^{n} s_k a_k = As \tag{3.3}$$

The goal of ICA is to find the source signals s using information stored in x. After matrix A is estimated, the the matrix inverse $\boldsymbol{W}$ ($\boldsymbol{W} = \boldsymbol{A}^{-1}$) can be used to calculate the original independent source signals by:

$$\boldsymbol{s} = \boldsymbol{W}x \tag{3.4}$$

The matrix $\boldsymbol{W}$ can be found using the FastICA algorithm of Hyvarinen and Oja [8].

3.3.1.3 Kernel Principal Component Analysis (KPCA)

Kernel principal component analysis (KPCA) extends traditional PCA to nonlinear data spaces. Instead of directly taking eigenvalue decomposition of the covariance matrix like PCA, KPCA takes a dataset with nonlinear features that PCA fails to preserve and projects them to a higher-dimensional space where they vary linearly. The high-dimensional space is called the *feature space* (F). KPCA first assumes that the data have been transformed nonlinearly using a nonlinear mapping function, $\Phi(x)$. Conventional PCA is then performed in the feature space to achieve the transformations for DR [9]. As with other nonlinear DR methods, the major weakness of **KPCA** is the high computational cost of calculating a feature matrix (or kernel matrix for **KPCA**). To determine the number of PCs, in accordance with Scholkopf et al. [10], the cut-off method using the average eigenvalue is used, which only includes PCs with eigenvalues above the average.

3.3.2 Neighbor Graphs

Neighbor graphs methods for DR generate a connected graph between the variables in the original high-dimensional space. Then, the algorithm looks for the topological manifold that minimizes the reconstruction error while preserving the relationship among the variables from the high-dimensional space to the low-dimensional embedding [4]. Two representative methods for the neighbor graphs approach are t-distributed stochastic neighbor embedding (t-SNE) [5] and uniform manifold approximation and projection (UMAP) [4].

3.3.2.1 Distributed Stochastic Neighbor Embedding (t-SNE)

The t-SNE is a nonlinear DR technique developed from stochastic neighbor embedding (SNE) [4]. Rather than using a linear operation to project the high-dimensional data into a low-dimensional space, t-SNE draws the low-dimensional projection by multiple iterations to minimize the probability differences (Kullback-Leibler divergence) between the two spaces.

It uses a student t-distribution with a heavy tailed probability distribution to solve the crowding problem found in the original SNE method [5].

Consider two pints, i and j, denote the probability distribution in the original space as:

$$p_{ij} = \frac{\exp\left(\frac{\|x_i - x_j\|^2}{2\sigma_i^2}\right)}{\sum_{k \neq l} \exp\left(-\frac{\|x_k - x_l\|^2}{2\sigma_i^2}\right)} \tag{3.5}$$

and the distribution in the latent space as q_{ij}

$$q_{ij} = \frac{\left(1 + \|y_i - y_j\|^2\right)^{-1}}{\sum_{k \neq l}\left(1 + \|y_k - y_l\|^2\right)^{-1}} \tag{3.6}$$

The cost function that minimizes the Kullback-Leibler divergences between high-dimensional space and the latent space is given as:

$$C = KL(P \mid Q) = \sum_i \sum_j p_{ij} \log\left(\frac{p_{ij}}{q_{ij}}\right) \tag{3.7}$$

Using the gradient descent method to optimize the cost function, the distribution in the original space can be expressed on the low-dimensional map. The gradient is given by:

$$\frac{\delta C}{\delta y_i} = 4\sum_j \left(p_{ij} - q_{ij}\right)\left(y_i - y_j\right)\left(1 + \left\|y_i - y_j\right\|\right)^{-1} \tag{3.8}$$

An important user-defined parameter in t-SNE is *perplexity* that adjusts attention between local and global aspects of your data. Looking at the numerator in Equation 3.5, the variance depends on the Gaussian distribution and the number of points surrounding the center of it. This is where perplexity is introduced. Perplexity is the target number of neighbors for our central point. Basically, the higher the perplexity is, the higher the value of variance. In other words, it is a guess about the number of close neighbors each point has. The perplexity parameter has a complex effect on the resulting representations. The performance of t-SNE is robust to changes in the perplexity, and typical values are between 5 and 50. In most cases, the default value of the perplexity parameter in t-SNE is 30. Generally, when reducing the perplexity parameter, local variations dominate and for larger values the clusters tend to merge.

3.3.2.2 Uniform Manifold Approximation and Projection (UMAP)

UMAP [4] is also a nonlinear DR method, which was inspired by t-SNE and Laplacian Eigenmaps [11]. UMAP aims to find a transformation function for mapping high-dimensional data onto a low-dimensional plane. This is a clear distinction between t-SNE and UMAP as t-SNE directly optimizes an embedding. The steps for finding the transformation function are as follows: First, a weighted graph is computed for the high-dimensional data W^H that models the local relationship of neighboring points. Second, a weighted graph is computed for a low-dimensional representation W^L similarly to the high-dimensional neighborhood. The low-dimensional representation is then determined by minimizing the cross-entropy, given by:

$$Loss^{UMAP} = \sum_{i \neq j} W_{ij}^H \log\left(\frac{W_{ij}^H}{W_{ij}^L}\right) + \left(1 - W_{ij}^H\right)\log\left(\frac{1 - W_{ij}^H}{1 - W_{ij}^L}\right) \tag{3.9}$$

The first term provides an attractive force between the points spans whenever there is a large weight associated to the high-dimensional case. This is because this term will be minimized when W_{ij} is as small as possible, which will occur when the distance between the points is as small as possible. In contrast the second term provides a repulsive force between points because the term will be minimized by making W_{ij} as large as possible. On balance, this process of pull and push, mediated by the weights on edges of the topological representation of the high-dimensional data, will let the low-dimensional representation settle into a state that relatively accurately represents the overall topology of the source data.

UMAP has several hyperparameters that can have a significant impact on the resulting embedding and each of these parameters has a distinct effect:

n_neighbors: This parameter controls how UMAP balances local versus global structure in the data. This means that low values of n_neighbors will force UMAP to concentrate on very local structure (potentially to the detriment of the big picture), while large values will push UMAP to look at larger neighborhoods of each point when estimating the manifold structure of the data, losing fine detail structure for the sake of getting a broader view of the data. The default value of n_neighbors for UMAP is 15.

min_dist: The min_dist parameter controls how tightly UMAP is allowed to pack points together. It, quite literally, provides the minimum distance apart that points are allowed to be in the low-dimensional representation. This means that low values of min_dist will result in clumpier embeddings. This can be useful if you are interested in clustering, or in finer topological structure. Larger values of min_dist will prevent UMAP from packing points together and will focus on the preservation of the broad topological structure instead. The default value for min_dist is 0.1.

n_components: As is standard for many dimension reduction algorithms, UMAP provides an n_components parameter option that allows the user to determine the dimensionality of the reduced dimension space we will be embedding the data into. Unlike some other visualization algorithms such as t-SNE, UMAP scales well in the embedding dimension, so you can use it for more than just visualization in two or three dimensions.

Metric: The final UMAP parameter is the metric parameter. This controls how distance is computed in the ambient space of the input data. By default, UMAP supports a wide variety of metrics. Any of which can be specified by setting the metric name.

UMAP has several significant advantages over t-SNE. Firstly, it is faster than t-SNE, making it a more efficient choice for data analysis. Additionally, UMAP captures global structure better than t-SNE, providing a more comprehensive understanding of the data. Moreover, UMAP is not limited to visualization and can be used as a general-purpose DR technique for ML preprocessing, while t-SNE has limited use beyond visualization. Lastly, UMAP has a solid theoretical foundation as a manifold approximation technique, in contrast to t-SNE, which is primarily a heuristic for visualization.

3.3.3 Self-Organizing Maps (SOM)

A self-organizing map (SOM), also known as a Kohonen network [12, 13], is a neural network used to visualize complicated, high-dimensional data. SOMs may be described formally as a nonlinear, ordered, smooth mapping of high-dimensional input data manifolds onto the elements of a regular, low-dimensional array. It simultaneously performs vector quantization and topographical preservation while representing complex multivariate data on a two-dimensional grid. The visualization abilities of SOMs are extremely helpful for the analysis of unlabeled plant data and exploring the underlying patterns in it [14, 15, 16].

A SOM is a graph composed of an array of nodes connected in a rectangular or hexagonal grid. Each node, called a prototype vector, models the observation vectors in the data space. These prototypes have a representation in the input (observation) and output (latent) space. The closest prototype vector to a data point in the observed space, known as the Best Matching Unit (BMU), can represent that data point on the map. The BMU of *i*th data point, v_i, is found by finding the closest prototype vector, m_k, according to:

$$BMU_i = \arg\min(\partial(m_k, v_i)), \vee k \tag{3.10}$$

To implement the SOM procedure, first the map shape is selected. Then, the prototype vectors positions are embedded in the data space along a PCA hyperplane. Next, the map is trained to capture curves in the manifold. The positions of the vectors are updated using each data point from a training set according to the formula:

$$m_k(t+1) = m_k(t) + \alpha(t)h_{k,BMU_i}(v_i(t) - m_k(t)) \tag{3.11}$$

where t is the discrete time coordinate of the mapping steps and α is the monotonically decreasing learning rate. The scalar h_{k,BMU_i} denotes a neighborhood kernel function centered at the BMU. Data vectors are matched to the prototype vectors with the smallest Euclidean distance from the data vector of interest. A neighborhood kernel centered at $m_k(t)$ is usually chosen in the Gaussian form:

$$h_{k,BMU_i} = \exp\left(\frac{\left\| m_k - BMU_i \right\|^2}{2\sigma^2}\right) \tag{3.12}$$

$\sigma(t)$ denotes the monotonically decreasing width of the kernel that allows for a regular smoothing of the prototypes. The algorithm continues to make passes over the data updating the locations of the prototype vectors and terminates after some predefined number of time steps have passed or prototype updating becomes negligible [17].

A key advantage of using SOM in process monitoring comes from the insights it provides into high-dimensional structures within the data on the two-dimensional (2D) map. The visualizations discussed below allow us to see the shape of the data distribution, cluster borders, projection directions, and possible dependencies between the variables. Common visualizations discussed here include component planes, distance matrices, and projections as proposed by Kaski [18], Vesanto [19, 20], and Himberg [21].

- *Component planes*: A component plane displays the value of individual variables at each prototype vector in the map. Each component plane is associated to one variable in the observation space. The value of the variable at all map nodes is visualized using different colors that allow the user to visually identify possible dependencies between variables [22]. The dependencies between variables can be seen as similar patterns in identical locations on the component planes. In that sense, the SOM reduces the effect of noise and outliers in the observations and, therefore, may actually make any existing dependency clearer than in the original data.
- *Distance matrices*: Distance matrices show the pairwise distances between nodes within the map structure. Using the lower-dimensional structure of the SOM, colors are used to show high or low distances, which provide insight into the dominant clustering patterns present in the data. The most common visualization of the distance matrix is the U-matrix [23], which visualizes all pairwise distances between nodes in the map. In a U-matrix, the dominant clustering structure can be seen as areas with small distances separated by large distances. As in the basic U-matrix, visualization of the clusters can be greatly improved by augmenting the distance matrix with an additional entry between each prototype vector and each of its neighbors.
- *Projections*: Finally, projections onto the map are very important visualization tools. Highlighting the BMU of data points represents them on the map. Two important figures derived from projections are hit diagrams and trajectories. Hit diagrams project a collection of data points to show a region belonging to a group of data. Trajectories take a series of time and project them on to the map, connecting them with lines. The trajectory makes it possible to visually indicate the current state of the process and observe how that state has been reached.

Example 3.1

Our analysis is based on data sourced from the Tennessee Eastman Process (TEP), as described in Appendix B. Specifically, we used data from TEP-1, where the A/C feed ratio is varied, and the B composition remains constant (step change). This is a simple application that serves to illustrate some of the ideas discussed in this chapter. The results will become even more evident in the next chapter when DR is used in conjunction with clustering techniques.

In Figure 3.2, we present the temporal variation of the data, focusing on selected variables for clarity. The plot clearly indicates a perturbation at a certain point in the process, indicating a faulty condition. Figure 3.2a illustrates the time evolution of the A and C feeds, while Figure 3.2b shows how the control system adjusts the feed of A to the reactor to compensate for the disturbance. This adjustment affects the stripper level, which is maintained through the control system (Figure 3.2c). Finally, Figure 3.2d represents the increase in the stripper temperature due to a fixed amount of steam (with less incoming steam 4).

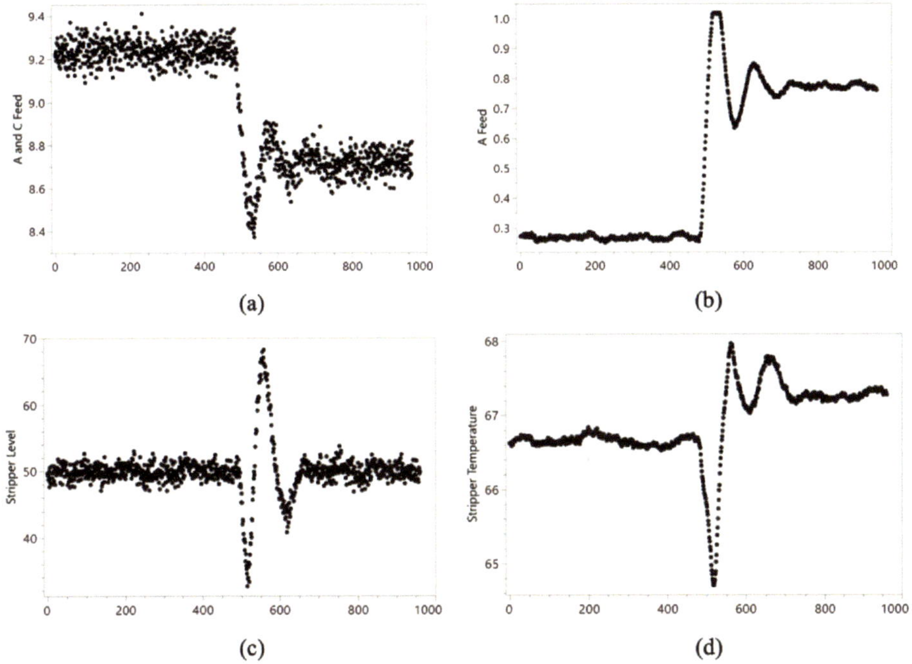

FIGURE 3.2 Process trend for the TEP plant: (a) A and C feed; (b) A feed; (c) stripper level and (d) stripper temperature.

Figures 3.3 and 3.4 illustrate dataset projections employing various DR techniques in both two and three dimensions, offering distinct perspectives of the dataset. From the time evolution plots, we expect to see at least two clusters in the DR projection. Notably, both t-SNE and UMAP capture this in a 2D plot, showcased in Figure 3.3. In contrast, the 2D representation using PCA shows only

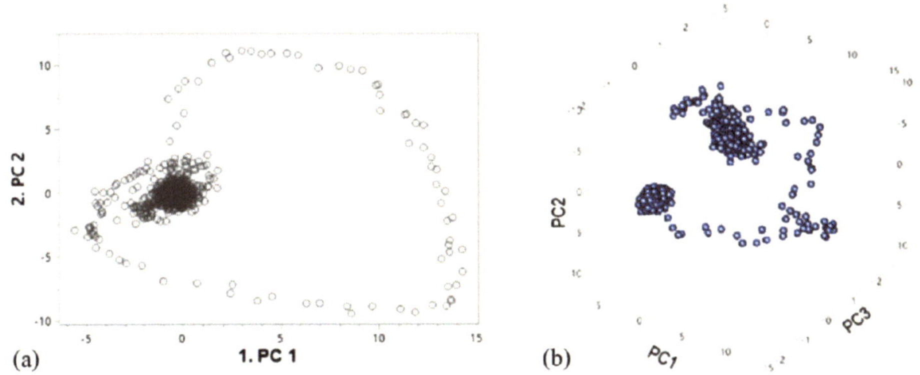

FIGURE 3.3 PCA projections of the data for the TEP process under fault 1: (a) 2D plot and (b) 3D plot.

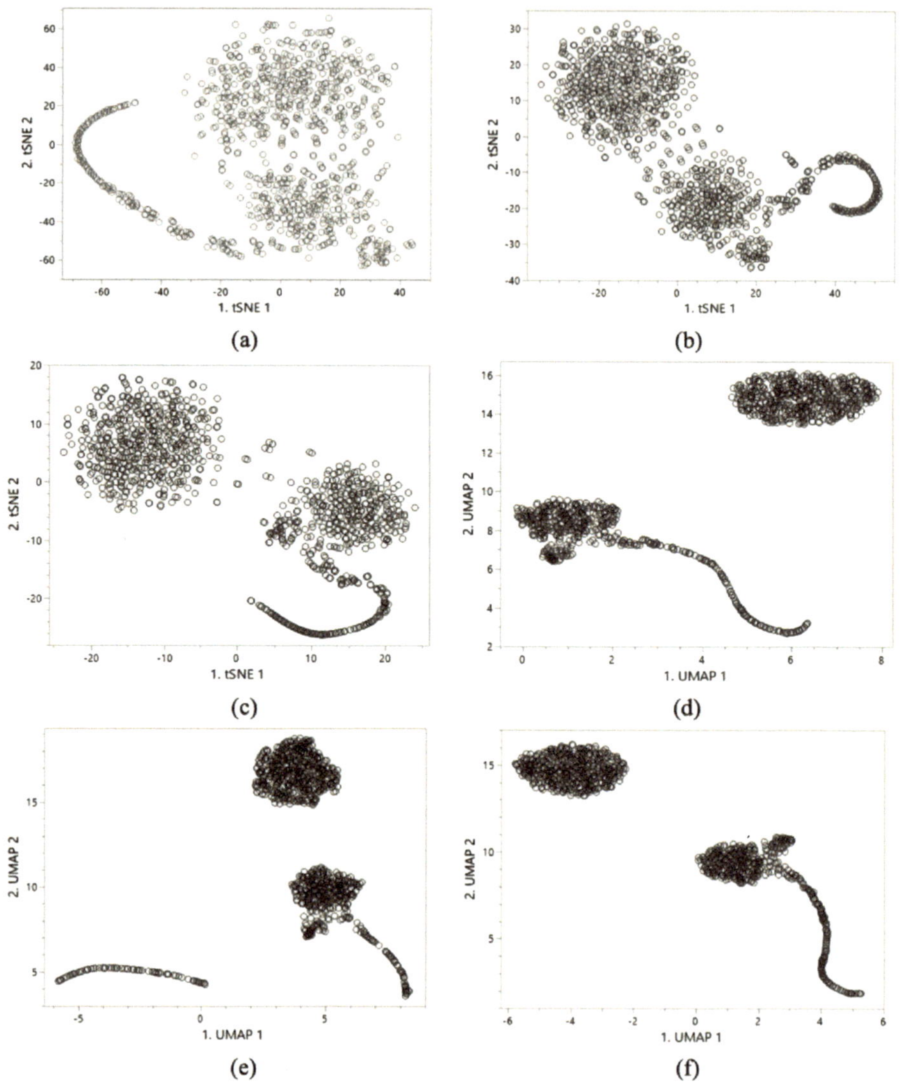

FIGURE 3.4 Projections of the data for the TEP process under fault 1 using alternative techniques. (a) t-SNE (perplexity 10); (b) t-SNE (perplexity 30); (c) t-SNE (perplexity 50); (d) UMAP (n_neighbor 15); (e) UMAP (n_neighbor 10); and (f) UMAP (n_neighbor 30).

one distinct cluster, leading to believe that after the fault the system returns to the original steady state. However, from the 3D representation, one can discern a clear distinction between operating regimes (pre- and post-faulty conditions). As it will be shown later, this behavior from PCA affects the data analysis when paired with clustering techniques, hampering proper classification.

In Figure 3.4, we present the results of the t-SNE DR technique, with a focus on the perplexity parameter. As the perplexity parameter increases, it leads to improved separation of regions within the data. Lower perplexity values prioritize capturing the global structure of the dataset, while higher values emphasize local behaviors. The two primary regions in the graph correspond to the data related to pre- and post-fault conditions. Additionally, the elongated region in the graph represents the transient process behavior resulting from the controller's actions.

UMAP, on the other hand, delivers comparable outcomes but excels in providing a distinctive differentiation between various process conditions. Also, in our experience working with industrial data, we have observed that UMAP exhibits less sensitivity to its parameters, as illustrated in the figure depicting the n_neighbors parameter.

Figure 3.5 illustrates the projections of the same dataset using an SOM. In Figure 3.5a, we present the SOM map, where various regions are distinguished

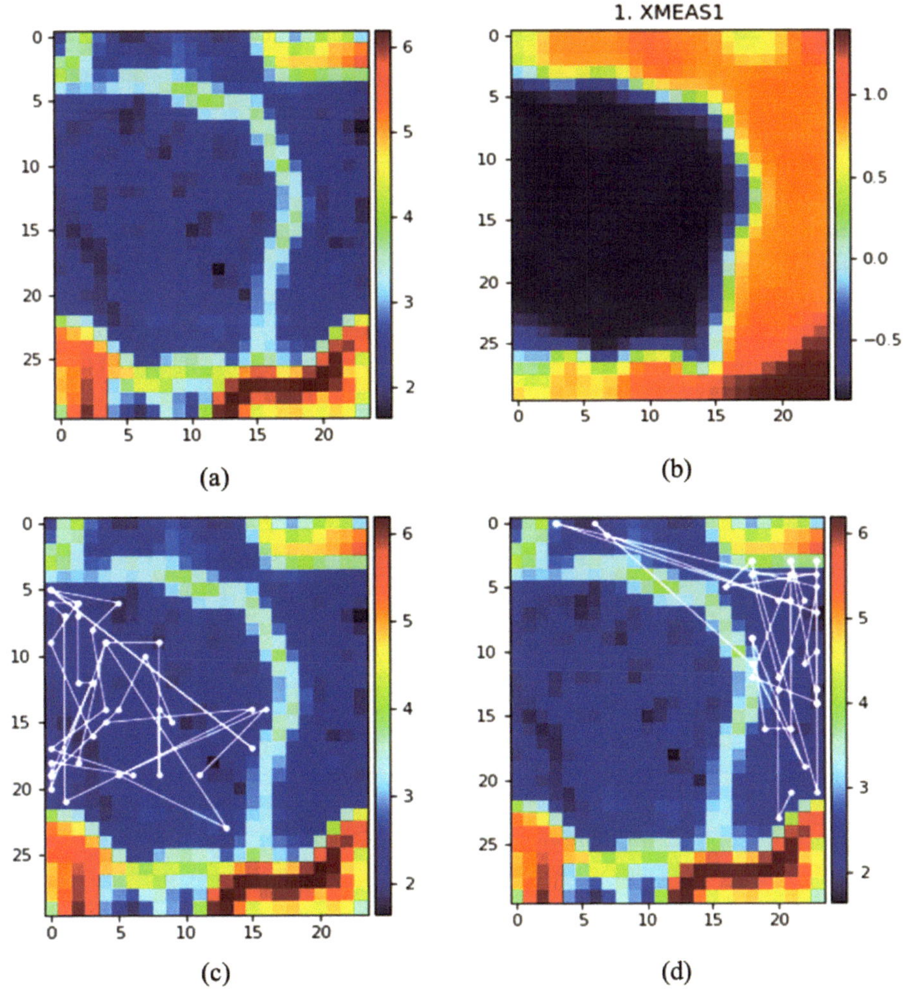

FIGURE 3.5 SOM results for the TEP plant. (a) SOM map; (b) component plane for a single variable and (c) and (d) process trajectories projected into the SOM map.

by different colors. Figure 3.5b showcases the component map, highlighting the strong correlation between the SOM map's shape and the relevant component, specifically the variable XMEAS1. This correlation, as discussed in the subsequent chapter, is pivotal for identifying the variables that contribute to operational changes. Moreover, Figure 3.5c and d demonstrates the trajectories of the data projected onto the map as the process evolves over time. Figure 3.4c represents the initial state of the process, while Figure 3.5d signifies the deviation from the initial condition due to the perturbation introduced along the way (A/C feed ratio step change).

3.4 GLOBAL AND LOCAL STRUCTURE PRESERVATION

The objective of DR is to decrease the number of features in a dataset to obtain a compressed representation that is useful for analysis, while maintaining as much of the original data structure as possible. Essentially, this means that data points that are far apart in the original high-dimensional space should also be far apart in the low-dimensional space (global structure), and likewise for data points that are close to each other (local structure). DR algorithms can be classified as primarily preserving either global or local data structure.

The loss of either crucial global or local information can have a significant impact on dimension reduction performance, which in turn affects process monitoring efficiency. The global structure defines the overall variance and outer shape of the process dataset, while the local structure describes the inner organization and detailed relationships between different data samples. Both aspects are essential in maintaining and describing industrial datasets, and the modeling performance of the monitoring method may be adversely affected if a process dataset loses significant global or local information. As a compromise, accurate DR should consider both local and global data information.

In PCAs, the first few components vector most of the global variance in the data [24]. Although the extracted PCs of the PCA method can retain most of the variation in the data, they can only capture the global structure of the process data. In other words, the detailed local structure information among the process data is ignored.

As mentioned, a newer class of DR techniques known as manifold learning has been developed in the pattern recognition area, including t-SNE [5] and UMAP [4]. One major drawback of several of these new methods is that they either were designed to preserve local structure (such as t-SNE) or struggle with balancing the preservation of both global and local structures (such as UMAP). Additionally, these methods can be sensitive to initialization, and more importantly, their hyperparameter's effects can be difficult to understand and tune.

The methods described above can be considered graph-based algorithms where nodes are observations, edges indicate similarity metrics between observations, and a loss function is defined and adjusted in order to find a low-dimensional embedding that is structurally similar to the high-dimensional graph. Recently, Wang et al. [25] identified a number of important principles needed for an "ideal" loss function by analyzing various cases and attempting to balance both global and local structure preservation. Following their findings, they proposed a new

DR algorithm called Pairwise Controlled Manifold Approximation and Projection (PaCMAP).

PaCMAP is a DR technique that uses a stochastic gradient descent algorithm, boasts effective and reliable performance with preserving local and global structure alike, and performs significantly faster than other contemporaneous DR methods. Of interest to note is that none of the previously described methods (PCA, t-SNE, and UMAP) are able to smoothly transition from local structure preservation to global structure preservation or vice versa through manipulation of parameters. PaCMAP, on the other hand, can and does so automatically by dynamically adjusting its own parameters over a series of iterations through three distinct phases [25]. Given the above-mentioned properties and its potential ability to describe chemical process data at both the global and local level, it is worth discussing PaCMAP in more detail.

3.4.1 The PaCMAP Algorithm

PaCMAP is able to smoothly transition from local structure preservation to global structure preservation by dynamically adjusting its own parameters over a series of iterations through three distinct phases [25]. PaCMAP optimizes the low-dimensional embedding by using three kinds of pairs of data points, neighbor pairs, mid-near pairs, and further pairs, in order to refine the global and local structure development. The input high-dimensional data is referred to as $\boldsymbol{X}$ and the projected low-dimensional output data is referred to as $\boldsymbol{Y}$.

The first step of the algorithm is to compile the sets of pairs. The neighbors, *NB*, are determined by pairing an observation a with its nearest n_{NB} neighbors according to the scaled distance between a pair of observations a and b as shown in Equation 3.13:

$$d_{ab}^2 = \frac{\left\|x_a - x_b\right\|^2}{\sigma_{ab}} \tag{3.13}$$

And $\sigma_{ab} = \sigma_a \sigma_b$, where σ_a is the average distance between observation a and its Euclidean nearest neighbors. The scaled distance is used during the initial pair selections to account for the fact that neighborhoods of points may have drastically different magnitudes. The default value for n_{NB} neighbors is 10. These near pairs will be strongly attracted to each other. The mid-near pairs, *MN*, are determined for observation a by sampling six observations and pairing a with the second closest of the six observations. The number of mid-near pairs that are assigned are defined by $n_{MN} = n_{NB} \cdot MN_{ratio}$, where the default value for MN_{ratio} is 0.5. These mid-near pairs will be weakly attracted to each other. The further pairs, FP, are determined for observation a by randomly sampling non-neighbor observations. The number of further pairs that are assigned are defined by $n_{FP} = n_{NB} \cdot FP_{ratio}$, where the default value for FP_{ratio} is 2. These further pairs will repulse each other.

The PaCMAP algorithm then applies the initialization procedure selected to set the initial values of the low-dimensional embedding (initialization can be done randomly or with PCA). Once initialized, the values are optimized. The algorithm uses the Adam optimizer [26] to reduce the loss function developed. The overall loss function is defined in Equation 3.14:

$$
\begin{aligned}
&Loss^{PaCMAP} = \\
&w_{NB} \cdot \sum_{i,j} \frac{\tilde{d}_{ij}}{10 + \tilde{d}_{ij}} && (\textit{neighbors; strong attraction}) \\
&+ w_{MN} \cdot \sum_{i,k} \frac{\tilde{d}_{ik}}{10,000 + \tilde{d}_{ik}} && (\textit{mid-near pairs; weak attraction}) \qquad (3.14) \\
&+ w_{FP} \cdot \sum_{i,l} \frac{1}{1 + \tilde{d}_{il}} && (\textit{further pairs; repulsion})
\end{aligned}
$$

where i and j are neighbors, i and k are mid-near pairs, i and l are further pairs, w refers to the weights of all the pairwise interactions that change over the course of iterations, and distances are defined by Equation 3.15:

$$\tilde{d}_{ab} = d_{ab}^2 + 1 = \|y_a - y_b\|^2 + 1 \tag{3.15}$$

The aim is to construct proper global and local structures by strongly attracting neighbor pairs, weakly attracting mid-near pairs, and repulsing further pairs. The PaCMAP algorithm's optimization is divided into three stages. The initial stage focuses on improving the global structure of the point embeddings. To achieve this, mid-near pairs are heavily weighted and the weighting is gradually decreased as the stage nears completion. In the second stage, the optimization shifts to improving the local structure while still maintaining the global structure from the first stage. The emphasis is on neighbor pairs, with the weighting increased. The third and final stage is dedicated to refining the local structure without attempting to adjust the global structure. The emphasis is on weighting the further pairs and neighbor pairs, with a zero weight assigned to the mid-near pairs. The first stage is completed by default in 100 iterations using the Adam optimizer, and the second stage after 200 iterations. The default total number of iterations is 450, allowing for the third phase to fully develop and refine the local structure, which has a comparatively large number of iterations.

Example 3.2

Let us revisit the dataset corresponding to the TEP process mentioned in Example 3.1. Figure 3.6 depicts the dataset's projections using PaCMAP, which results in a 2D plot. It is worth noting that the projection using PaCMAP starts by emphasizing the global structure of the point embeddings. As shown in Figure 3.7, the evolution of the loss function is presented as a function of the iteration numbers. The initial stage of the algorithm focuses on improving the global structure of the point embeddings, which is achieved by giving heavier

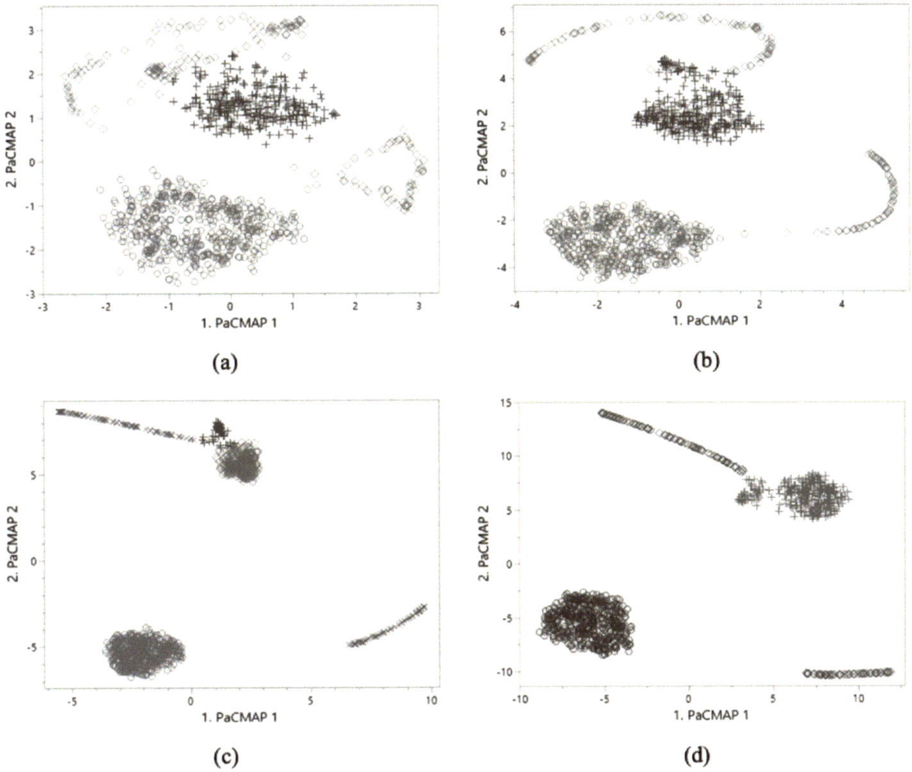

FIGURE 3.6 Projections of the data for the TEP process under fault 1 using PaCMAP under a number of iterations: (a) iteration 50; (b) iteration 100; (c) iteration 200; and (d) iteration 300.

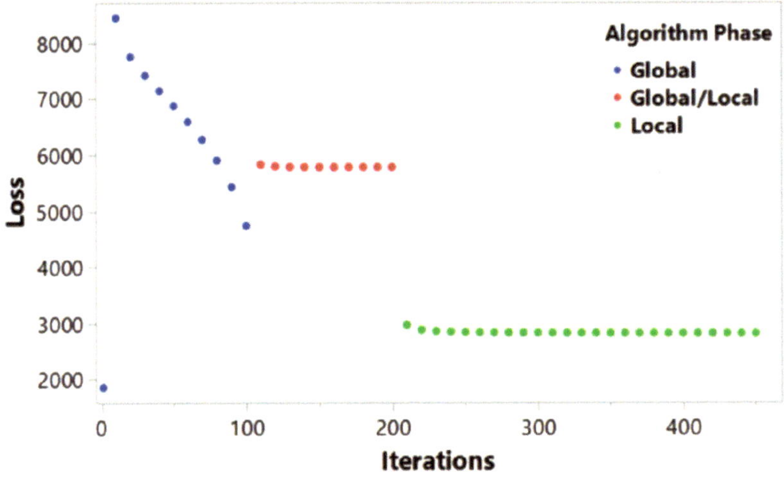

FIGURE 3.7 Loss function for PaCMAP.

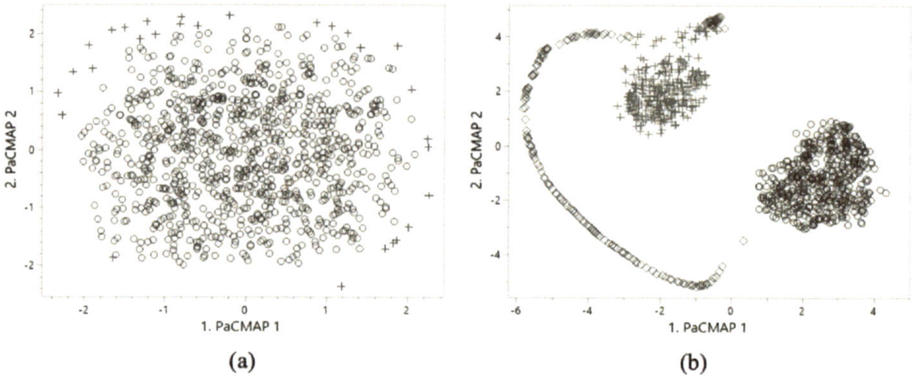

FIGURE 3.8 PaCMAP with random initialization: (a) iteration 5 and (b) iterations 100.

weights to mid-near pairs. The weightage is gradually decreased as the stage nears completion. As the iteration number increases, the optimization shifts toward improving the local structure, while still maintaining the global structure from the first stage. In this specific application, the algorithm had already captured the data structure with a good balance between local and global process behaviors at around 100 iterations. In our experience, while dealing with industrial data, most of the data structure is captured within the first stage of the algorithm. PaCMAP offers great flexibility as we can control the resolution level of the method by selecting the number of iterations. Additionally, PaCMAP has two initialization options: PCA or Random. Figure 3.8 demonstrates the impact of using random initialization for PaCMAP, while the previous results were obtained with PCA initialization. As previously mentioned, the algorithm is not sensitive to initialization, as both PCA and Random initialization produce similar results after approximately 100 iterations.

3.5 CONCLUSIONS

DR algorithms have successfully identified changes and different operational states in simulated data. However, neighbor graphs DR methods, such as t-SNE, UMAP, and especially PaCMAP, have demonstrated a superior ability to identify global and local changes in the process compared to matrix factorization methods like PCA.

In comparing t-SNE and UMAP, our analysis reveals that both techniques exhibit similar visualization capabilities. Nevertheless, UMAP has a clear advantage over t-SNE in terms of runtime. While both UMAP and t-SNE yield somewhat comparable results, the increased speed and improved preservation of global structure make UMAP a more effective tool for visualizing high-dimensional chemical process data. Furthermore, once a model has been created using historical information, incoming data points can be seamlessly passed through the same UMAP model for online process monitoring and fault detection systems.

We conducted an in-depth analysis of PaCMAP, which revealed that for chemical process data, adjusting the number of iterations allows for a smooth transition from a global data analysis to a local one. PaCMAP was evaluated against popular methods like PCA and t-SNE to illustrate its robustness and distinctive suitability for chemical process data analysis. While comparing these algorithms can be challenging due to their differing operational mechanisms, the results indicate that PaCMAP provides users with a uniquely simple and effective approach to apply deep and meaningful analyses to process data.

Additionally, we introduced SOMs among the DR tools. A key advantage of using SOM in the analysis of process data lies in the insights it offers into high-dimensional structures within the data, presented on a 2D map. The visualization capabilities of SOMs prove extremely valuable for analyzing unlabeled plant data and uncovering underlying patterns. Common visualizations, including component planes, distance matrices, and projections, become particularly important when employing data clustering in the subsequent chapter.

It is important to note that no DR technique is perfect. However, by developing a solid understanding of how these algorithms function and how to manage their parameters, one can use these powerful tools more effectively and efficiently to visualize and extract knowledge from large, high-dimensional datasets in the context of process data.

REFERENCES

1. Mervis J. (2012), U.S. Science policy: Agencies rally to tackle big data. Science, 335(6077), p. 22.
2. Gardner M., Bieker J. (2000), Data mining solves tough semiconductor manufacturing problems. KDD '00 Proceedings of the Sixth ACM SIGKDD International Conference on Knowledge Discovery and Data Mining, pp. 376–383.
3. Russell E. L., Chiang L. H., Braatz R. D. (2000), Data-Driven Techniques for Fault Detection and Diagnosis in Chemical Processes. Springer.
4. McInnes L., Healy J., Melville J. (2018), UMAP-Uniform Manifold Approximation and Projection for Dimension Reduction, arXiv, p. 1802.03426.
5. Van der Maaten L., Hinton G. (2008), Visualizing data using t-SNE. Journal of Machine Learning Research, 9(Nov), pp. 2579–2605.
6. Hotelling H. (1933), Analysis of a complex of statistical variables into principal components. Journal of Educational Psychology, 24(6), pp. 417–441.
7. Lee J., Qin S. J., Lee I. (2006), Fault detection and diagnosis based on modified independent component analysis. AIChE Journal, 52(10), pp. 3501–3514.
8. Hyvarinen A., Oja E. (2000), Independent component analysis: Algorithms and applications. Neural Networks, 13(4–5), pp. 411–430.
9. Lee J., Yoo C. K., Choi S. W., Vanrolleghem P. A., Lee I. (2004), Nonlinear process monitoring using kernel principal component analysis. Chemical Engineering Science, 59, pp. 223–234.
10. Scholkopf B., Smola A., Muller K. R. (1999), Kernel principal component analysis. Advances in Kernel Methods – Support Vector Learning. pp. 327–352, MIT Press.
11. Belkin M., Niyogi P. (2003), Laplacian eigenmaps for dimensionality reduction and data representation. Neural Computation, 15(6), pp. 1373–1396.
12. Kohonen T., Oja E., Simula O., Visa A., Kangas J. (1996), Engineering applications of the self-organizing map. Proceedings of the IEEE, 84(10), pp. 1358–1384.
13. Kohonen T. (2000), Self-Organizing Maps. Springer-Verlag.
14. Kohonen T. (2012), Self-Organizing Maps. Springer Science & Business Media.

15. Robertson G., Thomas M. C., Romagnoli J. A. (2015), Topological preservation techniques for nonlinear process monitoring. Computers & Chemical Engineering, 76, pp. 1–16.
16. Corona F., Mulas M., Baratti R., Romagnoli J. A. (2012), Data-derived analysis and inference for an industrial deethanizer. Industrial & Engineering Chemistry Research, 51(42), pp. 13732–13742.
17. Kohonen T., Somervuo P. (2002), How to make large self-organizing maps for nonvectorial data. Neural Networks, 15(47), pp. 945–952.
18. Kaski S. (1997), Data exploration using self-organizing maps. (Doctoral dissertation). http://users.ics.aalto.fi/sami/thesis/
19. Vesanto J., Ahola J. (1999), Hunting for correlations in data using the self-organizing map. Proceedings of the International ICSC Congress on Computational Intelligence Methods and Applications. Academic Press, pp. 279–285.
20. Vesanto J. (2002), Data exploration process based on the self-organizing map. (Doctoral dissertation). http://lib.tkk.fi/Diss/2002/isbn9512258978/
21. Himberg J. (2001), From insights to innovations: Data mining, visualization, and user interfaces. (Doctoral dissertation). http://lib.tkk.fi/Diss/2004/isbn9512273373/
22. Lampinen J., Kostiainen T. (2000), Self-organizing maps in data analysis – Notes on overfitting and overinterpretation. Proceedings of the European Symposium on Artificial Neural Networks, Bruges, pp. 239–244.
23. Ultsch A. (1993), Self-organizing neural networks for visualization and classification. In Opitz, O., Lausen, B., Klar, R. (Eds.), Information and Classification, pp. 307–313. Springer.
24. Yu J. (2012), Local and global principal component analysis for process monitoring. Journal of Process Control, 22(7), pp. 1358–1373.
25. Wang Y., Huang H., Rudin C., Shaposhnik Y. (2021), Understanding how dimension reduction tools work: An empirical approach to deciphering t-SNE, UMAP, TriMap, and PaCMAP for data visualization. Journal of Machine Learning Research, 22(201), pp. 1–73.
26. Kingma D., Ba J. (2015), Adam: A Method for Stochastic Optimization, arXiv:1412.6980 [cs.LG].

4 Clustering

4.1 INTRODUCTION

Chapter 3 introduced the notion of visualizing data and identifying patterns in it. Coupled with domain knowledge, that is, engineering expertise, these patterns can be associated to physical phenomenon taking place in the plant. These patterns, however, are not always as obvious as the examples presented up to this point, and often we need an algorithmic approach to capture them efficiently. In this endeavor, a set of unsupervised learning methods known as clustering becomes useful.

Clustering algorithms divide or partition data into natural groups. In our context, the notion of "natural" implies that the objects in a cluster should be internally similar to each other, but differ significantly from the objects in the other clusters [1]. The goal of data clustering is the unsupervised classification of data into groups or clusters that are useful and meaningful. Using data clustering, the key groups within a database can be isolated and connected to meaningful operating conditions in a plant.

4.2 CLUSTERING METHODS

One way of classifying clustering methods is as flat or hierarchical, based on the sequence the algorithm follows when forming the clusters. Flat clustering creates all clusters at the same time. Hierarchical clustering algorithms, on the other hand, find clusters one by one [2]. The hierarchical methods can be further divided into agglomerative and divisive algorithms, corresponding to bottom-up and top-down strategies. Agglomerative clustering algorithms merge clusters together one at a time to form a clustering tree that finally consists of a single cluster, the whole dataset. Divisive algorithms are similar but work in the opposite direction starting from a single cluster and dividing each cluster to sub-clusters until all instances are in a cluster of their own [2].

Clustering methods can also be classified as centroid-based or density-based, depending on how each cluster is built. Centroid-based clustering methods assume a certain distribution within the clusters and create new clusters around a point (i.e., a centroid) following this distribution. The number of clusters (centroids) is usually predefined, but it can also be part of an energy function [3]. In density-based clustering, no centroids are defined and no distribution of the data within the clusters is assumed. Density-based clusters are more versatile and powerful but require larger datasets to perform well [4]. In this discussion, the focus will be on the following two groups of cluster methods: flat centroid-based and hierarchical density-based.

DOI: 10.1201/9781003455905-5

4.2.1 Flat Centroid-Based Clustering

Flat centroid-based clustering algorithms find the splitting of the dataset that minimizes the sum of the distance between the members of a cluster and the centroid of said cluster. K-means is a commonly used flat clustering technique. In K-means, for example, K initial centroids are chosen corresponding to the number of clusters desired (the only hyperparameter for this method). Each point in the dataset is assigned to the closest centroid, and the centroid of each cluster is updated with each iteration based on the points assigned to the cluster [5].

4.2.1.1 K-Means Clustering

The fundamental concept of K-means clustering was developed independently by several researchers studying "optimal partitioning". MacQueen [6] coined the term "K-means" and provided the one of the earliest formal introductions of the K-means clustering process. In spite of its age, K-means is still one of the most widely used algorithms for clustering due to its simplicity and efficiency [7] and was recognized as one of the top 10 data mining algorithms by the Institute of Electrical and Electronic Engineers (IEEE).

For a d-dimensional spatial dataset, $X = \{x_i\}$, $i = 1, \ldots, n$, where n is the number of data points, K-means partitions the data into K clusters $C = \{c_k, k = 1, \ldots, K\}$ specified by the user. The K-means algorithm seeks to minimize the sum of squared error (SSE) between each mean μ_k of each cluster c_k given by:

$$SSE(C) = \sum_{K=1}^{K} \sum_{x_i \in c_k}^{n} \left\| x_i - \mu_k \right\|^2 \tag{4.1}$$

Minimizing the SSE objective function is an NP-hard problem; thus, K-means will converge to a local minimum. The main steps of K-means are shown in Algorithm 4.1. The K-means implementation used here randomly chooses k points from the dataset and set these points as the initial centroid of the specified number of clusters the algorithm must find [7].

Algorithm 4.1. K-Means Clustering

1. Given K, the number clusters.
2. $C \leftarrow c_k, \ldots, c_K$, the centroids in random positions.
3. Until stopping criteria is met:
   i. For each point in data, find the closest centroid, c_k.
   ii. Update C by computing the center for each cluster.

The implication of Algorithm 4.1 is that K-means not only seeks for spherical clusters in data [1], but it also clusters with roughly equal number of samples. This entails, of course, a nontrivial assumption about the data, hence K-means should never be blindly used to cluster any data without careful verification of the results. It should also be noted that as the number of clusters increases, the number of samples in clusters decreases, which makes this algorithm more sensitive to outliers [8].

4.2.2 Hierarchical Density-Based Clustering

Given a measure of proximity, density-based clustering looks for clusters of any shape and without need of initial centroids or the assumption of equal variance around a point (Gaussian sphere assumption). Furthermore, density-based clustering does not require every data point to be assigned to a cluster, allowing for data in sparse regions to be identified as noise. These features provide an advantage over partitional clustering, since fewer assumptions are made regarding the distribution of the data, which can become important when dealing with high-dimensional data [4].

Representative examples of density-based clustering algorithms are density-based spatial clustering of applications with noise (DBSCAN) [9] and hierarchical density-based spatial clustering of applications with noise (HDBSCAN). They will be briefly discussed next.

4.2.2.1 Density-Based Spatial Clustering of Applications with Noise (DBSCAN)

Density-based clustering locates regions of high density of any shape that are separated from one another by regions of low density [4]. DBSCAN divides all the data in a set into three categories based on their neighborhood: Core points making up the body of a cluster, border points in more diffuse regions of the data but near a few core points, and noisy points far from concentrated groups of data. DBSCAN is ideally suited to clustering chemical process data, since clusters in real datasets are rarely spherical, and often clouded by noise [5]. Removing this noise can focus attention on the densest clusters in the dataset.

For each core point within a cluster, the neighborhood of a given radius *Eps* must contain a minimum number of points (MinPts parameter), that is, the density of the point's neighborhood must exceed a threshold. The shape of the clusters found can depend on the distance metric selected. Unlike *K*-means, the clusters found by DBSCAN arise from a series of specific definitions. Each definition assumes the specification of a particular value of *Eps* and *MinPts*.

Definition 1 (Eps-neighborhood of a point):

The *Eps* neighborhood of a point *p* defined as:

$$N_{Eps}(p) = \{q \in D \mid dist(p,q) \leq Eps\}$$

The shape of this neighborhood is determined by the distance criterion used. Euclidian distance was used for the purpose of this book, but in practice any appropriate distance metric may be used. Appendix D shows a non-exhaustive list of distance metrics and their properties. A simple approach could require every point in the cluster to have at least *MinPts* in its neighborhood, but this would miss the border points on the outside of the cluster. Border points are within the *Eps* neighborhood of at least one core point, but a border point's neighborhood contains fewer points than *MinPts*.

Definition 2 (Directly density-reachable):

Point p is directly density-reachable (DDR) from point q if:

1. p is in $N_{Eps}(q)$
2. $N_{Eps}(q) \geq MinPts$ (i.e., q is a core point)

In other words, if q is a core point, other data points in its Epsilon neighborhood are said to be directly density reachable. Core points are DDR with each other. Border points are DDR from core points, but core points are not DDR from border points due to condition 2. Arising from the DDR definition is the concept of density reachability for groups of core points:

Definition 3 (Density reachable):

A point p is density reachable from a point q if there is a chain of points $p_1, \ldots, p_n$, where $p_n = q$ such that p_{i+1} is directly density reachable from p_i.

Two points are dimensionality reduction (DR) if there is a chain of core points DDR from each other connecting the two points. A border point can be density reachable from a core point, but because of the definition of DDR, no core points are DR from a border point. What finally defines clusters in DBSCAN is the concept of density connectivity:

Definition 4 (Density connected):

A point p is density connected to a point q if there is a point o such that both p and q are density reachable from o.

In other words, if both p and q are density reachable from a point o, they are said to be density connected. Therefore, clusters in DBSCAN are groups of density connected points. More formally:

Definition 5 (Cluster definition):

For database D, a cluster C is a non-empty set such that:

1. $\forall p,q$ if $p \in C$ and q is density reachable from p, then $q \in C$
2. $\forall p,q \in C$, p is density connected to q

Noise points are simply defined as data that are not associated with any cluster. The details of the DBSCAN algorithm are presented by Ester et al. [9].

To find clusters, the user must specify parameters *Eps* and *MinPts*. There is no simple way to determine the correct value of *Eps*, and Ester et al. [9] suggest a heuristic that uses the sorted *k-dist graph*, created by finding the distance to each point's k-th nearest neighbor, sorting the distances found into descending order, and plotting. *Eps* is determined by the distance value of the first "valley" of the sorted k-dist graph. In essence, the user is estimating the percentage of noise in the data.

4.2.2.2 Hierarchical Density-Based Spatial Clustering of Applications with Noise (HDBSCAN)

HDBSCAN [10] extends the DBSCAN algorithm by converting it into a hierarchical clustering algorithm to be more robust to cases with multiple scales of density. The HDBSCAN starts from the estimation of density based on k- nearest neighbor distance, which denotes as $core_k(x)$ for a data point x. To further distinguish data points in low-density regions, a new metric, mutual reachability distance is then defined as:

$$d_{m\ reach-k}(a,b) = \max\left(core_k(a),\ core_k(b),\ d(a,\ b)\right) \tag{4.2}$$

Using the mutual reachability distance, a spanning tree is used to construct the hierarchy of the dataset, where multiple density regions are identified. The HDBSCAN algorithm simply requires one parameter, *MinPts*, to define the minimum number of points to form a cluster. With a defined *MinPts* value, the hierarchy tree is pruned and merged to satisfy the requirement of minimum cluster size.

Example 4.1

Let's revisit the TEP process dataset that was introduced in Example 3.3. In addition to applying DR and projecting into a lower-dimensional space, clustering techniques will be used to classify the data. Figure 4.1 displays 2D projections of the data using various combinations of DR and DBSCAN clustering techniques. It's worth noting that all the methods – namely t-distributed stochastic neighbor embedding (t-SNE), uniform manifold approximation and projection (UMAP), and Pairwise Controlled Manifold Approximation and Projection (PaCMAP) – show excellent separability among classes (clusters). The only difference between them is that UMAP and PaCMAP subdivide the transient region between clusters (due to the action of the controller) into an additional region, showing a better characterization of the local process behavior. On the other hand, when using principal component analysis (PCA) for DR, the DBSCAN algorithm falls short in proper classification, illustrating the necessity of selecting an appropriate combination of techniques to fully and accurately classify the data.

Moving on to Figure 4.2, we can see the HDBSCAN clustering algorithm in action with two sets of min-sample parameters. While HDBSCAN's performance is similar to DBSCAN, it tends to add extra clusters that are otherwise part of the noise in DBSCAN. As for the min-sample parameter, increasing it generally leads to a decrease in the number of identified clusters.

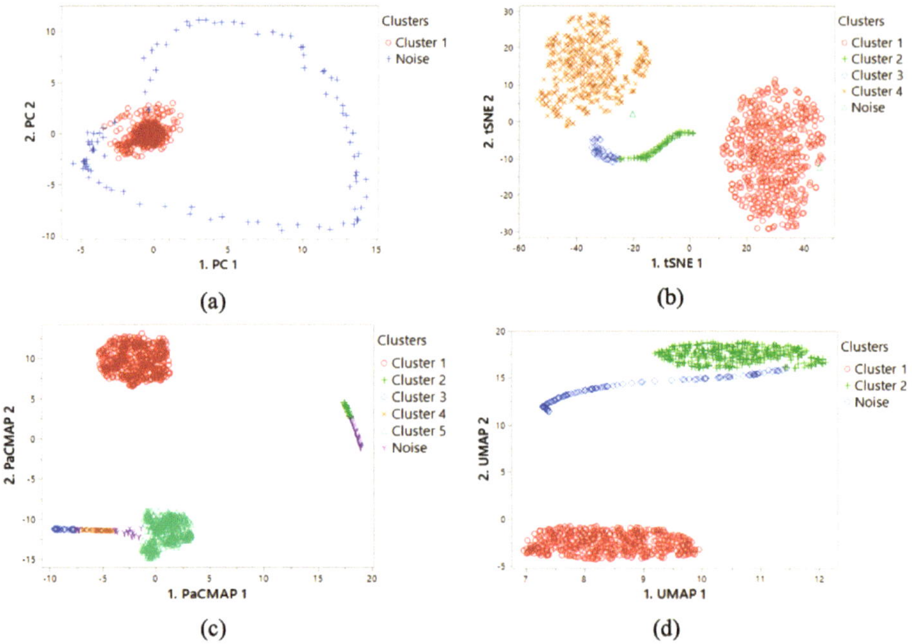

FIGURE 4.1 Clustering results for alternative combinations of DR-DBSCAN (min-sample 20): (a) PCA-DBSCAN, (b) t-SNE DBSCAN; and (c) PaCMAP-DBSCAN and UMAP-DBSCAN.

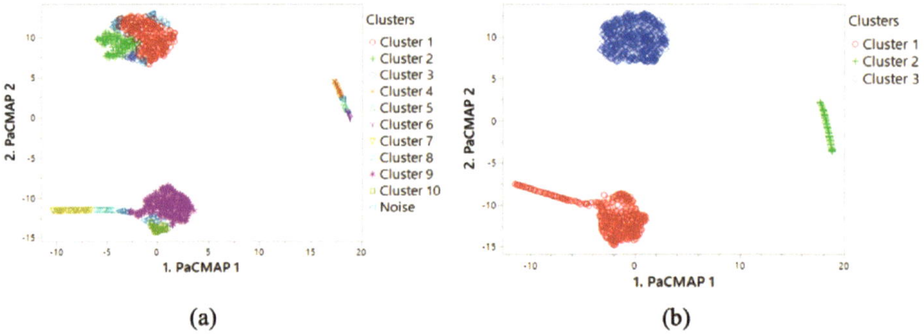

FIGURE 4.2 Clustering results for combination PaCMAP-HDBSCAN with alternative min-sample: (a) min-sample 10 and (b) min-sample 20.

Previously, it was demonstrated that the PCA-DBSCAN combination was ineffective in classifying the data compared to other combinations of DR and clustering techniques. To gain further understanding, we can examine the results of applying PCA DR in conjunction with alternative clustering techniques, as illustrated in Figure 4.3. Despite utilizing an alternative clustering technique, the PCA DR approach still fails to identify the two primary clusters as expected because of the disturbance. Also, we can associate the cluster formation with a specific

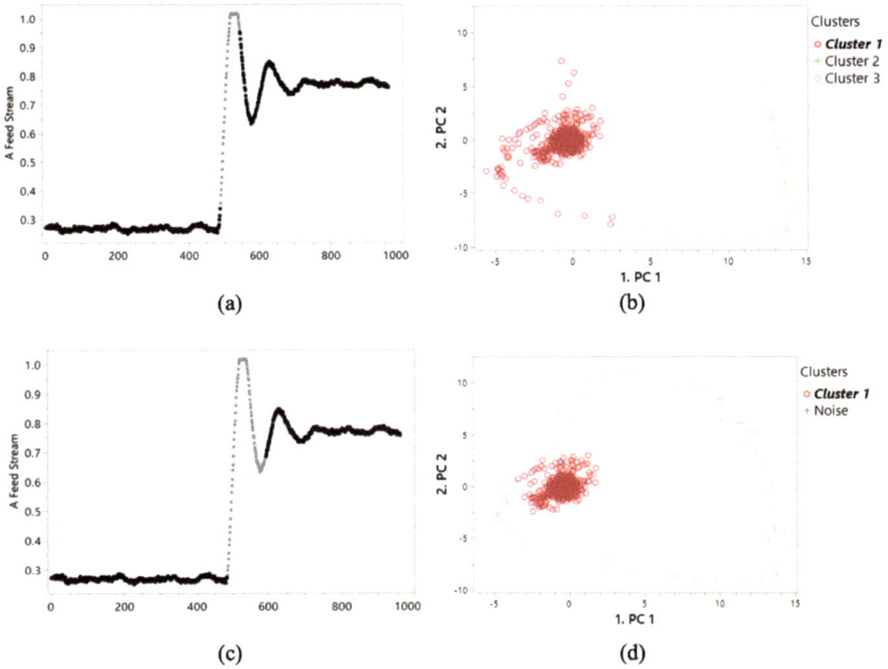

FIGURE 4.3 PCA dimensionality reduction combined with alternative clustering technique: (a) and (b) PCA-K-means (k = 3) and (c) and (d) PCA-DBSCAN.

period by projecting the clusters and highlighting the region in the time domain. This permits the identification of the exact period during which the process upset occurred. Further analysis is possible by zooming in on the specific period to investigate the incident.

Figure 4.4 depicts the projections of clusters onto the time evolution of the A feed stream, which were identified using the PaCMAP-DBSCAN combination. The clustering results reveal two primary regions that correspond to pre- and post-disturbance conditions in the process, as well as a third region that corresponds to the transient portion of the controller actions. This indicates that the PaCMAP approach can characterize not only the global but also the local behavior.

In addition, the SOM provides an excellent mapping of the various regions in the dataset, which enables the verification of clustering results for consistency by projecting the clusters onto the map. In Figures 4.5, the projections of the clusters are shown, indicating excellent agreement with the classification provided by the PaCMAP-DBSCAN combination and SOM.

Each DR and clustering technique had one or more parameters that needed to be specified. The determination of parameters has a substantial effect on clustering performance and the validity of the analysis and finding appropriate values can be a trial-and-error process. In the next section, unsupervised clustering metrics are discussed that can provide support in making this decision.

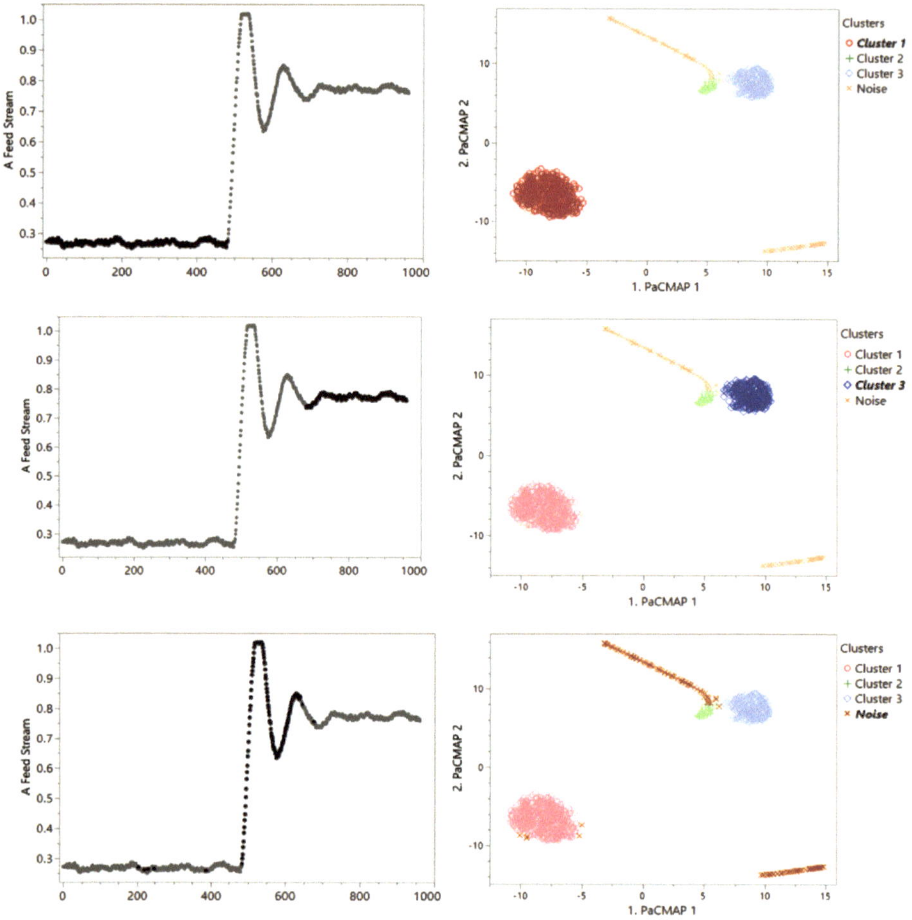

FIGURE 4.4 Clusters 1, 2 and noise cluster projected into the time evolution for PaCMAP-DBSCAN combination.

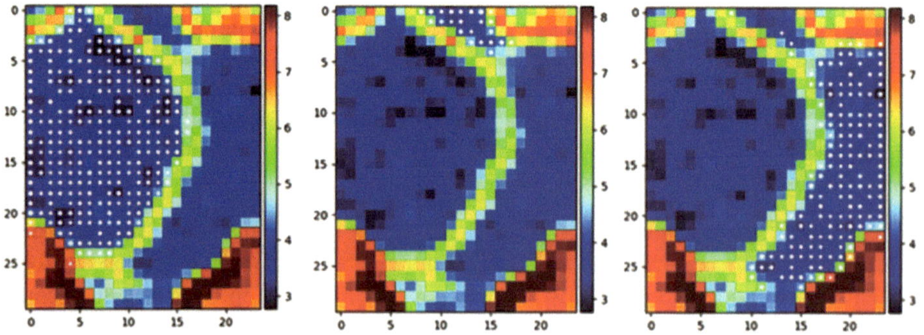

FIGURE 4.5 PaCMAP-DBSCAN cluster 1, 2, and 3 in Figure 4.4 projected into the SOM map.

4.3 UNSUPERVISED CLUSTERING METRICS

Unsupervised clustering evaluation metrics are key in the process monitoring workflow process. Depending on the dataset and method-specific parameters used to develop the clustering, the identified clusters may or may not reveal meaningful groupings of data. Therefore, evaluation metrics of the clustering result are often necessary to assess the quality of results at this stage before proceeding with analysis. Some trusted examples of these evaluation metrics are the Silhouette score [11], the Calinski-Harabasz Index [12], the Davies-Bouldin Index [13], and the Scatter Degree Between and Within Clusters (S_Dbw) validity index [14]. These metrics are briefly discussed next.

4.3.1 Silhouette Score

The Silhouette score [11] evaluates the compactness of individual clusters and the separation among clusters. In other words, it measures how similar an object is to its own cluster (cohesion) compared to other clusters (separation). For a set of data points that are to be clustered into k clusters $[c_1, \dots, c_k]$, two measures are calculated per each point shown in Equations 4.3 and 4.4. The average distance between data point i and all the other points in the same cluster as C_i, called a_i, is shown in Equation 4.3.

$$a_i = \frac{1}{|C_i| - 1} \sum d_{ij} \tag{4.3}$$

Where $|C_i|$ is the number of points in cluster i and d_{ij} is the distance between data points i and j within cluster i. The resulting a_i is a measure of how well data point i fits in the cluster to which it is assigned. Next, the average distance between data point i and all the other points in the nearest neighboring cluster to C_j, called b_i, is shown in Equation 4.4.

$$b_i = \min\left(\frac{1}{|C_j|}\right) \sum d_{ij} \tag{4.4}$$

Where $|C_j|$ is the number of points in cluster j and, again, d_{ij} is the distance between data points i and j within the cluster. The resulting b_i is the smallest mean distance of data point i to all points in any cluster of which point i is not a member. The most commonly used metric to evaluate these distances is the Euclidean distance.

Now, the Silhouette s_i is defined in Equation 4.5:

$$s_i = \begin{cases} 1 - a_i / b_i & if \quad a_i < b_i \\ 0 & if \quad a_i = b_i \\ b_i / a_i - 1 & if \quad a_i > b_i \end{cases} \tag{4.5}$$

Or, written as one formula:

$$s_i = \frac{b_i - a_i}{\max\{a_i, b_i\}} \tag{4.6}$$

Silhouette score values range from –1 to 1, with high scores indicating a data point is well-matched to its assigned cluster and poorly-matched to other clusters. A score of 0 means a point is on the boundary of two clusters. The overall Silhouette score for a clustering solution is calculated as the average Silhouette score over all the data points in the dataset. Therefore, the higher the Silhouette score, the better the clustering solution.

4.3.2 Calinski-Harabasz Index

The Calinski-Harabasz Index [12] is a ratio of the sum of the dispersion between-clusters and of the dispersion within-clusters for all clusters. This index is based on the premise that a good clustering result should have compact and well-separated clusters. Both aspects are measured by comparing the variance within each cluster to the variance between clusters.

First, the global centroid (centroid for the entire dataset) and the centroids for each cluster are calculated. Then, the between-cluster sum of squares is calculated as shown in Equation 4.7:

$$B_k = \sum_{k=1}^{K} n_k |c_k - c|^2 \tag{4.7}$$

Where n_k is the number of observations in cluster k, c_k is the centroid of cluster k, c is the global centroid, and K is the total number of clusters. Next, the within-cluster sum of squares is calculated as shown in Equation 4.8:

$$W_k = \sum_{k=1}^{K} \sum_{i=1}^{n_k} |X_{ik} - c_k|^2 \tag{4.8}$$

Where n_k is the number of observations in cluster k, c_k is the centroid of cluster k, X_{ik} is the i-th observation in cluster k, and K is the total number of clusters. Finally, the Calinski-Harabasz Index is in the form of a ratio of the separation over the cohesion, as shown in Equation 4.9:

$$CH_k = \frac{B_k}{\mathrm{W_k}} \cdot \frac{n - K}{K - 1} \tag{4.9}$$

Where N is the total number of data points, K is the total number of clusters, W_k is the within cluster variation shown in Equation 4.8, and B_k is the between cluster variation shown in Equation 4.7. A higher Calinski-Harabasz score suggests a model with better defined clusters; however, there is not a bounded range for this score.

For this reason, its usefulness comes from comparing the Calinski-Harabasz scores between different clustering iterations on the same dataset.

4.3.3 Davies-Bouldin Index

The Davies-Bouldin Index [13] captures the average similarity between clusters, where the similarity is a measure that compares the distance between clusters with the size of the clusters themselves. This index is also based on the premise that a good clustering result should have compact and well-separated clusters. These aspects are measured by comparing the average distance between data points within a cluster to the average distance between each cluster and its nearest neighbor cluster.

First, the centroids for each cluster are calculated. Then, for each pair of clusters i and j, the Euclidean distance between their centroids, denoted d_{ij}, and the average distance from each point in a cluster to its centroid, denoted S_i (and S_j), are calculated. Next, for each cluster i, the nearest neighbor cluster j is determined by searching for the smallest d_{ij}. The Davies-Bouldin Index is calculated for each cluster i as shown in Equation 4.10.

$$DBI_i = max\left(\frac{S_i - S_j}{d_{ij}}\right) \tag{4.10}$$

Finally, the overall Davies-Bouldin Index is calculated by averaging the DBI_i by the number of clusters k as shown in Equation 4.11.

$$DBI = \frac{1}{k}\sum_{i=1}^{k} DBI_i \tag{4.11}$$

While the Davies-Bouldin Index has no upper bound, zero is the lowest possible score. Values closer to zero indicate a better partition since this means the clusters are more compact and better separated.

4.3.4 Scatter Degree between and Within Clusters (S_Dbw) Validity Index

Finally, the Scatter Degree Between and Within Clusters (S_Dbw) validity index [14] is the sum of the average scattering for clusters and the inter-cluster density. For this validity index, a lower value indicates better clustering. Again, this measure is based on the concept that a good clustering result should have compact and well-separated clusters.

The S_Dbw index is determined by first calculating several measures for each cluster, capturing the inter-cluster density and the intra-cluster variance. The inter-cluster density is calculated as shown in Equation 4.12.

$$Dens\_bw = \frac{1}{k(k-1)}\sum_{i=1}^{k}\sum_{j=1}^{k}\frac{density(u_{ij})}{max\left\{density(v_i),\, density(v_j)\right\}} \tag{4.12}$$

Where k is the number of clusters, v_i and v_j are the centers of clusters C_i and C_j, u_{ij} is the middle point of the line segment defined by v_i and v_j, and $density(u)$ is defined in Equation 4.13.

$$density(u) = \sum_{i=1}^{n_{ij}} \left[f(x_m, u) = \begin{Bmatrix} 0 \text{ } if \text{ } d(x, u) > stdev \\ 1 \text{ } otherwise \end{Bmatrix} \right] \tag{4.13}$$

Where n_{ij} is the number of tuples belonging to clusters C_i and C_j. If a point belongs to the neighborhood of u, its distance from u is smaller than the average standard deviation of clusters, earning it a function value of 1 as shown in Equation 4.13. Next, the intra-cluster variance is calculated as shown in Equation 4.14.

$$Scat = \frac{1}{k} \sum_{i=1}^{k} \frac{\sigma(v_i)}{\sigma(S)} \tag{4.14}$$

Where $\sigma(S)$ is the variance of the dataset, and $\sigma(v_i)$ is the variance of cluster C_i. Finally, the validy index is defined as shown in Equation 4.15.

$$S_{Dbw} = Scatt + Dens\_bw \tag{4.15}$$

For this validity index, a lower value indicates better clustering, meaning the clusters are compact, well-separated, and generally regular in shape.

Since each of these methods considers different aspects of clustering to report a metric, only a combination of them would yield a comprehensive evaluation of the clustering approach.

Example 4.2

Next, we analyze the different cluster metrics for the TEP conditions of Examples 3.1 and 4.1. The results provided in Figure 4.6 correspond to the combination of PaCMAP-DBSCAN. In this figure, the four metrics are illustrated as function of the number of iterations in PaCMAP.

Figure 4.6 shows that all metrics coincide in terms of the number of iterations. Based on these metrics, the range of 100–200 iterations provides the most accurate information about clustering. This is supported by Figure 4.7, which provides a 2D projection of the clusters for different numbers of iterations. As shown in this figure, the algorithm identifies global events in the dataset as well as part of the local events within the 100- to 200-iteration range. This range provides a good compromise for characterizing both global and local structures. Beyond 200 iterations, the algorithm focuses on refining the local structure and identifying additional clusters during the transient process behavior between operating regimes, thereby focusing on local events as expected.

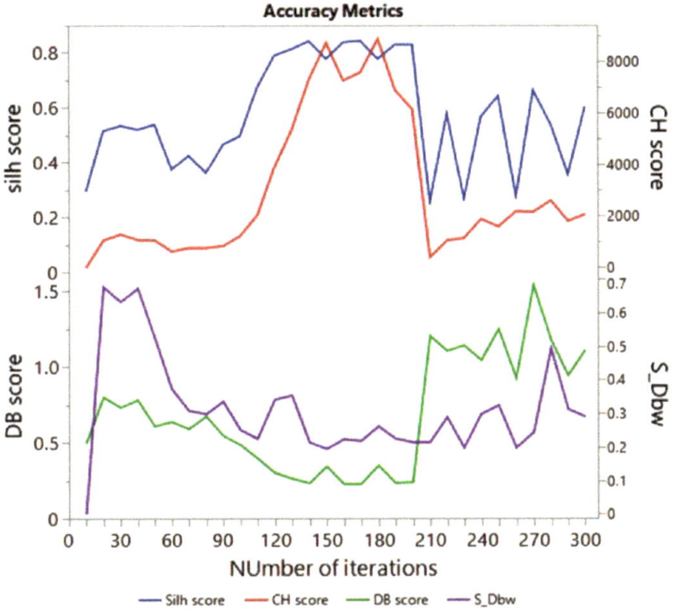

FIGURE 4.6 Clusters metrics for TEP-1 using PaCMAP-DBSCAN combination.

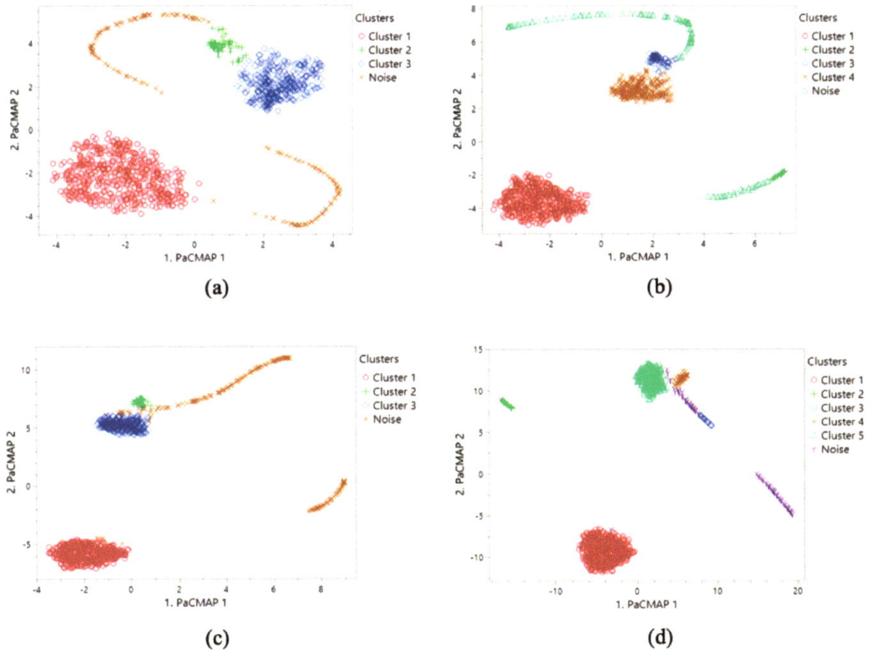

FIGURE 4.7 2D data projection for PaCMAP-DBSCAN combination as number of iterations: (a) 100 iterations, (b) 130 iterations, (c) 150 iterations, and (d) 450 iterations.

4.4 EXPLAINING THE CLUSTER FORMATION: SUBSPACE GREEDY SEARCH

Once different patterns are grouped into different clusters, verification of the formation of the clusters is the next step before using the datasets for supervised learning approaches. To verify the separation between different clusters and provide a meaningful explanation, a method called *subspace searching approach* can be used to find the most contributing combination of variables that cause the separation [15].

Theoretically, the most straightforward way to find the most contributing combination of variables is to define a score function that can measure disparity among dimensions equally. By comparing scores over combinations of variables across different dimension spaces, the cause of cluster formation can be identified. However, it is difficult to find a proper score function to measure the disparity equally across multiple dimensions due to the curse of dimensionality.

Another issue is computational complexity. For a measurement including d variables, the number of possible subspaces is 2^d. It is infeasible to score all subspaces and pick one with the highest score by brute-force searching. To address such issues, a greedy algorithm to search for the target subspace in multiple iterations, called subspace greedy search (SGS), can be applied [16], making it efficient in computation. Instead of brute-force searching that visits all subspaces, the greedy algorithm reaches the target subspace by searching only a few subspaces through multiple iterations. The search starts from the lowest dimension and ascends to higher dimensions until no possible subspaces are left. Each iteration corresponds to a certain dimension. In each iteration, subspaces are created in the same dimension by combinations of available variables, and variables in the subspaces that have high test scores are kept for the next iteration. One way of defining the test score is as the k-nearest neighbor distance (k-dist) between samples in two clusters for each subspace, which can fairly compare the disparity among different subspaces in the same dimensions.

Subspaces with high k-dist indicate the critical variables contributing to the clusters forming. Hence, according to the greedy algorithm, variables in the subspaces with high k-dist are kept for further testing and the rest are purged. The detailed implementation is summarized in Algorithm 4.2. Candidate list is designed to preserve the most contributing variables obtained in previous iterations.

Algorithm 4.2. Subspace Greedy Search

1: Candidate_list ← []
2: Local_best ← []
3: Variable_list ← all variables
4: d ← 1
5: while d < length(Variable_list) do:
6: Subspace_list ← all d-dimension combinations from Variable_list that contains Candidate_list
7: Calculate Score for Subspace in Subspace_list
8: Candidate list ← Subspace with maximum score
9: Purge Variable_list

10: Local_best.Add(Subspace with maximum score)
11: d ← d+1
12: Rank Variable in Local_best by frequency counting

A subspace list for each dimension is generated by a combination formula subject to the candidate variables. The setting of Candidate list can effectively reduce the number of combinations from ${}^{m}C_{n}$ to ${}^{m-d+1}C_{1}$ for each iteration, where m is the number of variables left in each iteration. The score is calculated by k-dist between data from different clusters at each subspace. The subspace with the highest score is recorded in a Local best list and set as candidate variables for the next iteration. Variables in the subspaces with scores above the average are kept for the next iteration, and the excluding variables are purged. After that, the dimension is increased by one for the next iteration. Once dimension d exceeds the number of variables left, the algorithm ends and variables in Candidate list are returned as the explanation between data from different clusters. The frequency of each variable in the Local best list can naturally rank their contribution to the disparity

Example 4.3

Let's return to the same dataset associated with the TEP process, as previously illustrated in Examples 3.3 and 4.1. By employing the SGS approach for cluster evaluation, we can pinpoint the primary factors contributing to cluster formation.

The results for UMAP-DBSCAN are presented in Figures 4.8 and 4.9. Figure 4.8 highlights the most significant variables, which in this instance are A feed (XMEAS1) and A and C feed (XMEAS4). These findings are further corroborated by the component plane plots in Figure 4.9, aligning accurately with the shape of the regions on the SOM map.

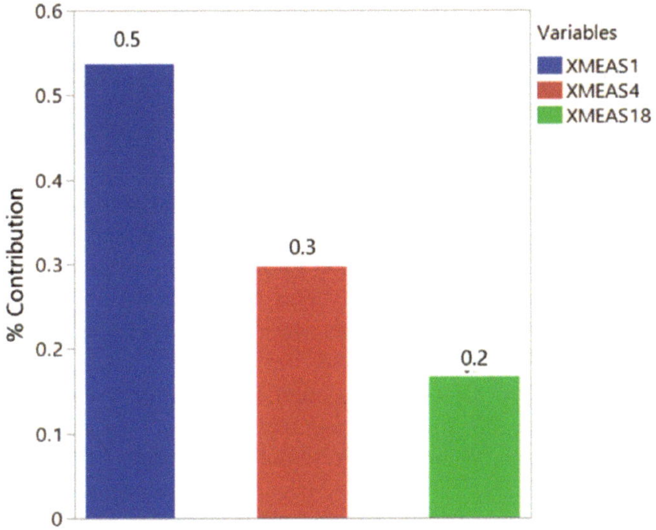

FIGURE 4.8 Variables contribution plot to the cluster's formation using.

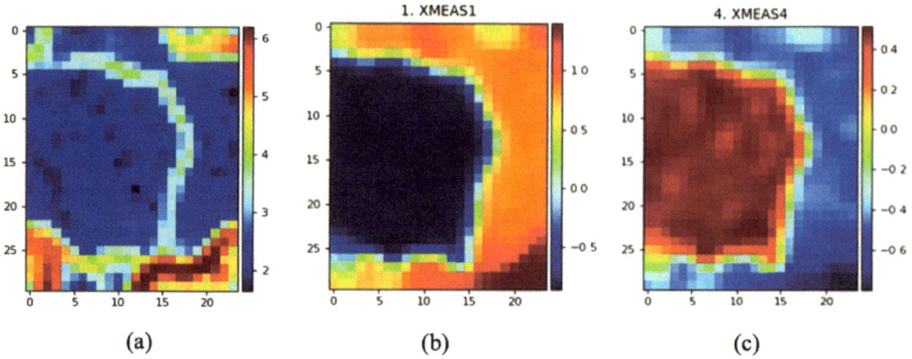

FIGURE 4.9 (a) SOM, (b) and component plane for variable XMEAS1 and (c) component plane for variable XMEAS4.

4.5 HYPERPARAMETER TUNING

Selection of both DR and clustering methods has a great impact on the outcome of the data mining and knowledge discovery procedure. In addition to this, each method may require user-defined parameters, such as n_neighbors in UMAP and min_samples in HDBSCAN, which are likely to be different for each use case. These user-defined parameters are often called hyperparameters and are critical not only for unsupervised learning but for machine learning (ML) applications in general. The process of finding the right set of hyperparameters is called *hyperparameter tuning* and its implementation for unsupervised learning will be the focus of this section.

A key challenge in the selection of DR and clustering methods, along with their hyperparameter tuning, is the fact that often we do not know how the data projection *should look like* because of the very nature of the unsupervised learning problem. Therefore, subject matter expertise is vital to find the right combination. When domain knowledge is available, a human expert can analyze the resulting data representation and assess whether it makes sense from the engineering and physical viewpoint. This, however, takes time and effort, and a more systematic approach is desirable. Now, even with a proper way of evaluating the results, there is still an extremely large number of possibilities to choose from, and this also requires an efficient strategy.

To help address these challenges, Webb et al. [17] designed an experiment to explore the behavior of DR and clustering ensembles using clustering metrics as evaluation applied on the Tennessee Eastman dataset. In the experiment, the task of finding the right ensemble was cast as an optimization problem (Chapter 7 provides a deeper discussion of optimization in the context of ML), and a genetic algorithm is used to solve the problem. In the implementation, the genetic algorithm [18] is set to find the combination of DR and clustering methods, with the corresponding hyperparameters that provides the better representation to identify the known clusters in the dataset. Three different clustering metrics were tracked to analyze their trends. The number of clusters was included in the optimization to prevent the algorithm to get stuck in the trivial solution of finding one cluster per data point. By analyzing the trends of the clustering metrics during the optimization, Silhouette score and, to a lesser extent, Davies-Boulding Index provide a good proxy for accuracy in the absence of knowledge about the data distribution.

4.6 UNSUPERVISED LEARNING COMPUTATIONAL FRAMEWORK (FASTMAN-JMP)

Practical applications of unsupervised learning often involve a combination of the methods and techniques introduced in Chapters 2–4. This poses the challenge of selecting not only a good method for each step, but also a good combination of methods for all the steps. This task can be daunting even for an experienced practitioner. To address this challenge, a computational framework was developed and implemented to execute the techniques outlined in this and preceding chapters. The FAST MANual Data Classification (FASTMAN) Python environment serves as a tool created by the Process Systems Engineering group at Louisiana State University (PSE@LSU). Its purpose is to facilitate the swift and straightforward application of pattern recognition techniques, drawn from data mining and unsupervised ML, for the analysis of historical datasets.

FASTMAN, designed to operate within Python, equips users with a suite of algorithms, data visualization tools, and variable contribution calculations. These components collectively aid users in efficiently validating clustering results for a specific dataset and the chosen hyperparameters for applied algorithms.

In tandem with FASTMAN, a recent addition to the toolkit is FASTMAN-JMP (Appendix A). This tool integrates the statistical data analysis software JMP with the FASTMAN Python environment. Packaged as an add-in using JMP Scripting Language (JSL), FASTMAN-JMP enhances the framework by providing a user-friendly and practical application interface. Upon the correct configuration of the user's Python environment, this combined framework seamlessly integrates powerful analysis methods.

FASTMAN-JMP simplifies the user experience by removing the necessity for a deep understanding of the complex implementation details of each provided method. This includes tasks such as data cleaning, normalization, sampling, DR, and data clustering. Serving as the offline model-building component in the process monitoring workflow, this tool streamlines the process. Within this workflow, clustering is executed on a historical database to train a model, which can subsequently be deployed for real-time process monitoring. For a comprehensive understanding of the FASTMAN framework and its capabilities, refer to Appendix A for a detailed explanation.

4.7 CONCLUSIONS

When analyzing clustering techniques, it becomes evident that hierarchical density-based clustering outperforms flat centroid-based clustering when dealing with chemical process data. HDBSCAN introduces a higher level of resolution in the analysis compared to DBSCAN. In most cases, DBSCAN strictly captures the new state, relegating the transient to the noise component or cluster. In contrast, HDBSCAN effectively reduces noise and incorporates a portion of the transient into the primary cluster. Generally, neighbor graphs and DR methods have proven adequate for capturing both local and global events. Combining t-SNE, UMAP, or PaCMAP with DBSCAN strikes a balance between resolution and cluster count reduction.

SGS can be used to identify the variables responsible for data distribution. The effectiveness of SGS was demonstrated through simulated data, where the algorithm accurately pinpointed the variables driving observed changes. In Chapter 5, a similar analysis will employ real data from the pyrolysis reactor to showcase the applicability of SGS in real industrial scenarios.

SOM, on the other hand, served as a tool for data visualization and clustering verification. Its capabilities in detecting data patterns and correlating original variables with overall data distribution were evident. Component plots, when used in conjunction with SGS, provided valuable insights into cluster formation.

The meaningfulness of identified clusters may vary depending on the dataset and method-specific parameters employed during clustering development. Consequently, evaluation metrics for the clustering results become essential to assess their quality before proceeding with the analysis. Each metric described in this chapter quantifies clustering accuracy differently and can occasionally yield contrasting conclusions. Therefore, it is advisable to use multiple methods in combination to compare clustering algorithms and select the best solution.

Achieving a high-quality clustering result requires a deep understanding of the process and familiarity with the data mining methods in use. Additionally, even when sticking to a single method, clustering results are sensitive to hyperparameter choices, making it crucial to consider these selections carefully. The employment of a multi-objective optimizer rooted in automatic ML can greatly assist in this endeavor.

REFERENCES

1. Tan P.N., Steinbach M., Kumar V., (2018), Introduction to Data Mining. 2 ed.. Pearson.
2. Jain A. K., Dubes R. C. (1988), Algorithms for Clustering Data. Prentice-Hall, Inc.
3. Buhmann J., Kühnel H. (1993), Complexity optimized data clustering by competitive neural networks. Neural Computation, 5(1), pp. 75–88.
4. McInnes L., Healy J., Astels S. (2017), hdbscan: Hierarchical density based clustering. The Journal of Open Source Software, 2(11), doi: 10.21105/joss.00205
5. Thomas M. C., Zhu W., Romagnoli J. A. (2018), Data mining and clustering in chemical process databases for monitoring and knowledge discovery. Journal of Process Control, 67, pp. 160–175.
6. MacQueen J. (1967). Some methods for classification and analysis of multivariate observations. Fifth Berkeley symposium on mathematical statistics and probability, Vol. 1, pp. 281–297, University of California Press, Berkeley.
7. Jain A. K. (2010), Data clustering: 50 years beyond K-means. Pattern Recognition Letters, 31, pp. 651–666.
8. Vesanto J., Alhoniemi E. (2000), Clustering of the self-organizing map. IEEE Transactions on Neural Networks, 11(3), pp. 586–600.
9. Ester M., Kriegel H. P., Sander J., Xu X., (1996), A density-based algorithm for discovering clusters in large spatial databases with noise. Proceedings of the Second International Conference on Knowledge Discovery and Data Mining. AAAI Press, pp. 226–231.

10. Campello R. J., Moulavi D., Zimek A., Sander J. (2015), Hierarchical density estimates for data clustering, visualization, and outlier detection. ACM Transactions on Knowledge Discovery from Data (TKDD), 10(1), p. 5.
11. Rousseeuw P. (1987), Silhouettes: A graphical aid to the interpretation and validation of cluster analysis. Journal of Computational and Applied Mathematics, 20, pp. 53–65.
12. Calinski T., Harabasz J. (1974), A dendrite method for cluster analysis. Communications in Statistics-Theory and Methods, 3(1), pp. 1–27.
13. Davies D., Bouldin D. (1979), A cluster separation measure. IEEE Transactions on Pattern Analysis and Machine Intelligence, PAMI-1(2), pp. 224–227.
14. Halkidi M., Vazirgiannis M., Batistakis Y. (2000), Quality scheme assessment in the clustering process. LNCS (LNAI), 1910, pp. 265–276.
15. Moulavi D., Jaskowiak P., Campello R., Zimek A., Sander J. (2014), Density-based clustering validation. Proceedings of the 14th SIAM International Conference on Data Mining (SDM), Society for Industrial and Applied Mathematics.
16. Zhu W., Sun W., Romagnoli J. A. (2018), Adaptive k-nearest-neighbor method for process monitoring. Industrial & Engineering Chemistry Research, 57, pp. 2574–2586.
17. Webb Z., Nnadili M., Seghers E., Briceno-Mena L., Romagnoli J. A. (2022), Optimization of multi-mode classification for process monitoring, Frontiers in Chemical Engineering, 4. doi: 10.3389/fceng.2022.900083
18. Deb K., Pratap A., Agarwal S., Meyarivan T. (2002), A fast and elitist multiobjective genetic algorithm: NSGA-II. IEEE Transactions on Evolutionary Computation, 6(2), pp. 182–197.

5 Unsupervised Learning Case Study

5.1 PROCESS DESCRIPTION

Olefins are among the most important building blocks in the chemical industry, primarily produced from the cracking of naphtha in pyrolysis reactors [1]. Because of the importance of these processes, ensuring efficient operation of pyrolysis reactors through process monitoring is key for both safety and productivity.

Pyrolysis reactors are placed in a fired furnace, where the energy required for the reaction is generated from fuel gas burning. In this study, data comes from an industrial scale furnace reactor, in which naphtha is cracked to ethylene. As shown in Figure 5.1, six coils of naphtha are mixed with steam before cracking in the furnace. The ethylene product is generated by cracking the mixture through the furnace at a high temperature (around 1000°C). Figure 5.2 shows 27 essential variables for this process and Table 5.1 their corresponding operating mean values. As a time-varying process, the major challenge for monitoring the pyrolysis reactor comes from normal drifting due to slow fouling in the tubes and shape-changing of cracking coils in a

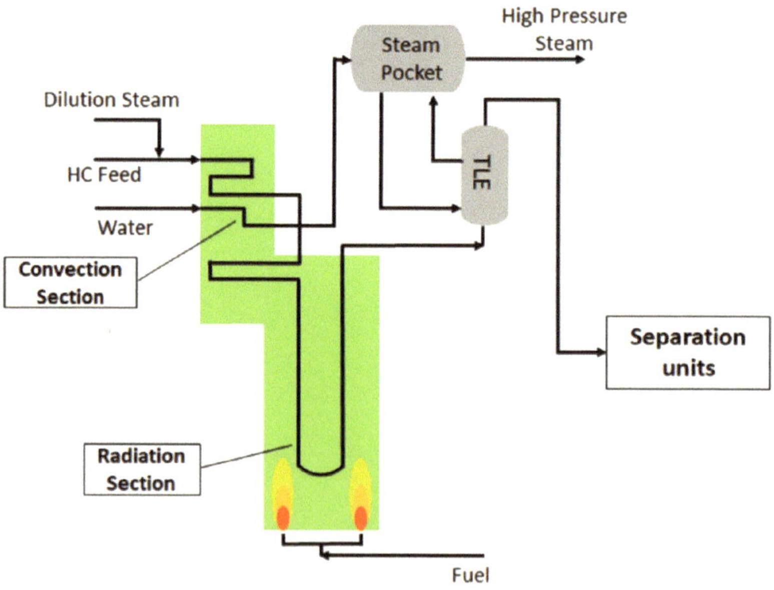

FIGURE 5.1 Schematic representation of the ethylene cracking unit.

DOI: 10.1201/9781003455905-6

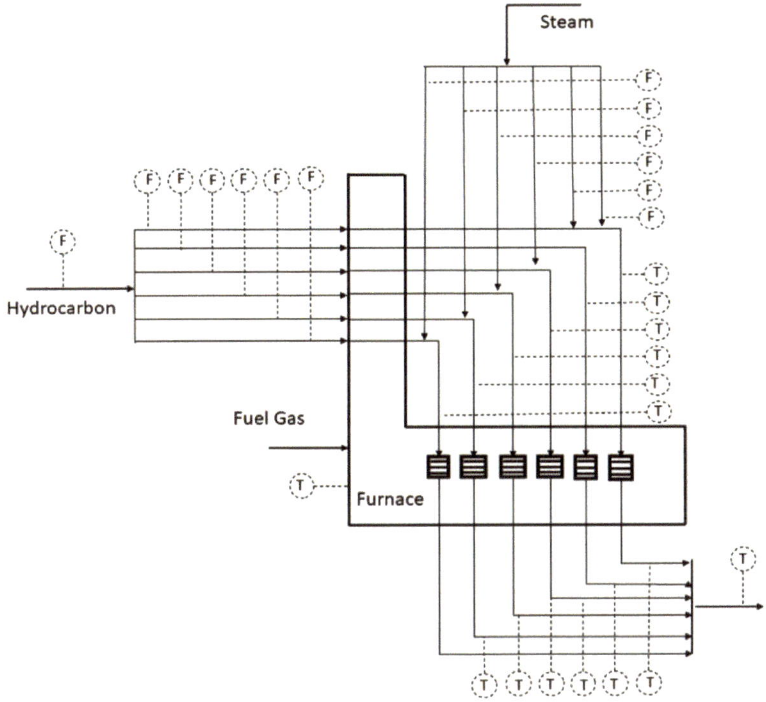

FIGURE 5.2 The furnace design scheme and associated measurements.

high-temperature environment. The operating cycle of one batch of the reaction is about 90 days, and afterward, it requires decoking for further operations.

Typically, process monitoring is based on the distributed control systems (DCS), where sensor measurements are visualized, and the status of the plant operations is observed through the information reported by the sensors. For many

TABLE 5.1
Furnace Process Variables Measurement with Their Mean Values

Coil	Hydrocarbon Flows	Steam Flows	Crossover Temperature		Cracking Temperature
1	8729	4285	611		832
2	8287	4464	617		837
3	8851	4361	621		835
4	8586	4272	611		836
5	8431	4320	618		837
6	8879	4317	622		838
Total hydrocarbon flow				52095	
Overall outlet temperature				838	
Furnace temperature				1026	

commercial DCS, the measurements visualization strategy is designed on individual variables, such that important measurements are expressed in a line chart showing its change over time or reported as instantaneous values on the process diagram. The faulty conditions are primarily determined by the violation of predefined thresholds for each measurement.

5.2 DATA ANALYSIS AND CLASSIFICATION

Figure 5.3 illustrates the time evolution of the hydrocarbon and steam flows respectively. The analysis was done on the total set of measurements; however, here only five are shown for simplicity. In our analysis, we restrict our discussion on these flowrates. The 2D principal component analysis (PCA) projection of the data and the resulting self-organizing map (SOM) are illustrated in Figure 5.4. Colors indicate the time dependencies, that is, blue beginning of the cycle and red end of the operating cycle. Clearly the unit is moving along the cycle through different conditions due to the tube cocking and the effect of the controllers to maintain constant pyrolysis temperature in the furnace.

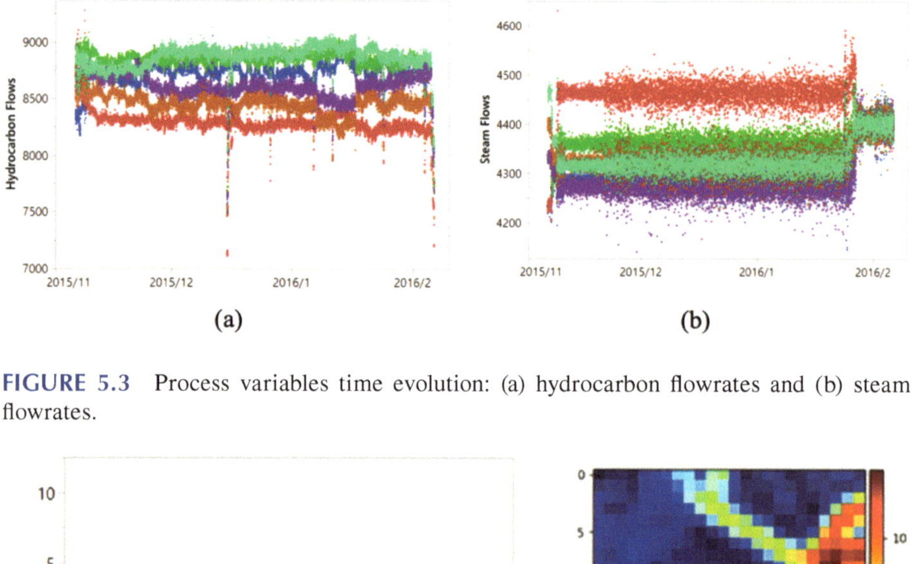

FIGURE 5.3 Process variables time evolution: (a) hydrocarbon flowrates and (b) steam flowrates.

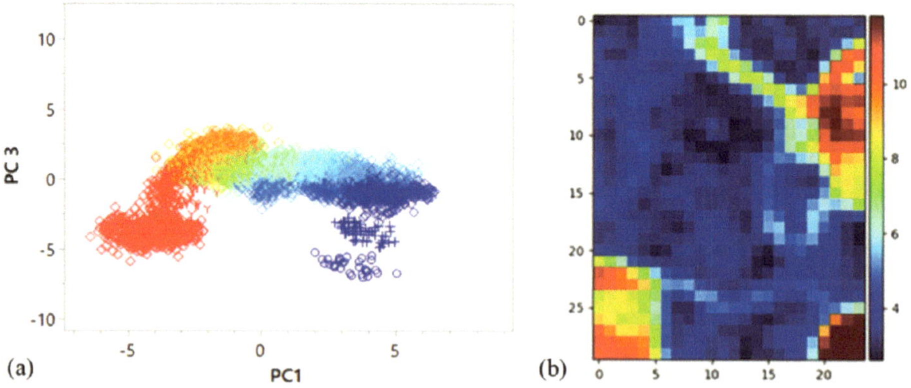

FIGURE 5.4 (a) 2D projection plot and (b) the SOM for pyrolysis data.

The SOM plot in Figure 5.4 can be used as a graphical illustration of the concepts discussed in Chapter 4 within the scope of industrial data. In this plot, as we discussed in previous chapters, observations that are very dissimilar to each other are revealed by the bright areas separating them (i.e., these regions illustrate global structure in the data). However, there are other regions visible that indicate more subtle, smaller separations between data, as indicated by the light blue lines. These observations are very similar to each other and indicate local structures in the data.

Based on the graph coloring as discussed, a user can identify regions that are characterized as global or local on the SOM. This information can then be used to validate the groupings of data that are identified after performing certain dimensionality reduction techniques by projecting these clusters on top of the SOM and comparing how these clusters overlap the SOM-identified structures.

5.3 PRINCIPAL COMPONENT ANALYSIS (PCA) ANALYSIS

5.3.1 Effect of Outliers

Figure 5.5 depicts the projection of the primary cluster, identified by PCA-DBSCAN (min_sample = 10) (where DBSCAN is density-based spatial clustering of applications with noise), onto the time evolution and the SOM. This combination effectively

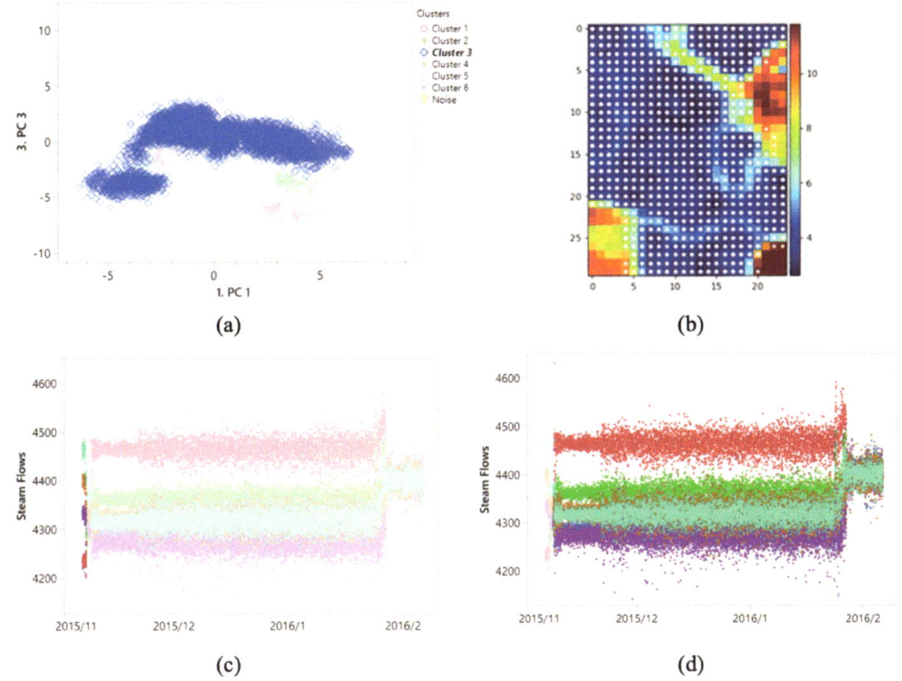

FIGURE 5.5 Projections of the main clusters as predicted by PCS-DBSCAN (*min_sample* 10) for raw data. (a) Projection of cluster 3 into a 2D space, (b) projection of cluster 3 into the SOM, (c) projection of cluster 3 into the time series, and (d) projection of cluster 1 into the time series.

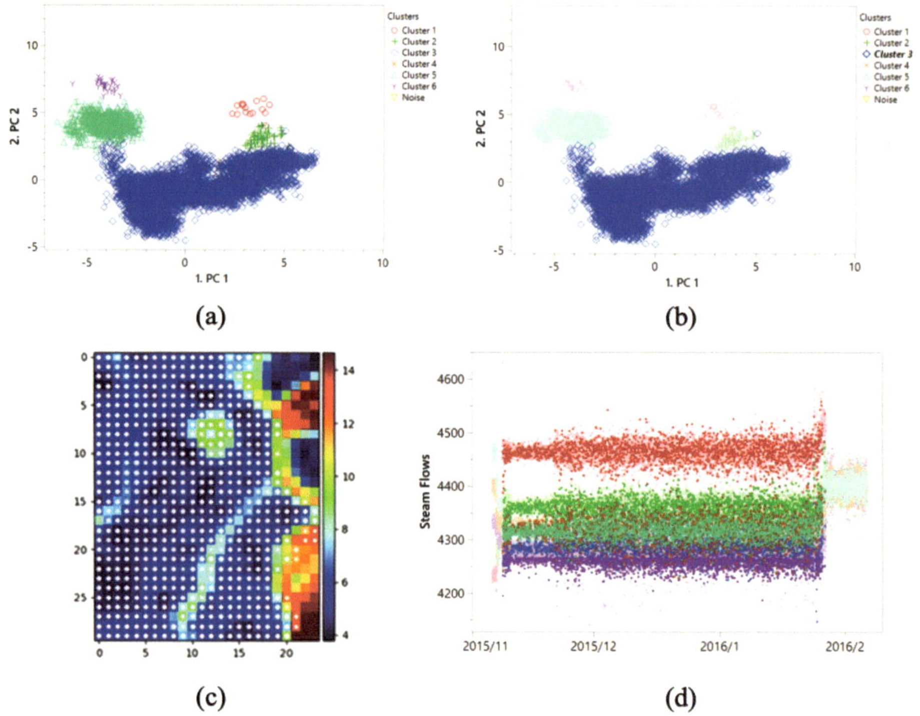

FIGURE 5.6 Projection of the main cluster as predicted by PCS-DBSCAN (min_sample = 10) for raw data after preprocessing and outlier elimination. (a) Projection of all data into a 2D space, (b) projection of cluster 3 into a 2D space, (c) projection of cluster 3 into the SOM, and (d) projection of cluster 3 into the time series.

characterizes most of the operating cycle as a single cluster. Nonetheless, upon examining the SOM, it becomes apparent that a few subregions (indicated by the light color separations) exist throughout the cycle.

Figure 5.6 shows the 2D PCA projection of the data and resulting SOM after eliminating outliers using k-nearest neighbors (k-NN) with $k = 10$. The elimination of outliers had a noticeable effect on the SOM structure and provided a clearer view of the data, which would have otherwise obscured the analysis. In Figure 5.6, the main cluster identified by PCA-DBSCAN (min_sample = 10) is projected into the time evolution and the SOM. Under these new conditions, the combination characterizes the single cluster previously identified into two clusters (Figure 5.6a), as confirmed by the SOM projection of cluster 3. Additionally, Figures 5.6c and d illustrate this issue when projecting the identified cluster 3 into the time evolution.

5.3.2 Effect of Sampling

In Chapter 2, we conducted an analysis of the sampling effect using only two of the flowrates. In this chapter, we extend our analysis to encompass all variables and

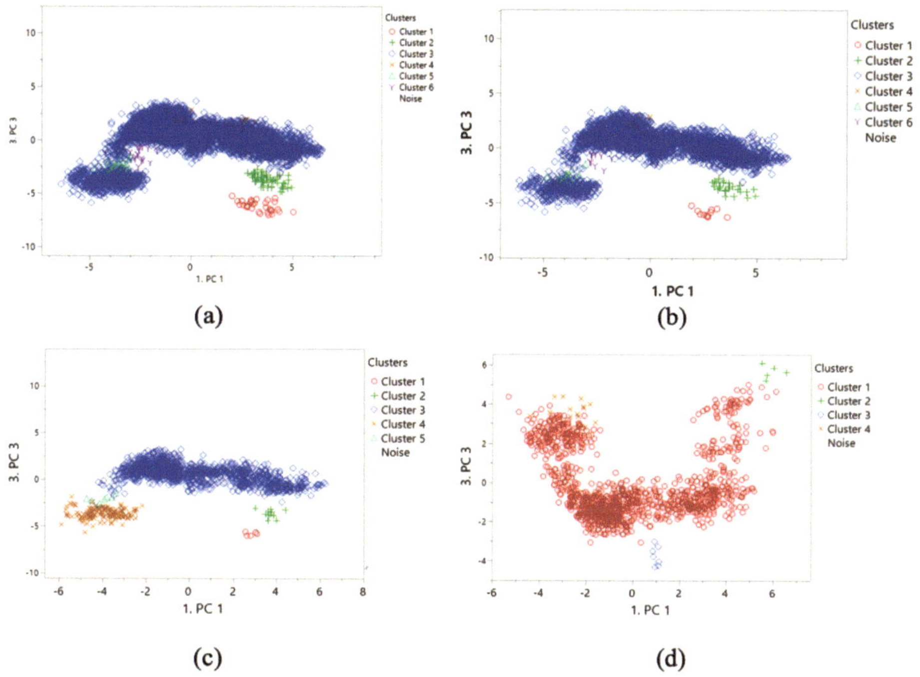

FIGURE 5.7 2D Projections results for alternative sampling techniques using PCA-DBSCAN (min_sample = 10). (a) No sampling, (b) random sampling with interval = 2, (c) random sampling with interval = 5, and (d) c-NN with $k = 5$.

consider their impact on data projection and classification. We investigate the consequences of sampling real plant data through random sampling and the Condensed Nearest Neighbor (c-NN) methods under PCA-DBSCAN dimensionality reduction (DR)/classification. In Figure 5.7, we depict the sampling effects for both c-NN and random sampling approaches with a fixed number of neighbors, $k = 5$ and varying intervals of two and five for random sampling.

In the case of the c-NN approach, the dataset size is reduced substantially, from 7800 to 1400 data points. Meanwhile, for random sampling, the reduction takes it from 7800 to 3900 and 1560 data points for intervals of two and five, respectively. Clearly, this reduction significantly influences the dataset's quality after projection and classification in the c-NN approach, as the primary events now appear more pronounced compared to the entire dataset. This observation aligns with the notion that this approach densely samples highly variable regions, potentially sacrificing some local effects in the process.

Conversely, random sampling maintains most of the original dataset's main properties. These findings are further corroborated in Figure 5.8, where SOM plots are provided. It is evident that minor changes occur in the case of random sampling as the sampling interval increases, with light blue lines gradually disappearing. In

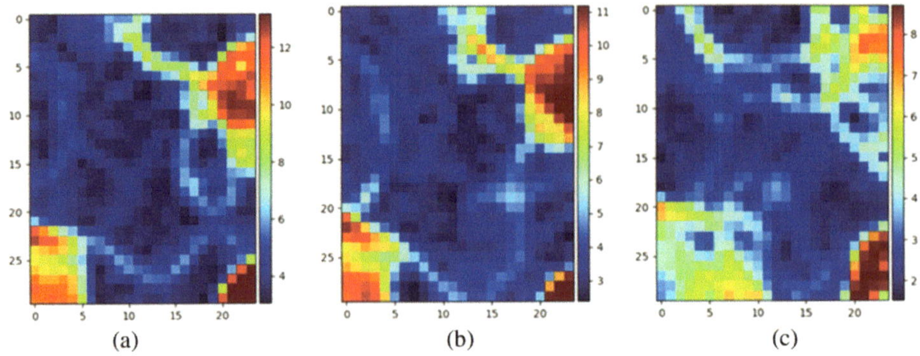

FIGURE 5.8 SOM results for alternative sampling techniques. (a) Random sampling with interval = 2, (b) random sampling with interval = 5, and (c) c-NN with $k = 5$.

contrast, with the c-NN approach, these changes are more pronounced, and some of the local effects are substantially lost.

Sampling can significantly enhance computational efficiency, but it requires careful handling to ensure that valuable information is not lost during the process. Moreover, when classifying industrial data, it is essential to adjust the min_sample parameter accordingly after sampling as the total data volume has undergone a change.

5.3.3 Effect of Clustering and Hyperparameters

In Figure 5.9, we observe the impact of reducing the min_samples parameter in DBSCAN from 10 to 5. The methodology, in this scenario, detects an extra region situated in the top right corner of the SOM, as well as additional regions during the startup and shut down conditions. The figure depicts the projections of these two primary clusters on both the time evolution plot and the SOM, as shown in Figure 5.7c, d, e, and f.

The next step is to analyze the impact of alternative clustering techniques in conjunction with PCA dimensionality reduction. Figure 5.10 displays the outcomes of using PCA/K-means algorithms ($k = 3$). However, the PCA/K-means combination does not yield accurate classification, as illustrated in the figures. Although it identifies a small region at the end of the operating cycle, it partitions the larger region into two clusters of similar sizes. To show the corresponding projections into the SOM, refer to Figure 5.11.

To compare the results, we performed a similar analysis using the PCA-HDBSCAN combination (where HDBSCAN is hierarchical density-based spatial clustering of applications with noise) with alternative min_sample parameters. The identified cluster's projection in both the PCA graph and the time evolution of the stem flowrates is illustrated in Figures 5.12 and 5.13. This combination yields

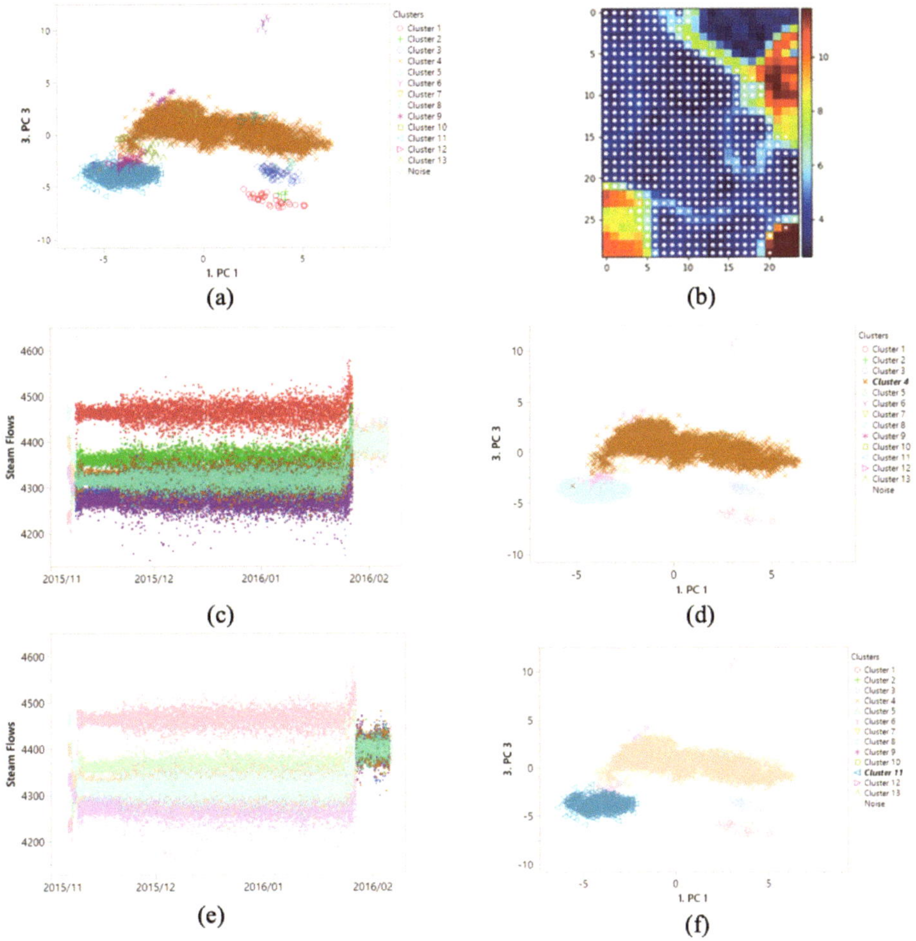

FIGURE 5.9 Projection of clusters PCA-DBSCAN (min_sample = 5). (a) Projection of all data onto the 2D space, (b) projection of cluster 4 onto SOM, (c) projection of cluster 4 onto the time series, (d) projection of cluster 4 onto the 2D space, (e) projection of cluster 11 onto the time series, and (f) projection of cluster 11 onto the 2D space.

comparable results to the PCA-DBSCAN combination, albeit with a reduction in the total number of small clusters.

As previously discussed, in PCA, the first few component vectors capture a significant portion of the overall variance present in the data. Consequently, PCA is mainly focused on capturing the global structure of the dataset. This makes it compelling to compare the performance of the Pairwise Controlled Manifold Approximation and Projection (PaCMAP) algorithm and PCA concerning characterizing both the global structure and local structure of the data.

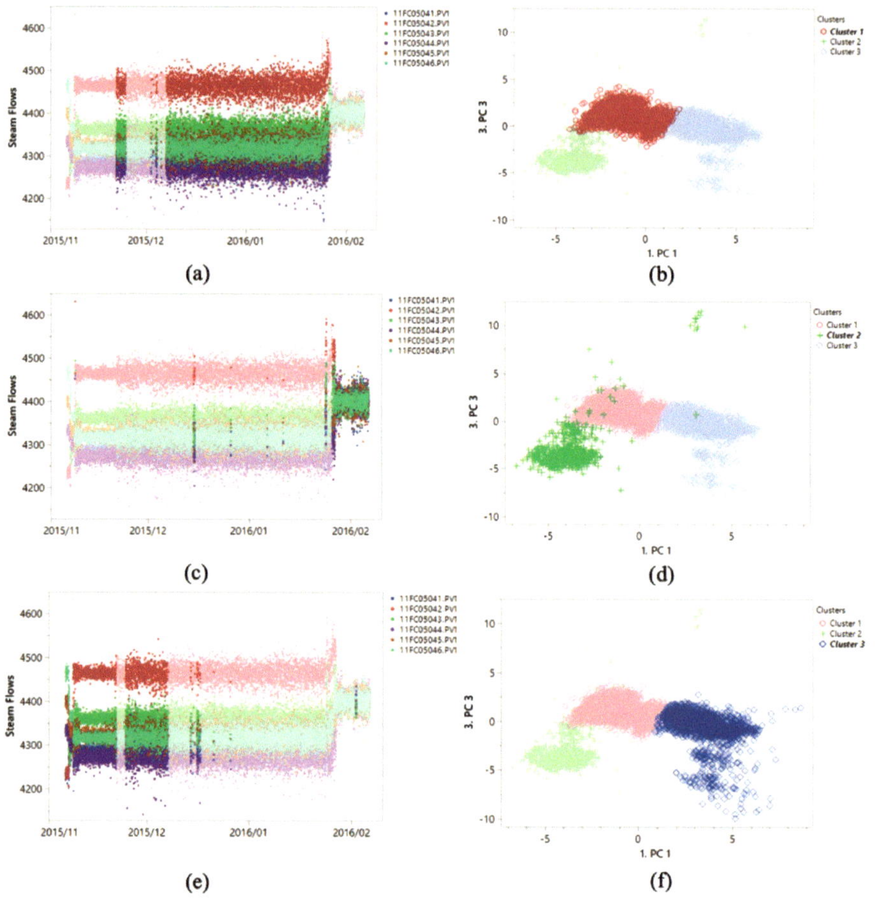

FIGURE 5.10 Projection of clusters PCA/K-means (k = 3). (a) Projection of cluster 1 onto the time series, (b) projection of cluster 1 onto the 2D space, (c) projection of cluster 2 onto the time series, (d) projection of cluster 2 onto the 2D space, (e) projection of cluster 3 onto the time series, and (f) projection of cluster 3 onto the 2D space.

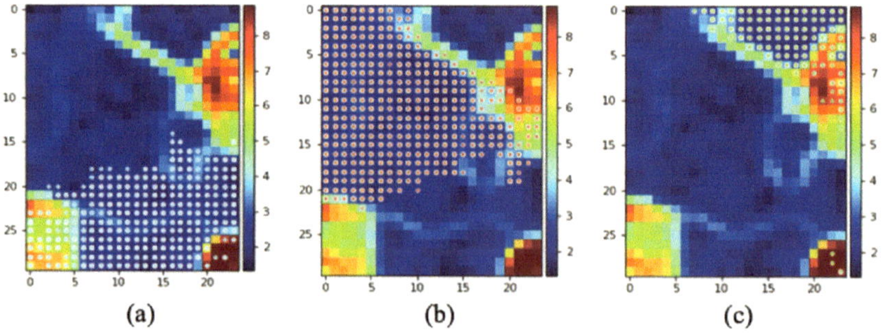

FIGURE 5.11 Cluster identified by PCA/K-means projection onto SOM. (a) Projection of cluster 1, (b) projection of cluster 2, and (c) projection of cluster 3.

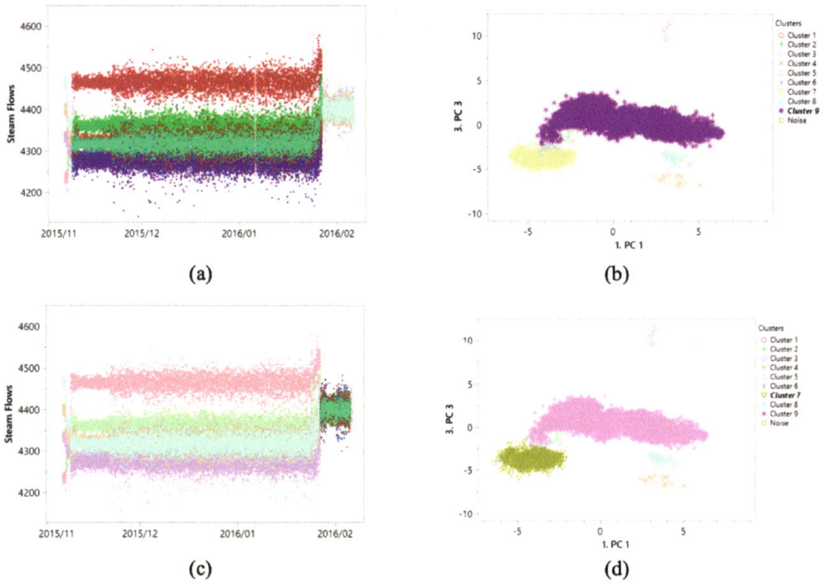

FIGURE 5.12 Projection of the identified cluster for PCA-HDBSCAN (min_sample = 10). (a) Projection of cluster 9 onto the time series, (b) projection of cluster 9 onto the 2D space, (c) projection of cluster 7 onto the time series, and (d) projection of cluster 7 onto the 2D space.

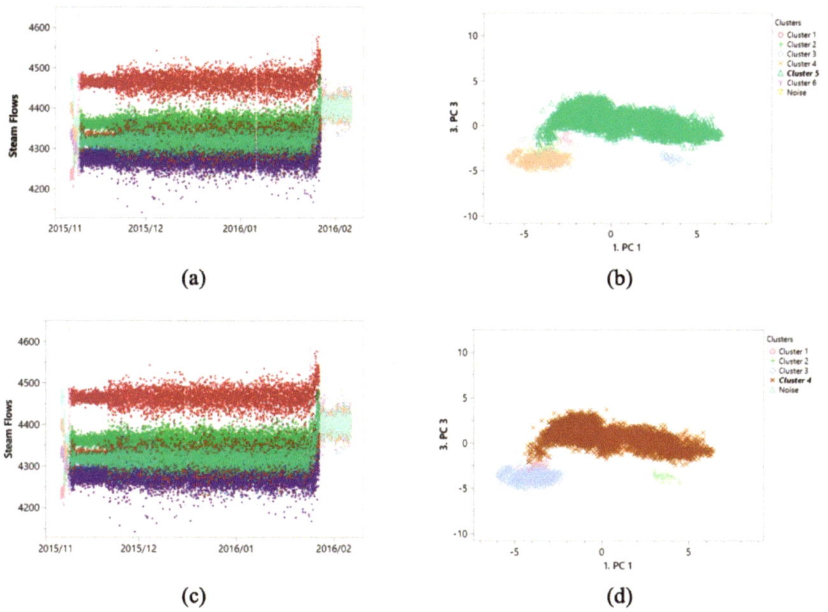

FIGURE 5.13 Effect of min-sample parameter on classification for PCA-HDBSCAN. Projection of the identified clusters: (a-b) min_sample = 20 and (c-d) min_sample = 30.

5.4 PAIRWISE CONTROLLED MANIFOLD APPROXIMATION AND PROJECTION (PaCMAP) ANALYSIS

Figure 5.14 illustrates the projection of the clusters identified by PaCMAP-DBSCAN (after 450 iterations) and the SOM projections (main clusters). Clearly, this combination, characterizes not only the main changes in the process conditions but also the local changes illustrated in the SOM by a number of subregions (light colors separations) throughout the process cycle. Projected into the SOM are clusters 3, 4 and 6 which in the PCA-DBSCAN combination were identified as single region.

Shown in Figure 5.15 is a graph of the value of the PaCMAP loss function for the industrial pyrolysis reactor dataset as a function of number of iterations. The loss function changes over the course of the PaCMAP algorithm stages based on the weights of neighbor pairs, mid-near points, and further points. In the second stage, there is a gradual change in the weighting of the mid-near pairs while the other weights remain static. However, it is clear from Figure 5.15 that the bulk of the loss function optimization occurs in the first stage for the tested industrial pyrolysis reactor dataset. The distinct variations in the graphed loss are due to the abrupt changes in the pairwise weightings in the loss function at the onset of the second and third optimization changes.

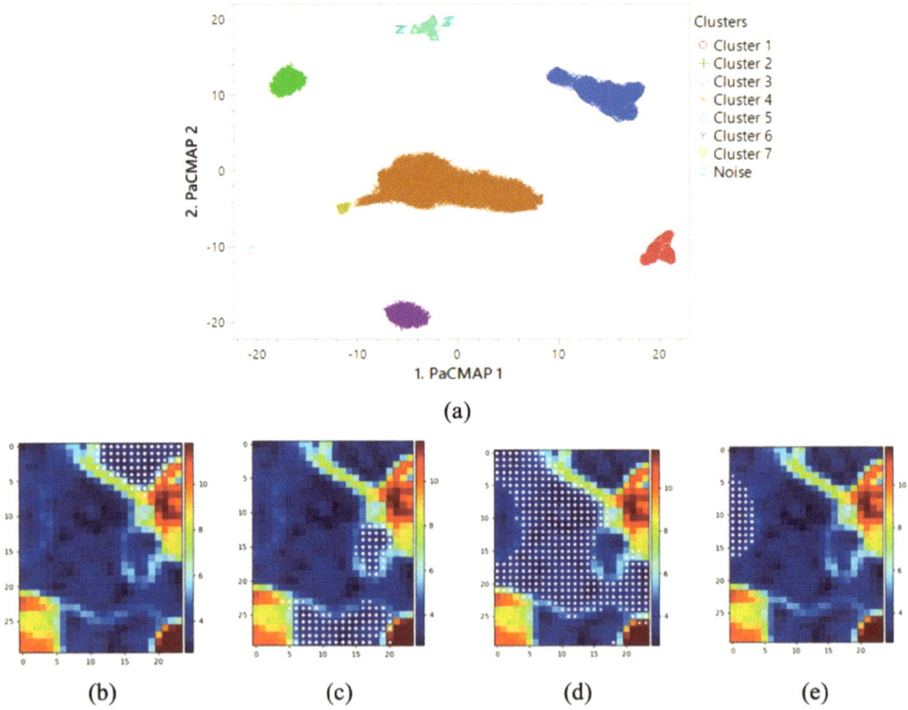

FIGURE 5.14 (a) 2D plot of PaCMAP results after 450 iterations (default) and (b, c, d, e) some clusters projected onto the SOM (min_sample = 70).

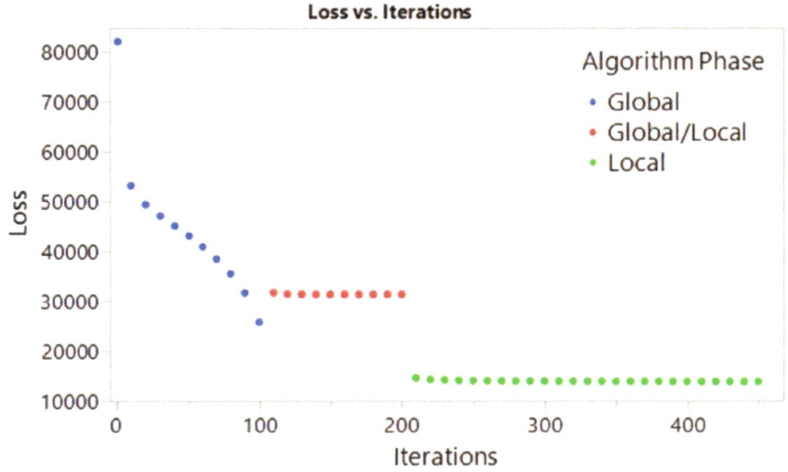

FIGURE 5.15 PaCMAP loss function evolution as function of iteration.

The performance of PaCMAP on the pyrolysis reactor dataset is analyzed in-depth in terms of the evolution of the projection with the number of iterations. Figure 5.16 illustrates the results of initial stages of the PaCMAP algorithm from 10 iterations with PCA initialization. This result is at the very beginning of entering the first stage

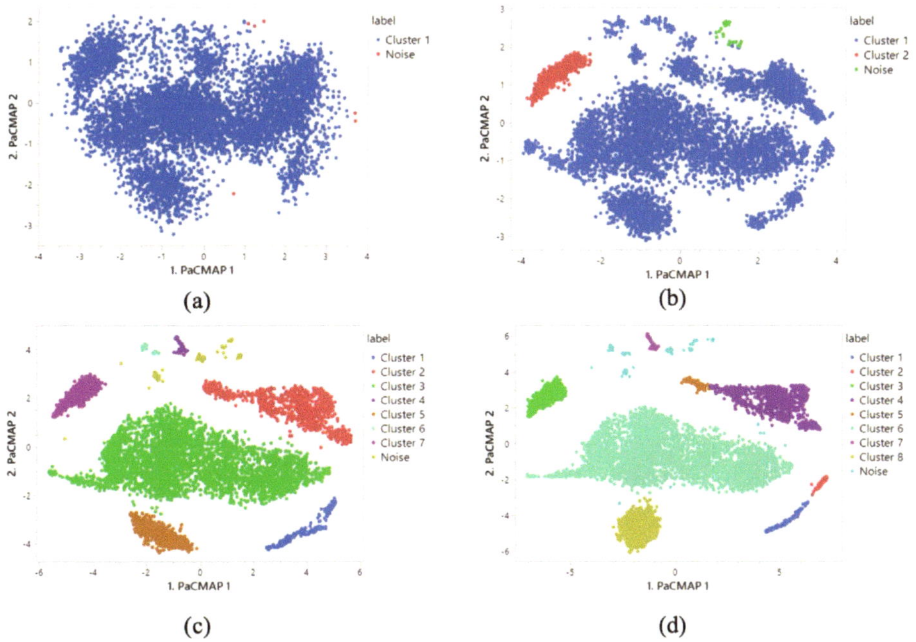

FIGURE 5.16 PaCMAP-DBSCAN projection as function of the number of iterations (PCA initialization). (a) 10, (b) 15, (c) 90, and (d) 100 iterations.

of the PaCMAP algorithm. There is only one cluster identified, categorizing most of the data into one large grouping. After 50 iterations of the PaCMAP algorithm, we are half-way point of the first stage in the PaCMAP algorithm. Now, there is an additional cluster identified than with only 10 iterations, but a large amount of the data is still categorized into a large global grouping.

After 100 iterations of the PaCMAP algorithm (marking the end of the first stage), there are even more clusters identified, and at this step there is already much local definition in the identified groupings, which is shown in the clusters projected on the SOM. "Cluster 1" and "Cluster 2" remain roughly the same (separated by bold SOM outlines) and "Cluster 8" is now the equivalent of "Cluster 4" that was achieved with 60 iterations. The significant difference in this step is that the large "Cluster 3" from Figure 5.17 has now been separated from one global cluster into five smaller clusters. "Cluster 3" through "Cluster 7" are now much more locally describing clusters, consistent with the light blue outlines on the SOM.

Together, the results in Figure 5.16 show that this dimensionality reduction method initially focuses on identifying the main regions, and at the end of the first stage of the algorithm (by default, around 100 iterations), the global structure of the data (i.e., the main process events) have essentially been identified. Again, in its three stages of optimization, PaCMAP initially emphasizes fixing global structure (by default, iterations 0–99), then gradually eases its global structure optimization and begins

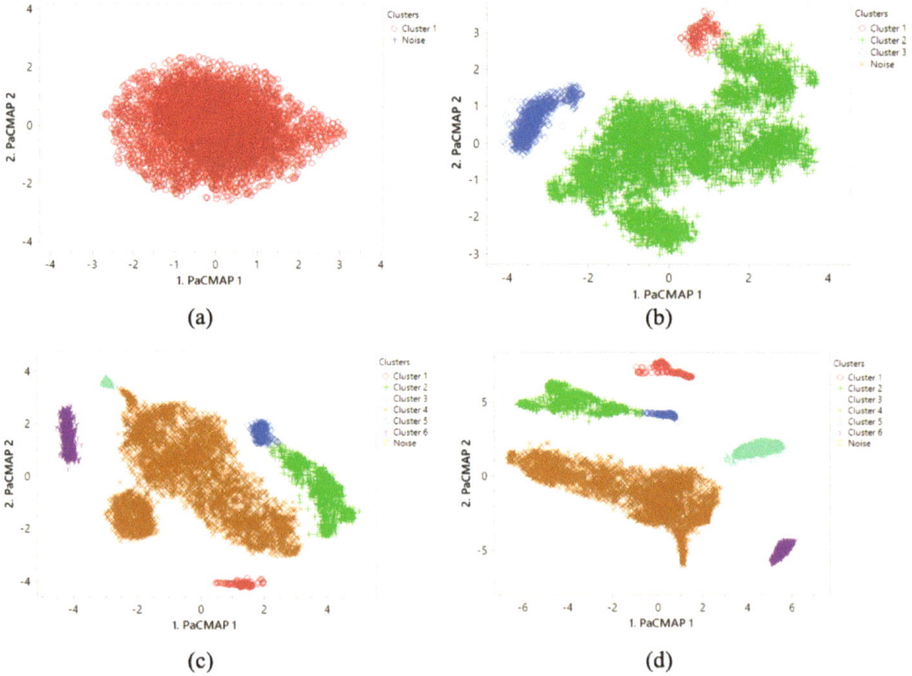

FIGURE 5.17 PaCMAP-DBSCAN projection as function of the number of iterations (random initialization). (a) 10, (b) 15, (c) 80, and (d) 105 iterations.

optimization of the local structure (by default, iterations 100–199). Finally, in the last stage, the algorithm just focuses on refining the local structure (by default, iterations 200 through 450).

Furthermore, PaCMAP allows the use of alternative initializations, that is, random or PCA initialization. It is also illustrative to analyze the results with both options. Figure 5.17 shows the results of iterating through the PaCMAP algorithm from 10 to 100 iterations using random initialization. Comparing the plots in Figure 5.16 with those in Figure 5.17, the initialization makes a considerable difference in the very beginning iterations of the PaCMAP algorithm, but it does not take many iterations for both initializations to achieve comparable results. At roughly 90 iterations, both runs show remarkably similar structure (mirror images flipped on an axis) and almost the same number of clusters. PaCMAP is fairly robust to the choice of initialization due to its mid-near points and dynamic graph structure.

Depending on the dataset and method-specific parameters used to develop the clustering, the identified clusters may or may not reveal meaningful groupings of data. Therefore, evaluation metrics of the clustering results are often necessary to assess the quality of results at this stage before proceeding with analysis. Figure 5.18

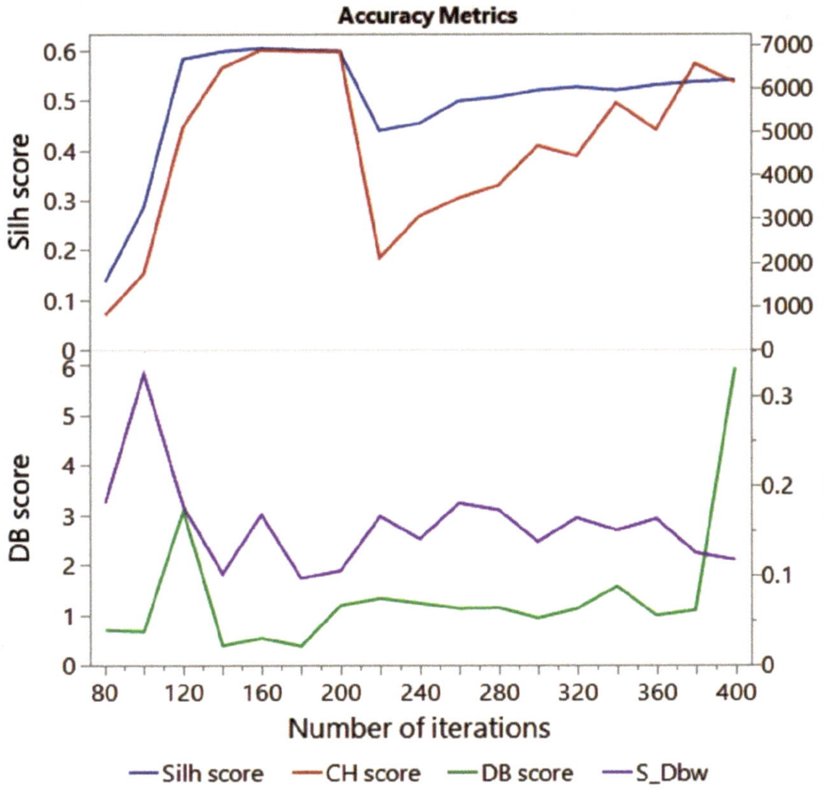

FIGURE 5.18 Clustering metrics evolution as function of the number of iterations for the PaCMAP-DBSCAN combination.

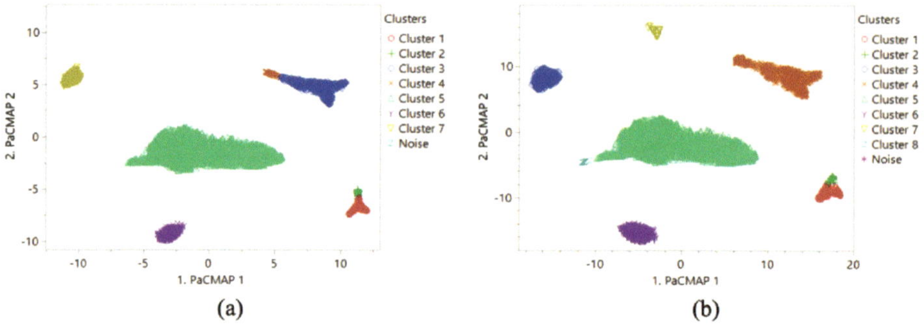

FIGURE 5.19 Clusters metrics as function of iteration numbers for PaCMAP: (a) clusters projected for *iteration* 150 and (b) clusters projected for *iteration* 320.

illustrates the clustering metrics evolution as function of the number of iterations for the PaCMP-DBSCAN combination.

As shown in the Figure 5.18, all metrics are coincident that in terms of "optimal" number of iterations. For the pyrolysis data, it seems that the ranges of 140–160 as well as the range 320–340 provide the most accurate information about the clustering information (based on these metrics). This is corroborated in Figure 5.19, where the clusters projection into 2D is provided for 150 and 320 iterations. In these ranges of iterations, the algorithm finds a good compromise between characterizing both global and local structures. Overall, however, the algorithm performs very well on almost the whole range, characterizing the main process operational status, differences are basically identifying or not small clusters indicating some very low/large values of operating variables or distinguishing those subtle, smaller separations between data, as indicated by the light blue lines in the SOM plot.

5.5 t-DISTRIBUTED STOCHASTIC NEIGHBOR EMBEDDING (t-SNE) ANALYSIS

Here the performance of the t-distributed stochastic neighbor embedding (t-SNE) algorithm is analyzed in terms of a key hyperparameter, the perplexity parameter. Figure 5.20 shows the projected pyrolysis reactor dataset using t-SNE dimensionality reduction with several choices for its perplexity parameter. The perplexity parameter is effectively a measure of the number of neighbors and is therefore a balance between the local and global aspects of the data. The default value for this parameter is 30. Adjusting this parameter, the embedding is changed based on the number of close neighbors to a locally appearing data point; thus, this method focuses mainly on developing the local structure of data and the global distances between separate clusters are not necessarily meaningful. The graph in Figure 5.18 covers a range of values for the perplexity from 5 to 250 using random initialization.

From the graph, it is evident that t-SNE is a locally focused dimensionality reduction technique. Even with very low values of perplexity (which should force the

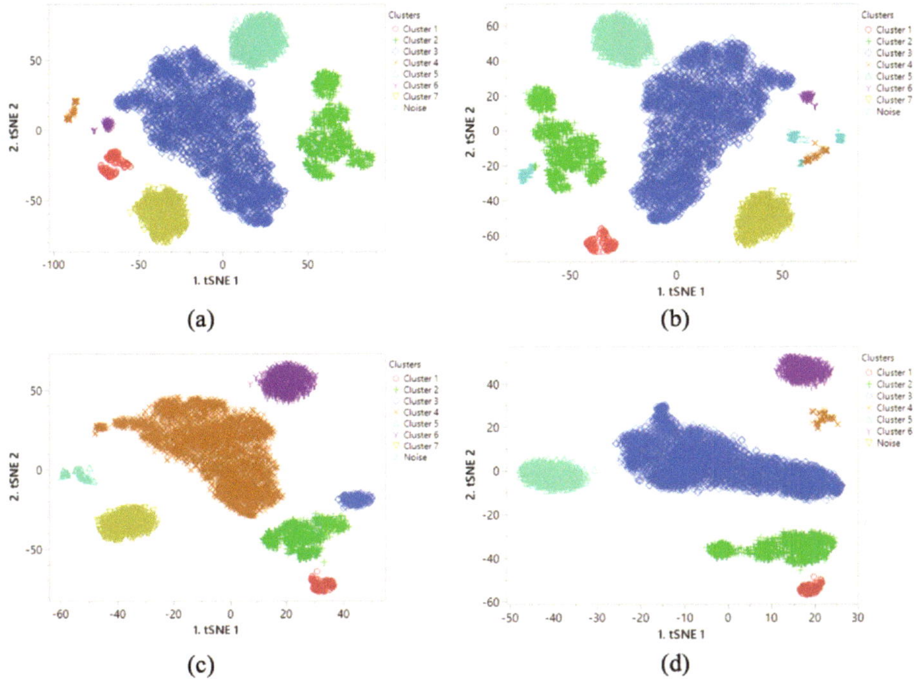

FIGURE 5.20 t-SNA-DBSCAN projection as function of the number of perplexity parameter; (a) 20, (b) 30, (c) 50, and (d) 100.

algorithm to focus more on global structure and less on local structure), the projected data shows clear groupings and separations of data.

5.6 CLUSTER ANALYSIS

In terms of chemical process data visualization and analysis, PaCMAP's flexibility to represent both global and local behavior in the same implementation enables the user to control the granularity and the scope, in time, of the analysis. Anomalous states can then be identified not only by their overall deviations but by their time scale as well, which provides plant operators with more information on how to deal with the fault or abnormal state.

For the pyrolysis dataset analyzed here, global effects are mainly reported by large process variable changes over a longer period of time (such as global clusters 3 and 4 on the order of months as shown in Figure 5.5), while local changes are detected on periods of time much shorter (such as local clusters visible in Figure 5.16).

The differences between global clusters should then be substantial changes or variables that affect a large part of the process. An example of a potential global event would be a major unit feed rate change, or even a steady, gradual drift of the feed rate since a change in an upstream variable such as the feed rate then affects the rest of the process and would be reflected elsewhere.

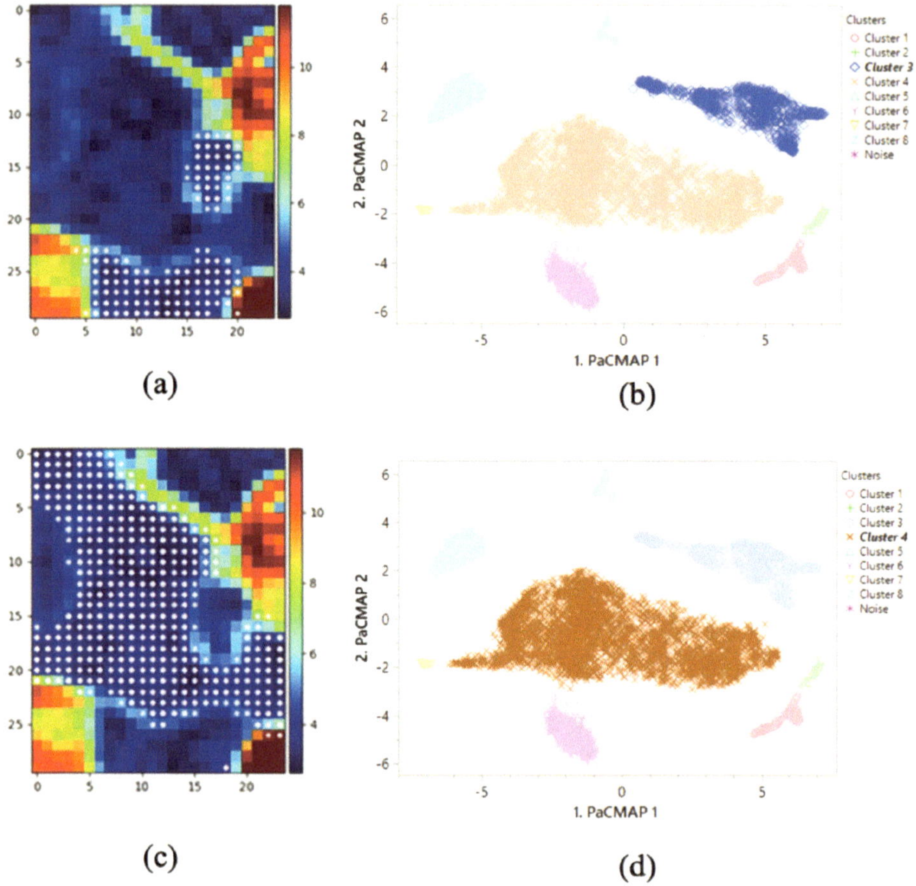

FIGURE 5.21 Cluster identified by PaCMAP-DBSCAN projection onto SOM. (a) Cluster 3 projected onto SOM, (b) cluster 3 projected onto the 2D space, (c) cluster 4 projected onto the SOM, and (d) cluster 4 projected onto the 2D space.

An analysis was performed using the PaCMAP-DBSCAN (iterations = 100; eps = 0.6; min_samples = 60) to analyze further the results of the feature extraction. Figure 5.21 illustrates the projection of some of the main clusters identified into the SOM. As can be seen, the large cluster identified previously using PCA is now subdivided into three subregions using PaCMAP (better resolution). These subregions correspond to local system behavior (a small number of hydrocarbon flows change significantly), as compared to the global changes discussed previously, for sharp changes in the steam flows. Figures 5.22–5.24 illustrate contributing variables identified by the SGS to the formation of the new clusters identified (cluster 4 and 5) as well as the component plane for the variables with higher contributions (hydrocarbon flows of coils 4 and 6). Figure 5.25 shows the time evolution of these variables indicating to portion of the data corresponding to the selected clusters.

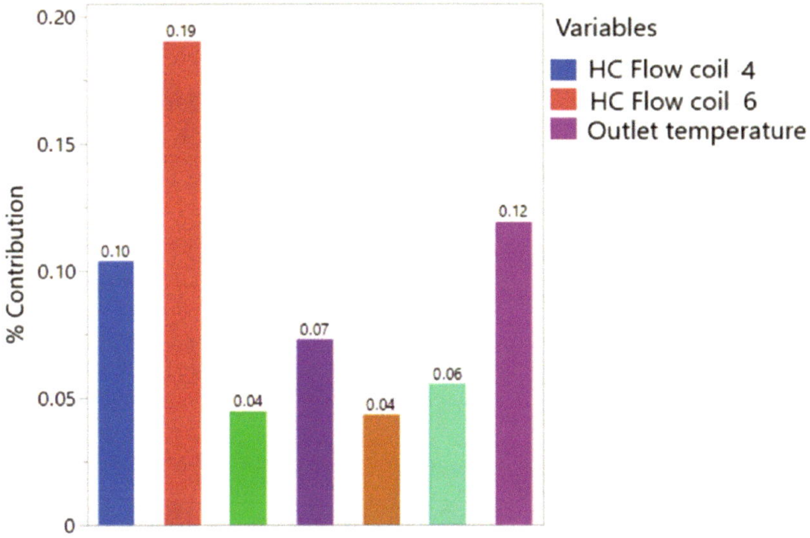

FIGURE 5.22 Contributing variables as identified by the subspace greedy search: contribution to formation of clusters 4 and 5.

Finally, it is possible to follow the process operation through the whole cycle using the project trajectories mapped into the SOM. Figure 5.25 illustrates the trajectories through different process stages. The trajectory projection over SOM makes it possible to visually indicate the current state of the process and observe how that state has been reached; it also enables the tracking of the different operation modes

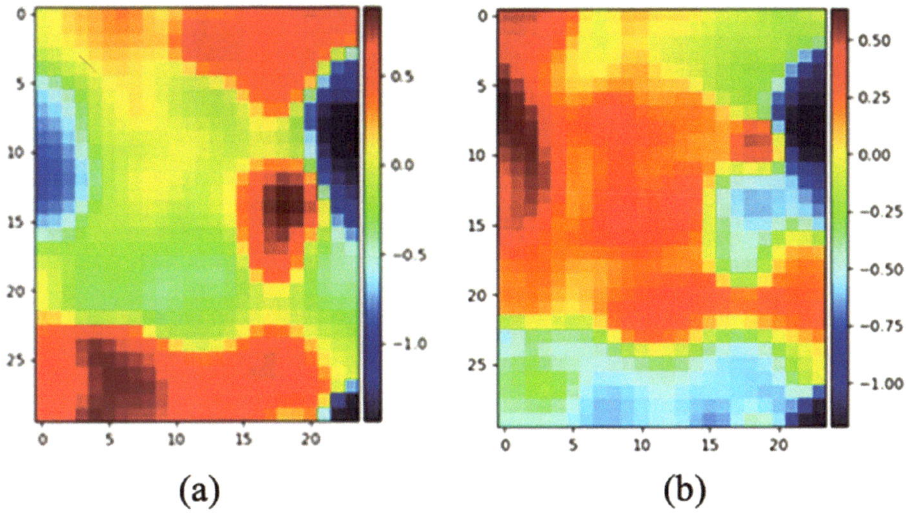

FIGURE 5.23 Contributing variables as identified by the component maps. (a) Hydrocarbon flow of coil 6 and (b) hydrocarbon flow of coil 4.

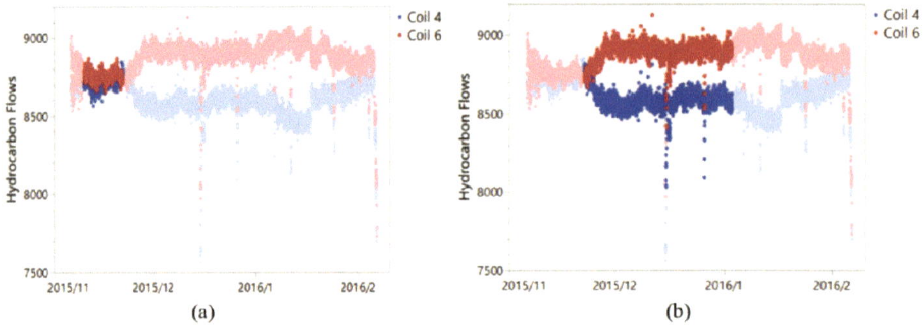

FIGURE 5.24 Time evolution of contributing variables as identified by the subspace greedy search. (a) Cluster 3 projected and (b) cluster 4 projected.

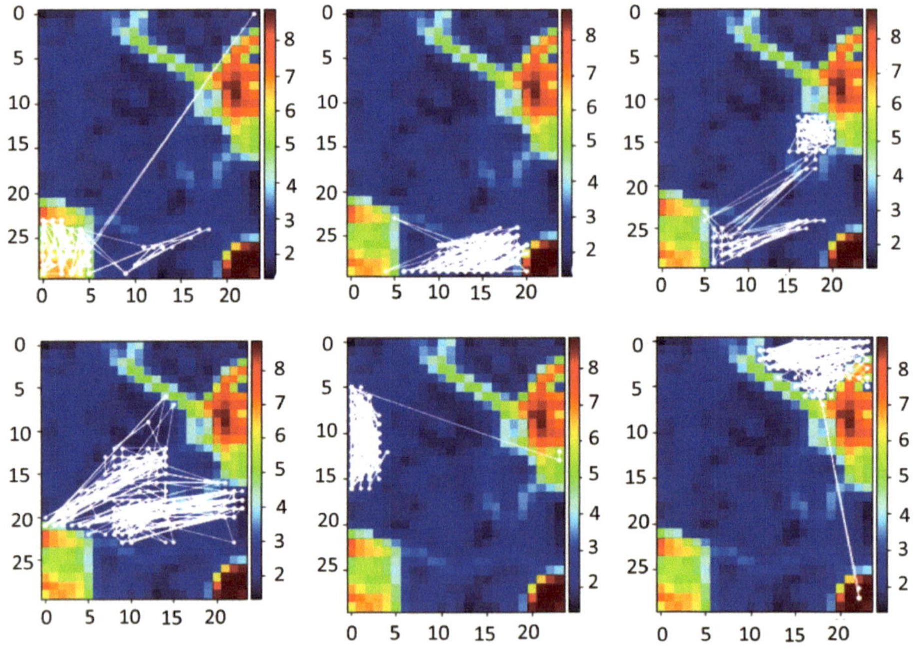

FIGURE 5.25 Process transitions trajectories for the pyrolysis reactor projected on the SOM.

of the plant. SOMs allow us to follow the data through transitions to different operating regions. The information provided in the previous analysis can be used now to develop an online monitoring strategy to identify normal and faulty conditions. The cluster information obtained about the drift of the process over time (i.e., the distribution of classes) can be saved as a new labeled dataset to be used later for an online monitoring implementation.

5.7 CONCLUSIONS

Unsupervised learning methods, namely dimensionality reduction and clustering can be used together to create a holistic visualization of process data from which we can draw insights about the behavior of the plant. In doing so, the engineer must make decisions regarding both data preprocessing and hyperparameter tuning. As shown in this example, selection of processing methods such as sampling and scaling can have a profound effect on the results. The same applies for hyperparameter tuning and even method selection for both dimensionality reduction and clustering techniques.

Method selection is a nontrivial task, and often requires experience and familiarity with the methods, the data, and the domain knowledge. Because of this, no hard guidelines can be provided for all cases. However, good practices such as testing multiple methods and visualizing the data with different techniques simultaneously (e.g., time series, clustered projection, and SOM), using systematic hyperparameter tuning (e.g., using an optimization approach), using model interpretation tools like SGS and always referring to domain knowledge can substantially ease the process of data mining.

Developments in Computer Science bring new methods that can facilitate the implementation of known principles. PaCMAP as an evolution of t-SNE in neighbor graph-based dimensionality reduction is a good example. PaCMAP's hyperparameters are fewer and more intuitive than those of t-distributed stochastic neighbor embedding (t-SNE) and uniform manifold approximation and projection (UMAP), and the algorithm has better out-of-the box performance. A balance between new approaches and well-known and proven methods is always the best way of making our analysis pipeline easier to implement while preserving the quality and interpretability of the analysis.

REFERENCE

1. Amghizar J., Vandewalle L. A., Van Geem K. M., Marin G. B. (2017), New trends in olefin production. Engineering, 3(2), pp. 171–178.

Part II

Supervised Learning

One or more data attributes could be assumed to be the causal result of interactions between other attributes. If this is the case, such attributes are designated as outputs, targets, or labels while the rest are called inputs or features. As it was discussed in Part I with Unsupervised Learning for Knowledge Discovery, this is not always the case. Part II of this book focuses on the case where the causal relationship between attributes is known or can be presumed, and outputs or targets are available. This is called supervised learning.

DOI: 10.1201/9781003455905-7

6 Concepts and Definitions

6.1 INTRODUCTION

There are two distinct types of problems in supervised learning: regression and classification. When the outputs are real numbers, they have a one-to-one relationship with the inputs, and the resulting model is called a *regression model*. On the other hand, if multiple combinations of inputs can produce the same output, the data is said to be divided in *classes*, and the resulting model is a *classification model* that seeks to assign the correct class to a given input.

Figure 6.1 illustrates the types of tasks in supervised learning. Common techniques used in the area include support vector machines (SVMs), decision trees (DTs), nearest neighbor, ensemble learning, and neural networks. These techniques have inspired great interest from industry and have already been used in many applications to improve product quality and process safety. Each of those will be introduced in this chapter, and in the subsequent chapters, several industrial applications using supervised learning algorithms will be discussed in detail.

6.2 SUPERVISED LEARNING AS A SEARCH PROBLEM

As a modeling approach, supervised learning can be understood as the task of finding a hypothesis that explains the behavior of a system in such a way that one is able to predict its behavior in an unseen scenario [1]. The process of finding this hypothesis, which is contained in the model's parameters, is called *training*. To train a model, three components are essential: examples, a performance metric, and a recursive method to automatically modify the model's parameters to improve the performance

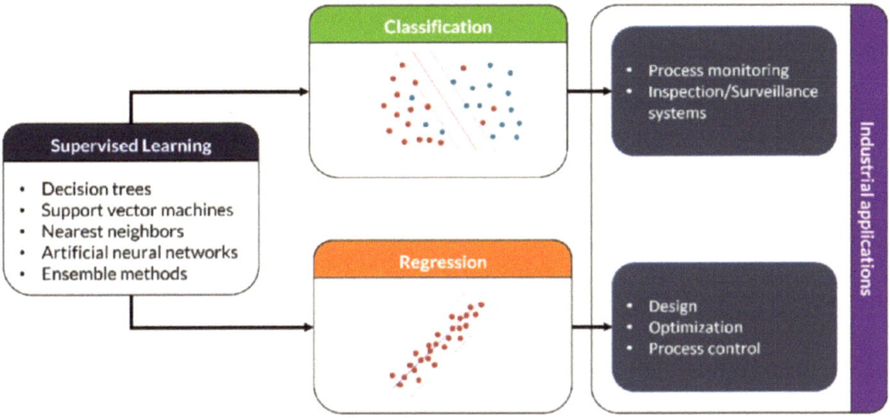

FIGURE 6.1 Schematic representation of supervised learning.

DOI: 10.1201/9781003455905-8

TABLE 6.1
Data for a Fault Detection Problem

Example	Pressure	Temperature	Humidity	Set-point	Fault
1	High	High	Normal	Same	Yes
2	High	High	High	Same	Yes
3	Low	Low	High	Change	No
4	High	High	High	Change	Yes

metric. In this section, a method-agnostic basis for training is presented, while the specifics for several methods are introduced in the next section.

Consider the data in Table 6.1. In a certain chemical process, four features or inputs are assumed to be good predictors of a faulty state. Therefore, we are interested in predicting whether the fault will occur in an arbitrary situation. This is equivalent to formulating a hypothesis about the relationship between the attributes and the target that is consistent with all the examples (instances) in the data.

The most general hypothesis one can propose is $\langle ?,?,?,? \rangle \rightarrow yes$, which means that every feature could take *any* value and the output would be "yes". This contradicts example 3, so the most general hypothesis is wrong. On the other hand, the most specific hypothesis is $\langle \varnothing, \varnothing, \varnothing, \varnothing \rangle \rightarrow yes$, which implies that there is no value an attribute can take for the output to "yes". Since some instances are labeled as "yes", we know that this is not the case. Now, given that both the most general and the most specific hypothesis are wrong, one can infer that the correct hypothesis lies somewhere in between. Furthermore, one can start with either hypothesis and use the data to modify it accordingly.

Let us start with the most specific hypothesis and apply it to the first instance. Since none of the $\varnothing$ in the current hypothesis is satisfied by this example, we replace it by the next more general hypothesis that fits the example, that is $\langle High, High, Normal, Same \rangle \rightarrow yes$. Next, we apply our new hypothesis to the second example. According to our hypothesis, humidity should be *Normal* for the output to be *yes*, which is contradicted by the current example. We then move to the next more general hypothesis that still agrees with the first example and accepts the new one, that is $\langle High, High, ?, Same \rangle \rightarrow yes$. The latest hypothesis also agrees with the third, so no changes are needed. From the fourth example, we learn that the last attribute, set-point, could take either value for the fault to be present. The new hypothesis is then $\langle High, High, ?, ? \rangle \rightarrow yes$. Since we have exhausted all the available data, the latest is our final hypothesis.

The previous example shows the intuition behind training in supervise learning. However, this procedure would be ineffective for large datasets, a higher number of features, or for a regression problem. In particular, at least three improvements are needed, namely (i) a way to automatically modify the current hypothesis to agree with new information, (ii) a way to deal with contradictions in the data, and (iii) a way to reduce the number of steps or iterations needed to incorporate all the information available in the data. The next section introduces some representative algorithms developed to deal with these challenges.

6.3 SUPERVISED LEARNING TECHNIQUES

There are many different classification techniques or classifiers, but some of the widely used ones include the following:

- DTs and random forest (RF)
- SVM
- Nearest neighbors
- Artificial neural networks (ANN)
- Ensemble methods

This section provides a brief description of each of these methods and their principles. More complex methods are available and new algorithms or implementations of the existing ones are being developed currently. Rather than providing an exhaustive list of all these methods, this chapter aims to provide the reader with a foundational understanding of supervised learning from which more sophisticated methods can be approached.

6.3.1 Decision Trees

The basic idea in the DT algorithm is to partition the dataset according to the values taken by the attributes, creating branches that are subsequently partitioned until all the instances in a branch (subset) have the same output or all the attributes have been exhausted [2]. Algorithm 6.1 shows a synthesis of the procedure. Each point at which a branch is created is called a decision node or just node, and the point where no more decisions are made is called a leaf node or just leaf. This strategy greatly improves the search, with the best possible scenario being that in which the data is divided in half every time. On the other extreme, the worst scenario is that in which an attribute only singles out one instance. This notion is formalized by defining entropy, a performance metric that reports how well a given attribute divides the dataset. For a binary classification, entropy can be expressed as follows:

$$Entropy(S) \equiv -p_{\oplus} \log_2 p_{\oplus} - p_{\ominus} \log_2 p_{\ominus}. \tag{6.1}$$

Here, $p_{\oplus}$ and $p_{\ominus}$ are the proportion of positive and negative examples in the sample S, respectively. From here, we can define a criterion to select the appropriate attribute to partition the data at any given point: select the attribute that provides the greatest reduction in entropy. This is called information gain and can be expressed as follows:

$$Gain(S, A) = Entropy(S) - \sum_{v \in Values(A)} \frac{|S_v|}{|S|} Entropy(S_v). \tag{6.2}$$

Algorithm 6.1. Decision tree.

1. Take A, the attribute that best classifies the instances in the dataset.
2. Use A to partition the dataset into subsets and create a branch for each subset.
3. For each branch:
   a. If all instances in a subset have the same output, create a leaf, and label it with the value of this output.
   b. Else, if all attributes have been used, create a leaf and label it with the value of this output.
   c. Else, repeat steps 1–3.

Consider again the data in Table 6.1. The first step in the DT algorithm is to select a partitioning attribute among pressure, temperature, humidity, and set-point. To do this, we can evaluate the *information gain* associated with the selection of each attribute.

The current value for entropy is as follows:

$$Entropy(S) = -\left(\frac{3}{4}\right)\log_2\left(\frac{3}{4}\right) - \left(\frac{1}{4}\right)\log_2\left(\frac{1}{4}\right) = 0.81.$$

Taking humidity, for example, the number of records in which the value of the attribute is normal is one, and it is positive (the value of the target is "yes"). On the other hand, the number of instances where the value of humidity is high is three, with two positive examples and one negative. This information is summarized below.

Humidity = Normal		Humidity = High	
1 positives	0 negatives	2 positives	1 negatives

From the information in Table 6.2, we can compute the information gain when humidity is selected as partitioning attribute. The entropy for the case of

TABLE 6.2
Attributes in the Flowmeter Data for Example 6.1

Attribute ID	Description
1	Flatness ration
2	Symmetry
3	Crossflow
4–11	Flow velocity in each of the eight paths
12–19	Speed of sound in all eight paths
20	Average speed of sound in all eight paths
21–36	Gain at both ends of each of the eight paths
37	Class attribute or health state of meter (0 or 1)

normal humidity is $-\left(\frac{1}{1}\right)\log_2\left(\frac{1}{1}\right)-0=0$ and for the case of high humidity is $-\left(\frac{2}{3}\right)\log_2\left(\frac{2}{3}\right)-\left(\frac{1}{3}\right)\log_2\left(\frac{1}{3}\right)=0.39$. The information gain is then:

$$Gain(S, \text{Humidity}) = 0.81 - \left[\left(\frac{1}{4}\right)\times 0 + \left(\frac{3}{4}\right)\times 0.39\right] = 0.5.$$

If, on the other hand, we take pressure as partitioning attribute, we notice that all examples with high pressure are labeled as positive, while all examples with low pressure are marked as negative. Therefore, the entropy for each case will be zero, and the information gain will be equal to the original 0.81. From this analysis, the best partitioning attribute is pressure. This result can be expressed as $\langle High,?,?,?\rangle \rightarrow yes$, and the procedure can be carried out again until a stopping criterion is reached.

The efficiency in the search makes DTs resilient to high-dimensionality problems. However, the greedy approach of selecting the partitioning attribute based only on the current distribution of data makes DTs susceptible to noisy data.

6.3.2 Support Vector Machines (SVMs)

The idea of SVM was first introduced by Drucker et al. [3] as an application of the concept of SVMs first proposed by Boser, Guyon and Vapnik [4]. An SVM is an algorithm that looks for the hyperplane that separates two linearly separable data classes maximizing the distance between the hyperplane and the classes (Figure 6.2). The line fairly separates the classes. This is what SVM essentially does – simple class separation.

If the data are not linearly separable, then a transformation, known as *kernel function*, is introduced to map the data into a new space (adding one dimension) in which it is linearly separable. This feature makes SVMs particularly flexible, as it

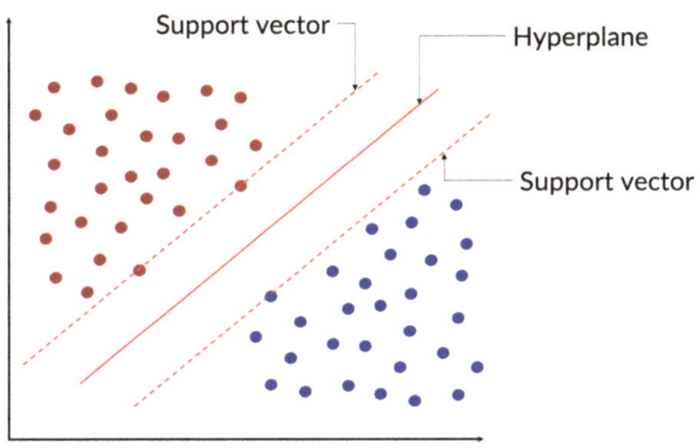

FIGURE 6.2 Schematic representation of a classification model using Support Vector Machines.

FIGURE 6.3 Kernel transformation.

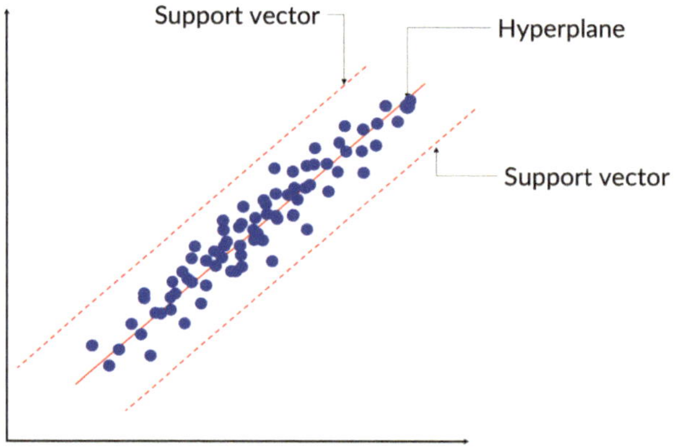

FIGURE 6.4 Support vector regression.

can be applied to a wide range of datasets. A kernel helps us find a hyperplane by projecting the data into a higher dimensionality in which it is now linearly separable (see Figure 6.3).

Support vector regression (SVR) uses the same principle as SVM but instead of finding the hyperplane that better separates the data, it seeks to find the support vectors (the two dashed red lines in Figure 6.4) that are closer to each other and that contain most of the data points. The SVR algorithm is discussed as follows.

Consider the set of training points $\{(\boldsymbol{x}_1, z_1), \ldots, (\boldsymbol{x}_l, z_l)\}$, where $\boldsymbol{x}_i \in R^n$ is the feature vector and $z_i \in R^1$ is the target output. Under the given parameters $C > 0$ and $\epsilon > 0$, the standard form of SVR is as follows:

$$\begin{cases} \min\limits_{w,\, b,\, \xi,\, \xi^*} \dfrac{1}{2} \boldsymbol{w}^T \boldsymbol{w} + C \sum\limits_{i=1}^{l} \xi_i + C \sum\limits_{i=1}^{l} \xi_i^* \\ \textit{subject to} \quad \boldsymbol{w}^T \phi(\boldsymbol{x}_i) + b - z_i \leq \epsilon + \xi_i. \\ \qquad z_i - \boldsymbol{w}^T \phi(\boldsymbol{x}_i) - b \leq \epsilon + \xi_i^* \\ \qquad \xi_i, \xi_i^* \geq 0,\ i = 1, \ldots, l \end{cases} \tag{6.3}$$

Here $\phi(\boldsymbol{x}_i) = \exp\left(-\gamma \|\boldsymbol{x} - \boldsymbol{x}'\|^2\right)$ with $\gamma = \dfrac{1}{n\ var(x)}$ is the kernel function, and $\boldsymbol{w}$ is the vector of weights. Usually, due to the high dimensionality of the $\boldsymbol{w}$, the following dual problem is solved:

$$\begin{cases} \min\limits_{\alpha,\alpha^*} \dfrac{1}{2}\left(\boldsymbol{\alpha} - \boldsymbol{\alpha}^*\right)^T \boldsymbol{Q}\left(\boldsymbol{\alpha} - \boldsymbol{\alpha}^*\right) + \epsilon \sum\limits_{i=1}^{l} (\alpha_i - \alpha_i^*) + \sum\limits_{i=1}^{l} \sigma_i (\alpha_i - \alpha_i^*) \\ \qquad subject\ to \quad \boldsymbol{e}^T\left(\boldsymbol{\alpha} - \boldsymbol{\alpha}^*\right) = 0 \\ \qquad 0 \le \alpha_i,\ \alpha_i^* \le C,\ i = 1, \ldots, l \end{cases}, \tag{6.4}$$

where

$$\boldsymbol{Q}_{ij} = \boldsymbol{K}\left(\boldsymbol{x}_i, \boldsymbol{x}_j\right) \equiv \phi(\boldsymbol{x}_i)^T \phi\left(\boldsymbol{x}_j\right). \tag{6.5}$$

After solving the dual problem, the approximate function is

$$z = \sum_{i=1}^{l} \left(-\alpha + \alpha_i^*\right) \boldsymbol{K}\left(\boldsymbol{x}_i, \boldsymbol{x}_j\right) + b. \tag{6.6}$$

SVM is a very robust and efficient method since it can capture complex functions from small datasets. For high-dimensional data however, SVM has shown to underperform relative to DT-based methods.

6.3.3 Adaptive *k*-Nearest Neighbors (Ak-NN)

This section presents a distance-based approach which is mainly utilized for monitoring processes and detecting faults as illustrated in Chapter 8. The method is designed to adaptively track time-varying process behaviors and extend the continuous normal operation regions [5]. Generally, detecting a fault involves determining that faulty data exhibits little similarity with normal data. Applying the *k*-NN rule, a fault is detected if the *k*-NN distance (*k*-dist) of the faulty data is greater than that of normal samples. To implement this approach, we use only normal samples to initialize the model and determine a threshold *k*-dist, $d\alpha$, for fault detection. After determining $d\alpha$ from training data, the model acquires the latest information from the process to monitor its time-varying behaviors.

The *k*-NN rule's computational complexity makes online implementation infeasible. Additionally, updating the model with consecutive data is unnecessary for a chemical process with slow time-varying behaviors since the incoming data often contains repetitive information. Therefore, we only update the model with prototypes that have relatively high similarity with the normal data and provide novel information to the model. This updating strategy reduces computational cost while maintaining prediction ability.

Next, we will formalize the selection of prototypes using the *k*-NN rule.

Definition 1 (Nearest neighbors' number of a point) For a data point x, the number of its neighbors in dataset D located within the k-th threshold distance d_α, denoted as N_x, is defined as $N_x = \left|\{q \in D \mid dist(x, q) \le d_\alpha\}\right|$.

According to Definition 1, a high similarity can be characterized by a large N_x value. On the other hand, a large N_x value also represents a data sample that is repetitive as the original set, which violates the purpose of updating the model adaptively from novel prototypes. To formulate the prototypes, a balance should be set between the high similarity and novelty. In terms of nearest neighbor number, a naive definition of the ideal prototypes is defined by the following criteria:

Definition 2 (Prototypes) A point x is considered as a prototype if $N_x = k$.

However, this definition is not robust enough. Since the degree of process variation can be uncertain, few points strictly obey the ideal definitions in practice. To match the degree of variation, and updating factor, p (p ∈ (0, 1)) is designed to tolerate Definition 2. Definitions of an ideal prototype and a fault are given as follows:

Definition 3 (Extended prototypes) A point x is considered as a prototype if $k(1-p) < Nx < k(1+p)$.

Definition 4 (Fault) A point x is defined as a fault if $Nx \le k(1-p)$.

Figure 6.5 demonstrates the definitions above, expressed as a k-dist comparison. As the sample in (a) completely drops within the threshold, the information it provides is repetitive of its neighbors. The sample in (b) is an ideal prototype. It has a high degree of similarity with original training samples and also contains novel information that is able to trace the possible variation. The sample shown in (c) shows a notable deviation from the dataset and is then classified as a fault.

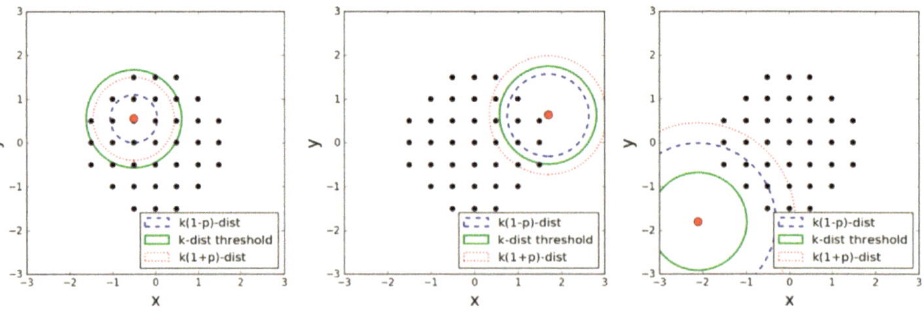

(a) (b) (c)

FIGURE 6.5 Three types of data positions are demonstrated ($k = 8$ and $p = 0.4$). (a) A repetitive data point. (b) An ideal prototype. (c) A fault.

The updating factor, p, is designed to govern the degree of process variation. The setting of the p value depends on the process. A large p value corresponds to the process with intense varying behaviors while a small p value is used for slow drifting in the process. A greater p value allows the model to extend its searching space in order to collect more qualified prototypes, while a smaller p value tightens the searching space. In practice, an accurate value of the updating factor can be estimated from calibration data that contains labels for each data sample. A reasonable p value can be determined from the overall consideration of different evaluation parameters.

In chemical processes, normal operating conditions are always in steady state or with a certain degree of slow drifting. Normal data samples always form dense clusters surrounding with noise samples or faulty samples [6]. Hence, to obtain the threshold distance from normal samples, the strategy used in density-based spatial clustering of applications with noise (DBSCAN) [7] and ordering points to identify the clustering structure (OPTICS) [8] is adopted. These clustering analysis algorithms use a sorted k-dist curve to characterize the distribution of clusters and obtain the threshold distance from the last "valley" point on the curve. As shown in the 2D demonstration in Figure 6.6, data samples inside the cluster (blue dots) exhibit a dense arrangement such that their k-dist are much lower than that of those outside or noise samples (red dots). The "valley" point is the natural separation between these two types of samples. In practice, since the sorted k-dist curve is always monotonically increasing, the "valley" point is approximated numerically as follows.

Definition 5 (Numerical approximation of valley point) Denote the sorted k-dist curve as $f(x)$. Set the domain of the $f(x)$ as [0, b], and the searching step as s ($s > 1$).

Since the "valley" points can be non-unique, in order to locate the last "valley" on the curve, the searching domain is constrained in [a, b] as the ending region of

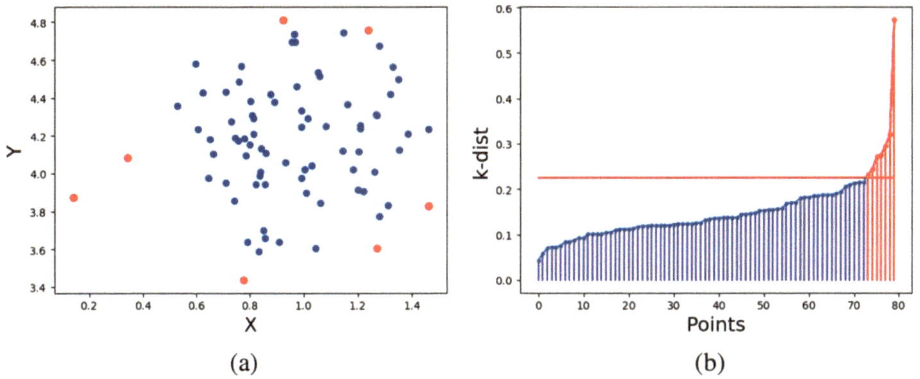

FIGURE 6.6 Demonstration of distance threshold determination by the proposed method. (a) A data cluster analogous to normal samples from steady state. (b) Its sorted k-dist curve.

the curve, where a ∈ (0, b). The "valley" point draws the separation between central dense regions and sparse boundary regions and is defined as follows:

$$x = min\left\{x \in [a,b] f\left(x + \frac{s}{2}\right) - f\left(x - \frac{s}{2}\right) s \geq f(b) - f(0) b\right\}. \tag{6.7}$$

6.3.4 Artificial Neural Network (ANN)

An ANN is a machine learning (ML) method in which the model is represented as nodes connected by edges and arranged in layers [9]. The basic unit of an ANN is a perceptron (see node 1, layer 1 in Figure 6.7). The output of each perceptron is computed as $o = activation\left(\Sigma_{i=0}^{N} x_i w_i\right)$, where w_i are the weights of the predictive model to be learned during training and *activation* is some function that regularizes the output of the perceptron.

ANNs are trained by calculating the predicted output for a given set of inputs, computing the error E between the result and the true value (target), and adjusting the weights accordingly. The weight between the nodes i and j, w_{ij}, is updated using the gradient descent algorithm:

$$w_{ij}^{updated} = w_{ij} + -lr \frac{\partial E}{\partial w_{ij}}. \tag{6.8}$$

Here, lr is the learning rate which regulates how much the weights change after each iteration and $\frac{\partial E}{\partial w_{ij}}$ is the partial derivative of the error between the predicted value and the target with respect to each weight. An extension of the gradient descent algorithm is the so-called mini-batch gradient descent (Eq. 5.6) which reduces the variance in the estimate of the gradient by processing a subset of instances (a batch)

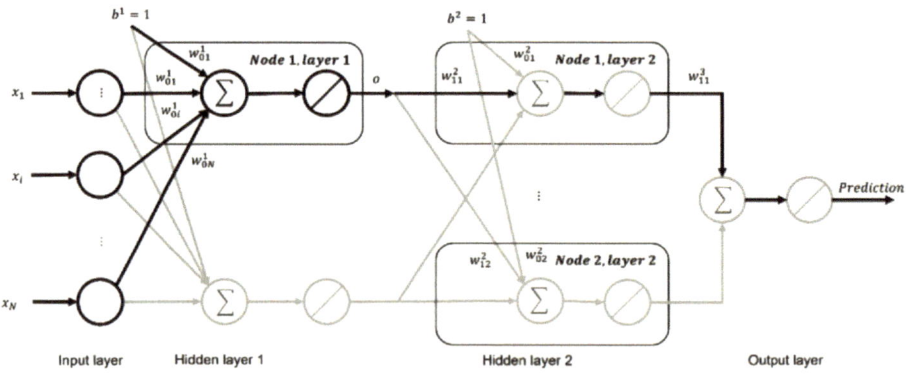

FIGURE 6.7 Simplified schematic of an artificial neural network.

at each iteration [10]. The weights' update equation for the mini-batch gradient descent algorithm is as follows:

$$w_{ij}^{updated} = w_{ij} + -lr\frac{1}{n}\sum_{m=n\times k}^{(k+1)\times n}\frac{\partial E}{\partial w_{ij}}, \tag{6.9}$$

where n is the batch size, $k = [1,\ N/k]$ is the number of batches for a dataset of size N. The number of nodes and layers and the batch size and learning rate are to be specified for each use case depending on the system to be modeled and the data available.

To train a neural network using gradient-based backpropagation, a significant bottleneck is to transit the gradient from the ending layers to the earlier layers. If the network has too many layers, the effect of the loss function on the first layers is almost negligible. This phenomenon is known as *gradient vanishing*. Some activation functions such as sigmoid [11] and tanh can cause gradient vanishing problems that constrain the depth of the network. To address the gradient vanishing problem, a number of activation functions have been proposed in the past few years. The rectified linear unit (ReLu) is one of the most popular function for neural networks [12] and is defined as follows:

$$f(x) = \max(0, x). \tag{6.10}$$

Many derivatives of the ReLu function have been proposed such as softplus [13] and leaky ReLu [14]. These functions provide stable gradients in the training process, as well as high computing effectiveness, which make the training of a neural network with many layers possible.

Since the topology of an ANN can be arranged in virtually any architecture, this method provides a very flexible platform for modeling. Also, the gradient descent algorithm provides a computationally efficient way for updating the model during training. However, ANNs still are susceptible to some shortcomings. While it is true that one can set up any architecture for the nodes and layers, the more complex the architecture (i.e., the larger the number of nodes or layers), the more data is needed for training. Bigger models are also prone to produce hypotheses that are highly specific to the data observed during training which gives them complications when trying to predict new, unseen data. This phenomenon, also called overfitting, will be discussed in detail later in this chapter. As mentioned before, another potential issue for ANNs is the vanishing gradient problem. When the number of layers is large, the influence of the output error on the update of the weights that are further away from the output layer may decrease fast, until the influence is not significant. A number of improvements over the basic fully connected ANN (the one presented here) have been made to address these challenges and will be introduced later in the book.

6.3.5 Ensemble Methods

A useful approach to overcome the limitations of individual ML methods is to train two or more models on the same data and then obtain the final prediction from a voting system. These strategies are referred to as ensemble methods and can be implemented in a wide variety of forms. A typical ensemble method for example are the so-called RF. RFs are DT-based methods in which instead of a single DT, a number of DTs (in the scale of hundreds) are trained and their outputs are averaged to produce a final prediction. This procedure helps overcome some of the issues inherent to DTs such as low resilience to noise. Ensemble methods could also be the mixture of different ML methods, such as ANNs, DTs, and SVM combined together, thus exploiting their respective benefits while compensating their shortcomings. One of the benefits of ensemble methods is the ability to measure the uncertainty associated with predictions. For example, if all the individual models in the ensemble make the same prediction for a given input, then it is reasonable to be confident about the prediction. On the other hand, if the individual models make drastically different predictions for the same input, this could be a sign that the prediction needs to be treated with caution. This measure of uncertainty is desirable in chemical engineering applications since most ML models are typically of a black-box nature and their behavior cannot be easily reasoned about. The underside of ensemble methods is that the number of parameters, that is, the complexity of the model, increases significantly. Additionally, the interpretation of ensemble methods is much more complicated if not intractable, in comparison with individual methods.

6.4 HYPERPARAMETERS

The concept of hyperparameters first introduced in Chapter 4 is revisited here in the context of SL. Hyperparameters can be thought of as specifications or constrains in the way the ML method searches for a valid hypothesis. Take for example an ANN. The mini-batch gradient descent algorithm outlined in Eq. 6.9 finds the best fitting set of weights, $W = \{w^0, \ldots, w^l, \ldots, w^L\}$, where $w^l \in \mathbb{R}^{n_i \times n_{i+1}}$, L is the number of layers in the ANN, and n_i is the number of nodes in the i-th layer. Although there are infinite possible combinations for W, the size and shape of the array – effectively the number of weights to adjust – is restricted by the predefined number of layers and nodes selected at the beginning of training. The algorithm is then constrained to find the best attainable solution with *this number of weights*. It may very well happen that any *useful* solution lies outside of this search space, and the ML method will never find a satisfactory model. The same logic applies for other specifications such as the learning rate, the batch size, or the activation function. In the same fashion as ANN, the other supervised learning methods presented in Section 5.3 also require hyperparameters to be defined by the user. The exhaustive list of all possible hyperparameters for each method is left to the reader and relevant references are provided at the end of this chapter. In most cases, a set of default hyperparameters is provided to the user in the code library or software in which the ML method is implemented.

Hyperparameters are essential to make the search for a suitable model tractable, but they also can make a good solution unreachable. The process of adjusting the hyperparameters for a given ML method so that a useful model is obtained is called tuning. Moreover, the task of finding a good set of hyperparameters for a given ML method is a search problem and can then be approached with a variety of search and optimization techniques as discussed in Chapters 7 and 8.

6.5 OVERFITTING AND REGULARIZATION

Intuitively, overfitting is the problem of a model adjusting too tightly to the training data to the point where it can no longer make good predictions for the testing data. This can be framed in the context of hypothesis search discussed before. Consider again the data in Table 6.1. Starting from the most specific hypothesis, data were used to modify the working hypothesis until it was able to predict the target variable for all the examples. Without the fourth example (supposed it was reserved for validation), the final hypothesis would be $\langle High, High, ?, Same \rangle \rightarrow yes$, requiring the last attribute to take the value "same" for the fault to be present. This hypothesis is too restrictive for the actual case and the model is overfitting. From a different perspective, one could artificially relax the hypothesis by picking one of the attributes at a time and evaluate the performance of the model on the validation data. The attribute that causes the accuracy to drop would be the one responsible for overfitting and therefore taken out from the model. The new model, which is a simpler one, generalizes better with the same amount of training data.

The previous scenario illustrates how overfitting can be understood either from the perspective of data availability or model complexity. In most applications, it is safe to assume that the data we are presented with is all the data available, and therefore, we will focus on model complexity to deal with overfitting. The principle of limiting model complexity to improve performance is known as regularization.

Regularization includes a variety of modifications to ML methods introduced to prevent the model from overfitting the training data. Given that each ML method handles training or learning in a different way, regularization strategies are model-specific. Here, we introduce three examples on how regularization looks like for different ML methods.

6.5.1 Decision Tree Pruning

Complexity in a decision-tree-based model is described by the size of the tree, that is, the number of branches (decision nodes) and its depth (levels between the root node and the leaves). Following the notion of reducing model complexity to prevent overfitting, pruning is a strategy in which a subtree rooted at a given node is removed and replaced by a leaf, which in turn is designated with the most common classification at below that node [1].

Although pruning can be performed as the tree is growing, pruning after a completed tree is obtained (post-pruning) is the most implemented strategy. The intuitive approach is to iteratively prune a subtree and evaluate the improvement in performance against validation data. This can be done recursively until a certain accuracy or model complexity criterion is met. For example, one could limit the depth of the DT to, say, five levels, forcing all branches beyond this depth to be pruned.

6.5.2 C and ϵ in Support Vector Machines

C and ϵ were introduced in Eq. 6.3 as part of the development of the SVR algorithm. Now, we turn attention to how these parameters contributed to regularizing model learning.

Perhaps the more intuitive of the two, ϵ, can be understood as the allowed distance between the hyperplane and the support vectors in Figure 6.5. A large enough value of ϵ will make all points fall within the region delimited by the support vectors, but the regression model would likely not be useful at all. On the other hand, forcing a very narrow margin between the hyperplane and the support vectors will force the hyperplane to pass through every single point, which would result in an overly complex model that will likely overfit. This is expressed mathematically in Eq. 6.3 showing ϵ as a relaxation term in the optimization constraints.

The C parameter works in close conjunction with ϵ. The ξ terms in Eq. 6.3 represent the distances between points beyond the support vector to it, and C allows the user to tell the algorithm how much these distances are penalized. A larger value of C translates into higher priority for the minimization of these distances (less regularization), thus forcing the model to "bend" the hyperplane to be close to the points outside of the region delimited by the support vectors. Conversely, a smaller value of C allows points beyond the support vectors to be far away without changing the hyperplane too much.

6.5.3 Dropout in Artificial Neural Networks

In ANNs, the learning rate hyperparameter (lr in Eq. 6.8) controls how much of the measured error is translated into updating the corresponding weight. Although this helps in moderating the influence of the training data over the final model, it does not change much about the model complexity. Hence, it is helpful to regularize learning once the topology of the model is known to be good enough.

Dropout is a strategy to regularize ANNs that can help reduce the complexity of the network by randomly selecting weights to be set to zero. The probability $\left(p \in [0, 1]\right)$ of a weight being zeroed (effectively removing its corresponding node) is a hyperparameter to be set by the user. Dropout can be implemented in a layer by layer basis, and typically input layers will have a lower dropout probability while a common value of 0.5 is often used for hidden layers.

6.6 METHOD SELECTION

The best attainable solution for a given method and a set of hyperparameters, even when the latter was optimized, might not be useful for an application. The different properties and characteristics of ML methods make them effective for certain situations and less convenient for others. Because of this, selecting the right method for the right application is an important task.

Although it is difficult to draw any hard guidelines on when to use an ML method over another, some basic intuitions can be considered. Often, the main driver on how well a method will perform is the data available for training. Characteristics of data, such as dimensionality, size, diversity, noise, and others, may influence model performance. Consider, for example, a DT-based approach. If data is noisy, the partitional and greedy nature of the algorithm could lead to selection of the wrong partitioning attributes or to high impurities in the leaf nodes. For ANN-based methods on the other hand, the backpropagation algorithm distributes the noise through the model, making it more resilient to noisy data. Alternatively, high-dimensional data will require a large number of weights in an ANN to be represented. This in turn sets the requirement of a big dataset to prevent overfitting. Such a problem is not present on DT-based methods since the search space decreases exponentially as the data gets partitioned.

Together with hyperparameter tuning, method selection is a search problem. A typical approach to find the right model is a nested optimization problem in which a subset of data is used to screen across multiple methods, optimizing the hyperparameter for each method and then comparing the best performing candidates. Once a method is selected, the rest of the available data is used to obtain the final model.

6.6.1 Example 6.1. Method Selection and Effect of Hyperparameters for State Classification on the Flowmeters Data

To illustrate some of the concepts introduced, we will use data from an eight-path liquid ultrasonic flowmeter to classify its state [15]. Data includes 87 records and 37 attributes as described in Table 6.2. Half of the records were used for training, and the remaining half was used for testing.

The correlation between the attributes and their distribution is shown in Figure 6.8. As observed, some attributes or features are highly correlated (dark green or dark red colors). This could indicate that not all the features are needed to find a good hypothesis that predicts the state of the flowmeter. This aspect will be discussed in the next section, and all attributes and instances will be used in this example.

Three different methods were used to perform the classification task: a DT, a SVM, and an ANN. All methods were implemented using the scikit-learn library in python [16], with the default hyperparameters unless otherwise noted. Figure 6.9 shows the results on the test set for the base case with default hyperparameters expressed as the receiver operating characteristic (ROC) curve. The ROC curve shows the relationship between the false positive rate (when the classifier assigns the wrong class) and the true positive rate. A random guessing would have a 0.5 probability of getting the classification wrong and show in the ROC curve as an identity line. As performance improves, the line approaches the upper left corner. A way of characterizing the results from a ROC curve is by calculating the area under the curve (AUC), with the best possible outcome being $AUC = 1$.

From the results in Figure 6.9, the ANN (MLPClassifier in the scikit-learn implementation) has the best performance followed by the SVM (SVC in the scikit-learn implementation) and with the DT in the third place. From this information, the obvious choice of method is the ANN. However, these results were obtained using little regularization, and we risk poor performance on new, unseen data. Therefore, we are interested in applying some regularization and assessing model performance again.

Table 6.3 shows the modified hyperparameters and Figure 6.10 shows the corresponding results. Performance of the ANN only decreased slightly with a great reduction in model complexity – from 100 to 10 neurons in the hidden layer. For the SVM, performance remained the same although C was reduced to a 10% of the default value. Meanwhile, performance of the DT showed a little improvement when pruning was in effect (depth with default parameters was 5). These simpler models are less prone to overfitting with comparable accuracy, which is desirable.

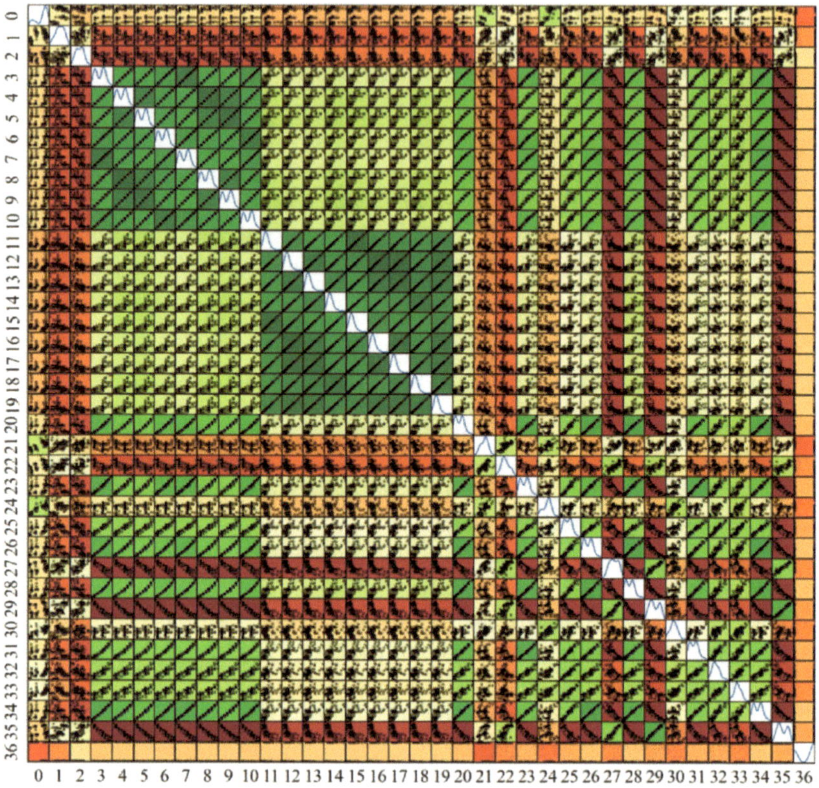

FIGURE 6.8 State diagnosis from a flowmeter. Numbers on axes represent feature identifiers. Darker red indicates a larger negative correlation. Darker green indicates a larger positive correlation. Elements in diagonal show frequency distribution of each feature.

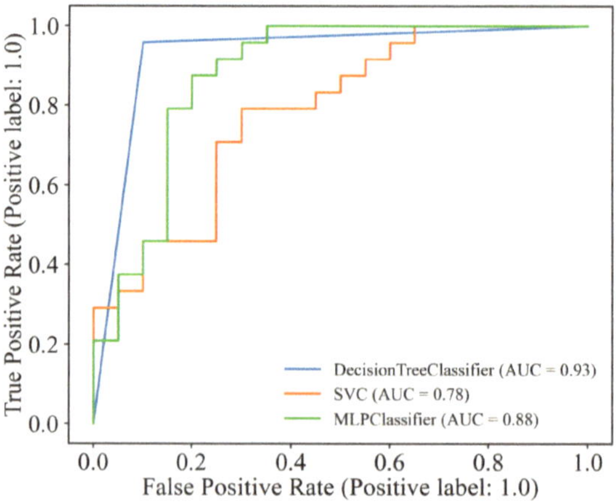

FIGURE 6.9 Receiver operating characteristic curve of different machine learning methods (default hyperparameters from scikit-learn) for classification on diagnosis data from a flowmeter.

TABLE 6.3
Hyperparameters for the ML Methods in Example 6.1

Method	Hyperparameters		
	Name	Default	Improved
Decision tree	Maximum depth	None	3
Support vector machine	C	1.0	5.0
Artificial neural network	Number of nodes per layer	100	10

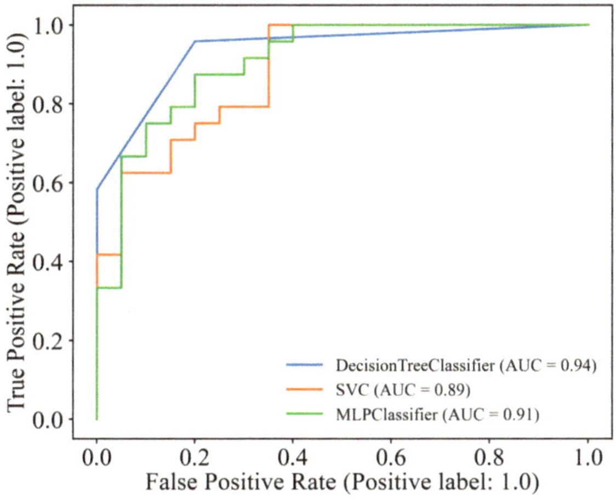

FIGURE 6.10 Receiver operating characteristic curve of different machine learning methods with the hyperparameters in Table 6.3 for classification on diagnosis data from a flowmeter.

6.7 FEATURE SELECTION

All features, that is, attributes in a dataset, might not be necessary for good model performance. Conversely, the necessary attributes to obtain a good model might not be readily available. Finally, some attributes may even hurt the model performance by introducing spurious information. The result of this is that before moving to select and optimize a ML method, a good practice is to have a look at the data and get a feel of the different attributes and their effect on the target variable.

As a general rule, the fewer features we have, the simpler the model we need and the easier the modeling task. One way of discarding less useful features is by looking at the correlation between these features and the target variable as well as the correlation between features (see Figure 6.7). If a feature has a low correlation with the target, then it is probably a good idea to discard it. On the other hand, if a feature is highly correlated to another one, domain knowledge could be used to select one of them and discard the other.

6.7.1 Example 6.2. Feature Selection for State Classification on the Flowmeters Data

To illustrate the effect of feature selection, we will use the same data used in Example 6.1. As stated before, there are highly correlated features in this dataset. Therefore, we will use feature correlation as a feature selection criterion. The result of the selection is illustrated in Figure 6.11. For pairs of features in which the Pearson correlation coefficient was greater than 0.9, one of the features was removed from the dataset.

Using the updated dataset, which now contains 11 features out of the original 37, we can train the models using the default hyperparameters. The results are summarized in Figure 6.12.

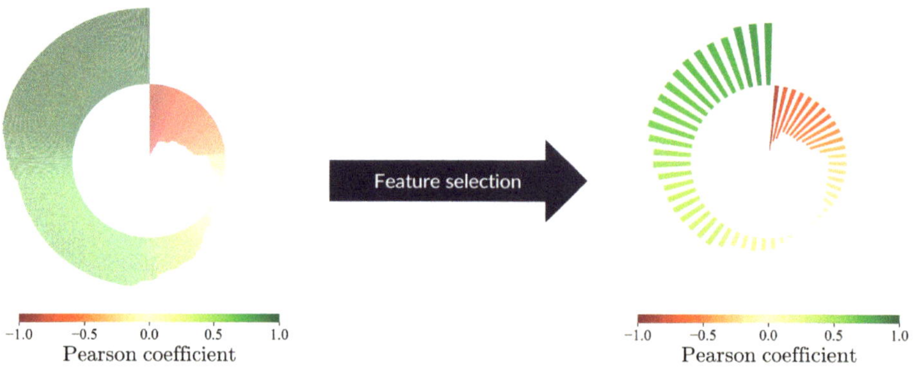

FIGURE 6.11 Pearson coefficient for features in the state diagnosis data from a flowmeter.

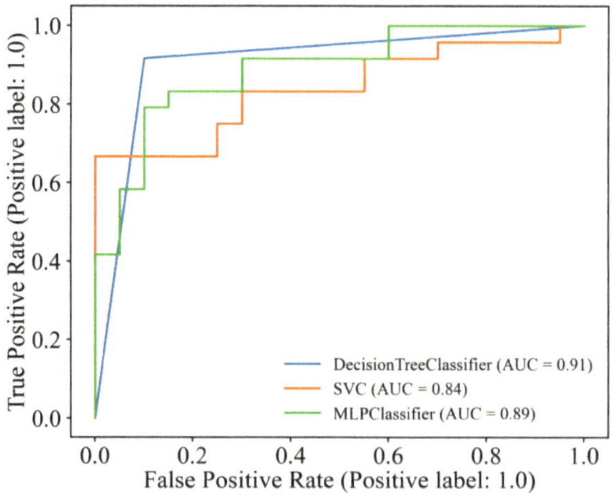

FIGURE 6.12 Receiver operating characteristic curve of different machine learning methods (default hyperparameters from scikit-learn) for classification on diagnosis data from a flowmeter using features selected by correlation.

After removing some to the features, performance of the SVM and ANN models is better than those trained on the full dataset (with default hyperparameters). This illustrates the effect of spurious features and how they can hurt model performance. This also serves as an example of how more data is only better if it is *good* data for the *right* ML method. Additionally, for the case of the DT model, performance is slightly worse, which showcases how flexible DT models can be. The DT model trained with the full dataset was able to exploit the information from data even when high correlations were present and when regularization constraints were in place.

6.8 FEATURE ENGINEERING

In some cases, features must be processed before they are passed as inputs to the ML model. This may be needed when the ML model is expecting tabular data but some of the features are in a different format. Consider for example a dataset in which one of the features is the molecular structure of a reactant which can be difficult to concatenate with the rest of features. Another reason for feature engineering is the ability to augment the information available for the model by performing calculations based on a feature but using a mechanistic model. For example, one could take a molecular structure and a surface to which said molecule is attached and run molecular dynamics simulations to obtain additional information about the energy state of the molecule or its orientation. Then, we pass these generated features to the ML model. This latter example falls into the realm of hybrid modeling, which will be introduced in the next chapter. Other applications of feature engineering are also detailed in subsequent chapters.

6.9 TRANSFER LEARNING

Transfer learning (TL) [17] is an ML technique derived from the notion of transfer of learning first proposed in the educational research context [18]. In TL, a data-driven model previously trained (general training) for a given task (source domain) is used as the base to build a model for a new task (target domain), with fewer data being required for the new training stage (task-specific training). Typically, the source and target domains are similar or closely related. As an illustrative example, a large database of labeled images of dogs (source domain) could be used for pretraining a learning agent to classify breeds, and then repurposed to classify images of cats (target domain). Formally, given a source domain $\mathcal{D}_s$, learning task $\mathcal{T}_s$, a target domain $\mathcal{D}_T$, and a learning task $\mathcal{T}_T$, TL aims to help the learning of the target predictive function $f_T(\cdot)$ for the target domain using the knowledge in $\mathcal{D}_s$ and $\mathcal{T}_s$, where $\mathcal{D}_s \neq \mathcal{D}_T$ and $\mathcal{T}_s \neq \mathcal{T}_T$. In chemical process applications, the source domain $\mathcal{D}_s$ and learning task $\mathcal{T}_s$ may come from simulations generated using a low-fidelity knowledge-based model. The target domain $\mathcal{D}_T$ and learning task $\mathcal{T}_T$ come from experimental or plant data.

A useful extension of TL is the so-called Few-Shot Learning (FSL) in which the task-specific training stage uses a very small amount of data (i.e., on the order of 1×10^1) [19]. By using FSL, new findings can be easily incorporated with little experimental cost. Furthermore, the information contained in a model for a given device can be readily transferred to a model for different, yet similar, device architecture [20].

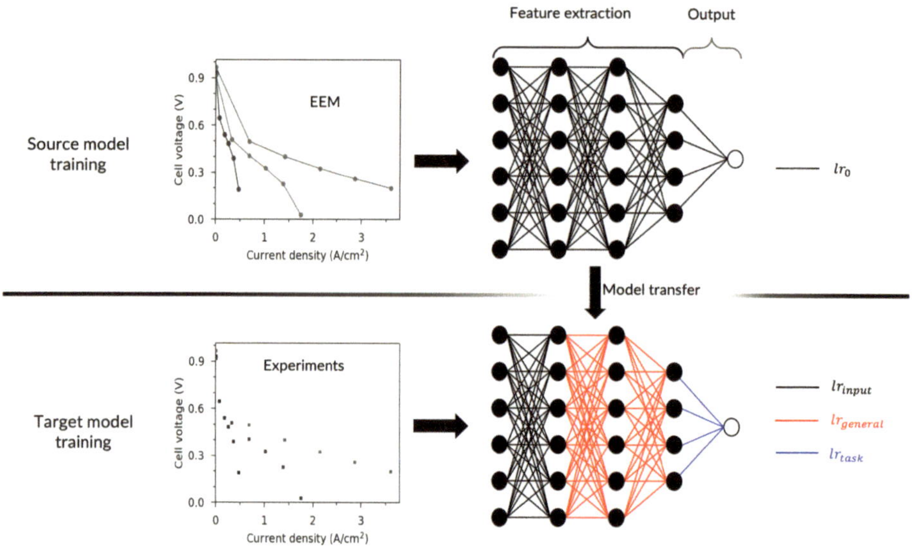

FIGURE 6.13 Schematic representation of the differential learning rate strategy for few-shot transfer learning. EEM = Explicit equation model.

FSL strategies are broadly based upon the notion of preserving information obtained through a first stage of training (i.e., a source model) and reusing it in some fashion to develop a target model. A practical way of implementing TL is through regularization. In the regularization approach, the target model has the same architecture as the source model (Figure 6.13), and the learned parameters (weights) are adjusted during a new training stage, but some restriction is applied to prevent the model from overfitting the new data [20]. In ANNs, for example, one form of regularization involves freezing most of the weights of the model and allowing only the weights of the last few layers of the model to be updated using the new data. While this approach works in some cases, figuring out which layers to freeze can be difficult. Another form of regularization, called differential learning rate, updates all weights during the task-specific training stage, but at a different rate (Figure 6.13), which helps control the influence of the new data over the model.

The problem of obtaining a good model via TL can then be divided in two stages: (i) find a good source model and (ii) find the best way to perform the task-specific training. The first stage involves finding the proper architecture and training conditions (number of hidden layers L, number of nodes N, the learning rate lr, and the batch size n) for the given source domain. The second stage requires finding a proper set of learning rates and batch size for a given target domain. Since the best combination of hyperparameters will be different for each application, there are not many guidelines as to what could work best. Because of this, obtaining the appropriate set of hyperparameters for a given application can become time-consuming. Although random search and grid search are widely used to explore combinations of hyperparameters, evolutionary algorithms have been shown to perform better for multi-objective hyperparameter optimization [21]. Chapter 8 will introduce more details on how to exploit TL in engineering applications.

6.10 CONCLUSIONS

The process of building a supervised learning model can be understood as that of finding the best hypothesis that explains the data available. In this process, characteristics of the data such as dimensionality (the number of features) and size of the dataset (number of examples available) are important as we must balance model complexity with data availability to prevent overfitting. The nature of the model architecture and its hyperparameters come into play as well as they guide and constrain the search for a good hypothesis. A highly complex architecture will likely have a large combinatorial space for the hyperparameters and require lots of data to converge to a good performance. On the other hand, a simpler architecture might be easier to tune, and require smaller datasets, but it might not be able to reach a satisfactory performance. In this chapter, we explored some representative supervised learning algorithms, along with important considerations in their implementation such as hyperparameter tuning and regularization. Also, data preprocessing methods including feature selection and feature engineering were presented to illustrate how transforming our data before the actual training step can lead to better results. Finally, we introduced the notion of TL and how can it be implemented in ANNs to alleviate the need for large datasets which can be scarce in some applications.

REFERENCES

1. Mitchell T. M. (1997), Machine Learning. McGraw-Hill, Inc.
2. Tan P.-N., Steinbach M., Kumar V. (2016), Introduction to Data Mining. Pearson Education India.
3. Drucker H., Burges C. J. C., Kaufman L., Smola A., Vapoik V. (1997), Support vector regression machines. Advances in Neural Information Processing Systems, 9, pp. 155–161.
4. Boser B., Guyon I., Vapnik V. (1992), A training algorithm of optimal margin classifiers. Proceedings of the Fifth Annual Workshop on Computational Learning Theory. Pittsburg.
5. Zhu W., Sun W., Romagnoli J. A. (2018), Adaptive k-nearest-neighbor method for process monitoring. Industrial & Engineering Chemistry Research, 57(7), pp. 2574–2586, doi: 10.1021/acs.iecr.7b03771
6. Thomas M. C., Zhu W., Romagnoli J. A. (2018), Data mining and clustering in chemical process databases for monitoring and knowledge discovery. Journal of Process Control, 67, pp. 160–175.
7. Ester M., Kriegel H. P., Sander J., Xu X. (1996), A density-based algorithm for discovering clusters in large spatial databases with noise. Proceedings of the Second International Conference on Knowledge Discovery and Data Mining. Portland, Oregon, United States.
8. Ankerst M., Breunig M. M., Kriegel H. P., Sander J. (199), OPTICS: Ordering points to identify the clustering structure. SIGMOD Record, 28(2), pp. 49–60, doi: 10.1145/304181.304187
9. Bishop C. M. (2006), Pattern Recognition and Machine Learning (Information Science and Statistics). Springer-Verlag.

10. Dekel O., Gilad-Bachrach R., Shamir O., Xiao L. (2012), Optimal distributed online prediction using mini-batches. Journal of Machine Learning Research, 13, pp. 165–202.
11. Narayan S. (1997), The generalized sigmoid activation function: Competitive supervised learning. Information Sciences, 99(1), pp. 69–82.
12. Hahnloser R. H. R., Sarpeshkar R., Mahowald M. A., Douglas R. J., Seung H. S. (2000), Digital selection and analogue amplification coexist in a cortex-inspired silicon circuit. Nature, 405(6789), pp. 947–951.
13. Nair V., Hinton G. E. (2010), Rectified linear units improve restricted Boltzmann machines. Proceedings of the 27th International Conference on Machine Learning (ICML-10). Haifa, Israel.
14. Maas A. L., Hannun A. Y., Ng A. Y. (2013), Rectifier nonlinearities improve neural network acoustic models. 30th International Conference on Machine Learning, Atlanta, Georgia: JMLR: W&CP.
15. Gyamfi K. S., Brusey J., Hunt A., Gaura E. (2018), Linear dimensionality reduction for classification via a sequential bayes error minimization with an application to flow meter diagnostics. Expert Systems with Applications, 91, pp. 252–262.
16. Pedregosa F., Varoquaux G. V., Gramfort A., Michel V., Thirion B. (2011), Scikit-learn: Machine learning in python. Journal of Machine Learning Research, 12, pp. 2825–2830.
17. Pan S. J., Yang Q. (2010), A survey on transfer learning. IEEE Transactions on Knowledge and Data Engineering, 22(10), pp. 1345–1359.
18. Perkins D. N., Salomon G. (1992), Transfer of learning. In Husen T., Postlethwaite T. N. (Eds), International Encyclopedia of Education. Pergamon Press.
19. Yang Q., Zhang Y., Dai W., Pan S. J. (2020), Transfer Learning. Cambridge University Press.
20. Zhu W., Braun B., Chiang L. H., Romagnoli J. A. (2021), Investigation of transfer learning for image classification and impact on training sample size. Chemometrics and Intelligent Laboratory Systems, 211, p. 104269.
21. Vishwakarma G., Haghighatlari M., Hachmann J. (2019), Towards autonomous machine learning in chemistry via evolutionary algorithms. ChemRxiv, doi: 10.26434/chemrxiv.9782387

7 Predictive Models

In the previous chapter, we discussed supervised learning methods, focusing on the computational aspects of their implementation. In this chapter, we introduce ideas and concepts related to the task of *modeling* in the context of engineering applications and show how supervised learning methods fit into the toolbox for model development. Rather than providing an exhaustive review of all the relevant topics and strategies, we introduce the key concepts and provide representative examples as well as references for further reading.

7.1 INTRODUCTION

Let us start with a definition of *model*. A model is a mathematical description of a physical system that can return information about a new state of the system based on the information currently available. The level of detail of the description can vary widely and must be chosen by the one building the model. A detailed description requires a more complex model, while a coarse description requires a simpler model. Simplicity is often preferred as simpler models are easier to build and use. However, a simpler model could potentially overlook important aspects of the physical systems and return inaccurate information. Also, since no model can fully capture the reality of the physical world, the scope of what the model is expected to capture must be defined. By defining *assumptions,* the scope of the model is reduced from all the possible events in the physical system to a subset of them. The model is expected to return valid information only if the assumptions are met. The more assumptions are defined, the narrower the scope of the model, and the less detailed and complex it needs to be. When considering assumptions and model complexity, as Ellie Arroway put it recalling Occam's Razor, "all things being equal, the simplest explanation tends to be the right one".

Another – and perhaps more practical – way of defining a model is as follows: a set of rules and parameters that map a set of inputs to a set of outputs. In some contexts, inputs are called *independent* variables while outputs are called *dependent* variables. Here, the more general terms inputs and outputs are adopted. *Parameters* are quantities that when combined with the *inputs* following the *rules* of the model produce the *outputs*. The degree of deviation of the model from the expected outputs is called *accuracy*.

One way of describing the complexity of a model is by the number of rules and parameters required to return an output. The higher the number of parameters or rules, the more effort is needed to calculate an output. Moreover, the number of inputs is related to the number of aspects of the physical aspects that are recognized by the model, while the number of outputs tells us how much information we can obtain about the state of the system.

DOI: 10.1201/9781003455905-9

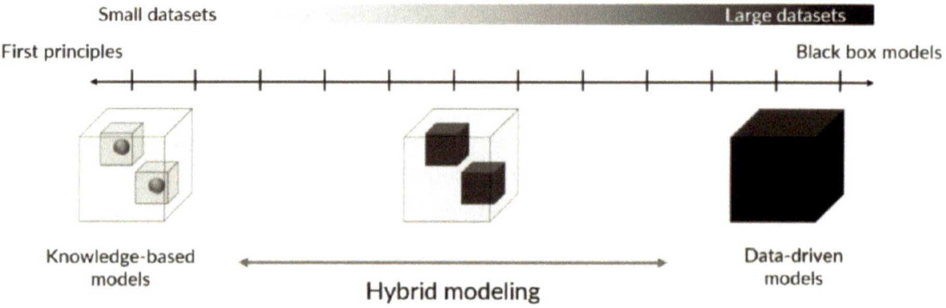

FIGURE 7.1 Spectrum of modeling approaches.

Given a set of outputs (measurements of quantities that represent the state of the system) and a set of inputs (measurements of quantities that influence the state of the system), the task of building a model is that of finding a set of rules and parameters that can return the expected outputs. It is in this procedure that challenges might arise. For example, the necessary measurements for both outputs and inputs might not have adequate resolution (think of frequency for events over time) or not be available at all. Also, the actual causal relationships in the physical system might be intractable, which will render some rules inaccessible. In such cases, assumptions can be defined to make the problem tractable and the model simpler.

The cost of simplifying a model can be represented by two distinct outcomes: (i) the accuracy of the model will be less than that of the model without the assumptions and (ii) some parameters of the model will not have physical meaning or will lump together more than one causal relationship, which obscures their interpretation as physical realities. The latter can be used to describe the spectrum of modeling strategies going from first principles (far left in Figure 7.1), in which all parameters have physical meaning, to data-driven modeling strategies (far right in Figure 7.1), in which none of the parameters bear physical meaning.

7.2 KNOWLEDGE-BASED MODELING

In engineering applications, knowledge-based models can be built when in-depth understanding of the laws of physics governing a certain phenomenon is available. For example, when a ball is thrown up in the air, the time for it to reach the ground can be predicted from Newton's laws of movement as $t = (v_0 \sin\theta)/g$. The model for the movement of the ball includes only one rule (one equation), two inputs (initial velocity, v_0, and angle at which the ball is projected, θ), and one parameter (gravitational acceleration, g) which has also physical meaning. As showcased by this example, knowledge-based models are powerful as they can have high accuracy with a few parameters and rules.

As powerful as it is, the model for the ball tells us nothing about the characteristics of the ball or its surroundings. This information is contained in the implicit assumption that these aspects of the physical system do not affect its behavior. If one were to ask whether the time to reach the ground would change if the ball had a different color or texture, or if the wind blows in a certain direction, the model would not be able to provide relevant answers. The model is now too simple for the information we intend to capture, and more rules (equations) and parameters need to be included in the model. As more information is needed from the model, its complexity will increase until it becomes intractable.

7.3 DATA-DRIVEN MODELING

When the laws of physics governing a phenomenon are not well known, one could infer the rules by observing the system. Using the same example from the previous section, let us assume that Newton's laws of movement are not available. An experiment can be set up in which a ball is thrown with different velocities and at different angles, and its position (height and length) is recorded at different times. The resulting records, that is, data, can be fitted to a mathematical function, say a sinusoidal wave $t = v_0 A sin(\theta)$. For a certain value of the parameter A, the function can predict the time it takes for the ball to reach the ground. Note, however, that unlike g in the knowledge-based model, A does not have a physical meaning. One might be tempted to conclude, by inspection, that $A = 1/g$, but for the data-driven model, A is capturing not only the gravitational effects but also *all other* effects that occurred during the experiments (characteristics of the ball and conditions of wind).

The data-driven modeling strategy gives us a simple description of a complex system. The cost of this simplicity is the physical interpretation of the model. One cannot be certain as to which aspects of the experiment will affect A and to which degree and, because of this, the model can only be used under the specific conditions of the experiment. A similar exercise can be repeated but with more inputs, that is, changing more aspects of the experiments. The model will then require more parameters and rules that will take it further away from physical interpretation. As more information is needed from the model, its opacity increases until it becomes a black box that while able to predict the state of the system cannot be interpreted from the physical point of view which limits its use.

7.4 SURROGATE MODELING

In surrogate modeling, simulated data from a complex model are used to build a simpler model that still represents the input-output relationships of the original model [1]. The new model would have some desired characteristics like low computational cost, high smoothness, or better explainability and is therefore better suited than the original for a given application, namely optimization [1]. Of special interest are machine learning (ML)-based surrogate models in which an explicit equations model (e.g., a system of differential equations) is used to generate simulated data that can in turn be used to train an ML model like those introduced in Chapter 6. Due to their intrinsic characteristics, ML models are easier to obtain than explicit equations models and provide great flexibility for different applications.

7.5 HYBRID MODELING

Hybrid modeling is a broad term used to describe the combination of different models and model synthesis methods. This notion is fundamental in Chemical Engineering practice, where sizing, optimization, and control methods rely on the offline estimation of empirical parameters that are then coupled into mechanistic models. A clear example of this is the sizing of chemical reactors. The mechanistic basis of the model is given by the mass and energy balances, while the kinetics of the chemical reactions are described by a set of parameters obtained from laboratory-scale experiments. In *Modeling in Chemical Engineering* [2], Octave Levenspiel refers to this, together with the boundary layer concept, the film model for heat transfer, and others, as some of the most influential advances in Chemical Engineering. Hybrid modeling as a concept provides a framework to analyze and tune the complexity of a model in response to the task at hand and the information available (see Figure 7.1). The examples mentioned above from typical Chemical Engineering practice would be located near the far left of the axis in Figure 7.1, with most of the system's behavior being captured by the mechanistic basis and only some aspects of it being described through data-driven models.

A working definition of hybrid modeling can be drawn from Falkenhainer and Forbus's *compositional modeling* [3]. Given a scenario description, the physical structure S and a (possibly empty) set of statements about its behavior (initial conditions, steady-state, etc.), a domain theory Th, consisting of domain model fragments $\left\{m_1, \ldots, m_n\right\}$ and a set of rules constraining their use, and a query about the scenario's behavior, the task is to obtain the most useful, coherent scenario model $M = \left\{m_i\sigma_i, \ldots, m_k\sigma_k\right\}$ for answering the query, where σ_i is a binding between the variables in m_i and objects in S, and each instance $m_i\sigma_i$ is unique.

Early implementations of hybrid modeling date back to the early 1990s and focused mainly in the introduction of empirical or data-driven models to fill gaps in mechanistic or knowledge-based models [4]. This could be conceptualized as moving to the right in the axis in Figure 7.1 as the scope of the data-driven elements grows. Despite the evolution of hybrid modeling and developments in data-driven modeling (e.g., ML), the prevalent idea is that of using data-driven models to improve mechanistic models, leaving the alternative, that is, using domain knowledge to improve data-driven models unexplored. The reader is encouraged to explore the works of Sansana et al. [4] and Bradley et al. [5] for an in-depth analysis of hybrid modeling strategies.

7.6 SEARCH AND OPTIMIZATION

Models, be it knowledge-based, data-driven, or hybrid, need to be found. That is, the modeler's task is to search for an adequate description of the system under study while accounting for certain limitations, such as information availability. Similar ideas were introduced in Chapter 4 when we discussed hyperparameter tuning. In this chapter, a formal definition of optimization is introduced and its relevance for modeling is illustrated by discussing the topic of parameter estimation. Further applications of optimization for modeling are shown in Chapter 8.

Given a set of functions $F = \left\{f_1, f_2, \ldots, f_n\right\}$ and a set of constraints $G = \left\{g_1, g_2, \ldots, g_m\right\}$ defined over a set of decision variables $x = \left\{x_1, x_2, \ldots, x_k\right\}$, an optimization problem can be defined to seek for the values of x that maximize (or minimize) the values assumed by the functions in F, subject to the constraints in G [6].

To solve the optimization problem, an optimization algorithm (OA) is utilized. One way of classifying OAs is as gradient based or gradient free. Gradient-based OAs rely on computing the gradient of the objective function and guide the search for the optimal values toward the direction of greater improvement (the steepest change in the gradient). To do this, the objective function must be continuous and twice differentiable, and when these requirements are not met, gradient-based OAs cannot be employed. Gradient free, on the other hand, exploit some structure or characteristic of the optimization problem to reduce the computational cost of exploring the search space without computing the gradient, at the expense of finding solutions that may not be optimal but are close to optimality. Since optimization problems in engineering applications can involve objective functions that are not continuous and/or differentiable, gradient-free OAs are often employed [7].

7.7 PARAMETER ESTIMATION

For a generic function $y = f(x_1, \ldots, x_n, a_1, \ldots, a_m)$, where the inputs $\{x_1, \ldots, x_n\}$ are mapped to y through some operations involving the coefficients $\{a_1, \ldots, a_m\}$, the latter are known as parameters of f. Mathematical models for physical systems will likely contain parameters that are to be fitted or estimated for the model to accurately predict the behavior of the system. Ideally, for knowledge-based models, these parameters will carry physical meaning, which implies that their values are linked to a physical property of the system or its components. However, in most cases, one or more parameters will not have physical meaning, or it will not be obvious.

Once a model is proposed, the next task is then to find the values of its parameters for a given system configuration. A wide variety of methods have been proposed for parameter estimation. As an example, the Jaya algorithm, proposed by Rao, is presented here [8]. The objective function for the Jaya optimization is the minimization of the standard error between the predicted values and the actual values:

$$\min\left(F = \sqrt{\frac{\Sigma_{k=1}^{K}\left(\hat{E}_k - E_k\right)^2}{k(k-1)}} \right). \tag{7.1}$$

The procedure starts sampling an initial solution population following **Algorithm 7.1**. Then, the best and worst solutions are selected, and the parameters of the solutions in the current population are updated such that

$$M_1'[d,j] = M_1'[d,j] + a_1\left(g_{best}[j] - \left|M_1'[d,j]\right|\right) - a_2\left(g_{worst}[j] - \left|M_1'[d,j]\right|\right), \tag{7.2}$$

where $M_1'[d,j]$ is the updated j-th parameter of the d-th solution, $g_{best}[j]$ and $g_{worst}[j]$ are the j-th parameter of the best and worst solution, respectively, and a_1 and a_2 are two random uniformly distributed numbers in $(0,r)$. The value r can be considered as a hyperparameter.

Algorithm 7.1. Sampling method for parameter estimation.

Input: Given J, $D \in \mathbb{R}$; $z \in \mathbb{R}^J$; u
Output: $\hat{S}_j$, $\hat{S}_{Tj}$, $j = 1, \ldots, J$

Initialization
1: Generate $D \times J$ matrices $\boldsymbol{M}_1$, $\boldsymbol{M}_2$, and $\boldsymbol{M}_3$ with entries $z_{d,j}$ varying randomly $z_j \pm u\%$
Sampling and resampling
2: for $m \leftarrow [1, 2]$ do
3: for $d \leftarrow [1, \ldots, D]$ do
4: $g_m[d] \leftarrow g(M_m[d, :])$ $\{M_m[d, :]$ is the d-th row of matrix M_m and $g_m[d]$ is the d-th entry of the vector $g_m\}$
5: if ($g_m[d]$ is not a feasible value) then
6: while ($g_m[d]$ not a feasible value) do
7: Use the first not used row k of M_3
$g_m[d] \leftarrow g(M_3[k, :])$
8: end while
9: $M'_m[d, :] \leftarrow M_3[k, :]$
10: end if
11: end for

The objective function is evaluated again for the updated solutions, then is compared against the current best solution, and if it is better, the solution is kept, otherwise is dropped. The algorithm finishes when the new best solution satisfies the stopping criterion or when the maximum number of iterations is reached (see **Algorithm 7.2**).

Algorithm 7.2. Jaya parameter estimation.

Input: Given $\boldsymbol{M}'_1[d, :]$, $d = 1, \ldots, D$; k, tol; it_{max}; $r_1, r_2 \in (0,1)$; $F = f(z)$, $\hat{S}_j$, $j = 1, \ldots, J$
Output: $z[j]$, $j = 1, \ldots J$

Initialization
1: $\boldsymbol{g}_{best} \leftarrow init > tol$
2: while $\boldsymbol{g}_{best} > tol$ or $it < it_{max}$ do
3: for $d \leftarrow [1, \ldots, D]$ do
4: $\boldsymbol{g}[d] = f(M'_1[d, :])$
5: end for
6: $\boldsymbol{g}_{best} = \min(\boldsymbol{g})$
7: $\boldsymbol{g}_{worst} = \max(\boldsymbol{g})$
8: Using the k parameters with highest $\hat{S}_j$
9: for $j \leftarrow [1, \ldots, k]$ do
10: for $d \leftarrow [1, \ldots, D]$ do
11: Update solutions using (E31)

end for
end for
$it \leftarrow it + 1$
end while
return $\boldsymbol{M}_1'[d, j]$

7.8 MODEL INTERPRETATION

Model interpretation is the procedure by which a model is systematically analyzed to establish the influence of its input variables or its parameters over the predicted values. Depending on the model, the interpretation could be straightforward or require a more involved procedure. Consider, for example, a simple linear regression model $y = mx + b$. The influence of the input x on the output y is given by the parameter m. The greater the m, the greater the influence x will have on the model output, and the sign of m will dictate the direction on that influence. This information can be extracted by inspection from the model. Consider now an artificial neural network with two layers, each one with three nodes. The expression for such model is as follows:

$$y = act^3\left(\boldsymbol{W}^3\left(act^2\left(\boldsymbol{W}^2\left(act^1\left(\boldsymbol{W}^1 x + b^1\right)\right) + b^2\right)\right) + b^3\right), \tag{7.3}$$

where $\boldsymbol{W}^i \in \mathbb{R}^{3\times 3}$ is the matrix of weights (parameters) for the i-th layer. Although still tractable, the interpretation is much more complicated than that of the linear model. Because of this, several methods have been proposed for the efficient and accurate interpretation of data-driven models.

7.8.1 Sensitivity Analysis

7.8.1.1 Sobol' Sensitivity Analysis

Sobol' method for global sensitivity analysis (GSA) was first introduced in 2001 [9]. The method proposes the series expansion of a function $G = g\left(z_1, \ldots, z_j, \ldots, z_J\right)$ into orthogonal terms of increasing dimensions with mutually independent input variables:

$$G = g_0 + \sum_{j=1}^{J} g_j(z_j) + \sum_{1 \le j < b \le J} g_{jb}(z_j, z_{jb}) + \cdots + g_{1,2,\ldots,J}\left(z_1, \ldots, z_J\right), \tag{7.4}$$

where the index j denotes the variable of interest, b another variable, and J is the total number of evaluated variables. Each term in G has quadratic integrability over the domain of existence, where g_0 is a constant. The decomposition of the variance of G is given by:

$$V(G) = \sum_{j=1}^{J} V_j + \sum_{1 \le j < b \le J} V_{jb} + \cdots + V_{1,\ldots,j,\ldots,J}, \tag{7.5}$$

where V_j, V_{jb}, and $V_{1,\ldots,j,\ldots,J}$ are the individual variances of $g_j(z_j)$, $g_{jb}(z_j, z_{jb})$, and $g_{1,2,\ldots,J}(z_1, \ldots, z_J)$, respectively.

The calculated variances can be used to compute the sensitivity indices as

$$\hat{S}_j = \frac{\hat{V}_j}{\hat{V}}, \tag{7.6}$$

$$\hat{S}_{Tj} = 1 - \frac{\hat{V}_{-j}}{\hat{V}}, \tag{7.7}$$

where $\hat{V}_j$ and $\hat{V}_{-j}$ are the sum of all variances terms that include and exclude z_j, respectively.

This sensitivity indices provide a way to classify variables according to their individual contribution to the changes in the overall variance of the function. Furthermore, the sensitivity indices can indicate whether the model is additive (interactions are present) or not. If $\hat{S}_j < \hat{S}_{Tj}$, then the model is nonadditive and if $\hat{S}_j = \hat{S}_{Tj}$, the model is additive [10].

The standard Sobol' method has been subject of several modifications to improve its performance [11, 12]. In this work, we adopted the method proposed by Salas et al. in which a third sampling matrix is used as a contingency for the case of unfeasible solutions (see Algorithm 7.3) [13].

A description of the GSA algorithm is shown on the next page. For this implementation, two objective functions were used – one for a parameter sensitivity analysis, G_1, and another for variable sensitivity analysis, G_2:

$$G_1 = \sqrt{\frac{\Sigma_{k=1}^{K}\left(E - E_{exp}\right)^2}{k(k-1)}}, \tag{7.8}$$

$$G_2 = \max\left(\frac{EI}{A}\right). \tag{7.9}$$

Algorithm 7.3: Sobol' global sensitivity analysis.

Input: Given J, $D \in \mathbb{R}$; $z \in \mathbb{R}^J$; u; $G = g(z)$
Output: $\hat{S}_j$, $\hat{S}_{Tj}$, $j = 1, \ldots, J$

Initialization
1: Generate $D \times J$ matrices $\boldsymbol{M}_1$, $\boldsymbol{M}_2$, and $\boldsymbol{M}_3$ with entries $z_{d,j}$ varying randomly $z_j \pm u\%$
Sampling and resampling
2: for $m \leftarrow [1, 2]$ do
3: for $d \leftarrow [1, \ldots, D]$ do

$g_m[d] \leftarrow g(M_m[d,:])$ {$M_m[d,:]$ is the d-th row of matrix M_m and $g_m[d]$ is the d-th entry of the vector g_m}
if ($g_m[d]$ is not a feasible value) then
while ($g_m[d]$ not a feasible value) do
Use the first not used row k of M_3
$g_m[d] \leftarrow g(M_3[k,:])$
end while
$M'_m[d,:] \leftarrow M_3[k,:]$
end if
end for
Output varying the input variables

$$\hat{g}_0 \leftarrow \frac{1}{2D}\sum_{d=1}^{D}(g_{1_d} + g_{2_d})$$

for $j \leftarrow [1,\ldots,J]$ do
$N_j \leftarrow M'_2$; $N[:,j] \leftarrow M'_1[:,j]$ {$N_j[:,j]$ and $M'_1[:,j]$ are the j-th columns of the matrices N_j and M'_1 respectively}
$N_{T,j} \leftarrow M'_1$; $N_{T,j}[:,j] \leftarrow M'_2[:,j]$
for $d \leftarrow [1,\ldots,D]$ do
$g'_{1,j}[d] \leftarrow g(N_j[d])$
if ($g'_1[d]$ is not a feasible value) then
$g'_{1,j}[d] \leftarrow \hat{g}_0$
end if
$g'_{2,j}[d] \leftarrow g(N_{T,j}[d])$
if ($g'_{2,j}[d]$ is not a feasible value) then
$g'_{2,j}[d] \leftarrow \hat{g}_0$
end if
end if
end for
First and global sensitivity indices

$$\hat{V} \leftarrow \frac{1}{2D}\sum_{d=1}^{D}(g_{1_d}^2 + g_{1_d}^2) - \hat{g}_0^2$$

for $j \leftarrow [1,\ldots,J]$ do

$$\gamma_j^2 \leftarrow \frac{1}{2D}\sum_{d=1}^{N}(g_1[d]g_2[d] + g'_{1,j}[d]g'_{2,j}[d])$$

$$\hat{V}_j \leftarrow \frac{1}{2D}(g_1 \cdot g'_{2,j} + g_2 \cdot g'_{1,j}) - \gamma_j^2$$

$$\hat{V}_{-j} \leftarrow \frac{1}{2D}(g_1 \cdot g'_{1,j} + g_2 \cdot g'_{2,j}) - \gamma_j^2$$

$\hat{S}_j \leftarrow \frac{\hat{V}_j}{\hat{V}}$

$\hat{S}_{Tj} \leftarrow 1 - \frac{\hat{V}_{-j}}{\hat{V}}$

end for

return $\hat{S}_j$, $\hat{S}_{Tj}$, $j = 1, \ldots, J$

7.8.1.2 Subspace Greedy Search

The subspace greedy search (SGS) algorithm seeks the combination of the most contributing subspace of variables which causes the separation between the clusters of data. To do this, SGS searches from the lowest dimension and moves to higher dimensions until no possible subspaces are left. A testing score, defined as the k-distance between clusters for each subspace, allows for comparison between different subspaces at the same dimension. The subspaces with the highest scores are kept for further testing and the rest are discarded. Further details about the SGS implementation can be found in Zhu et al. [14].

7.8.2 Model Explanations

Extracting model explanations from ML models can be difficult, especially for complex models. To deal with this, Lundberg and Lee [15] proposed Shapley additive explanation (SHAP) values to express the importance of variables. SHAP rests on the idea of obtaining an explanation model g for each individual prediction, that is a linear combination of the binary variables $z' \in \{0, 1\}^M$, where M is the number of input features. As shown in Eq. 9.8, the final prediction will then be the additive result of the individual influences $\phi_i \in \mathbb{R}$.

$$g(x') = \phi_0 + \sum_{i=1}^{M} \phi_i x_i'. \tag{7.10}$$

Figure 7.2 illustrates the idea of the explanation model. The prediction starts at zero when no information is available. Then, the predicted value moves according to the influence of each input, given by ϕ_i. To calculate the SHAP values for each input, an explanation model with all the inputs is evaluated and compared to the evaluations of all possible explanation models that do not include the variable of interest. The final value for a given input is then a weighted average of all the differences.

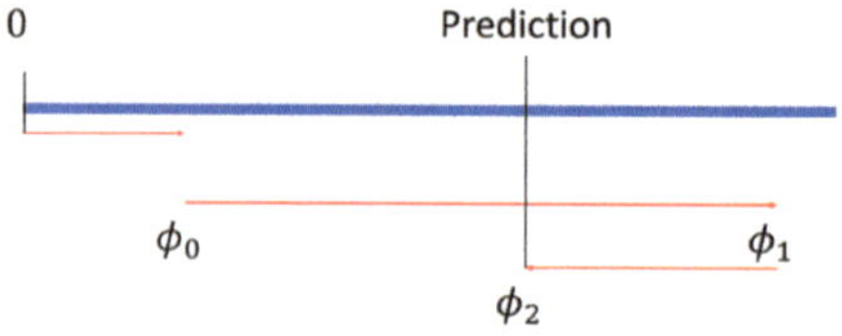

FIGURE 7.2 Shapley additive explanation values attribute to each feature the change in the expected model prediction when conditioning on that feature.

7.9 EXAMPLE: HIGH-TEMPERATURE PEM FUEL CELL: HYBRID MODELING

7.9.1 Hybrid Modeling

The formulation of a hybrid model of an experimental high-temperature polymer electrolyte membrane fuel cell (HT-PEMFC) is an excellent candidate to illustrate some of the concepts discussed in this and the previous chapters. A detailed description of the system is given in Appendix C and more information about the study can be found elsewhere [16]. The goal is to obtain a model capable of predicting the total fuel cell potential, E, at a given current density (i) given a set of operating conditions, cell design, and materials' specifications. The key challenge, however, is the scarce availability of cell potential data at different operating conditions for a variety of cell designs and materials. Even though there are data available for materials' characterization, performance of these materials in actual fuel cells at different operating conditions is not available. Hence, we are interested in finding a way of combining these data so that we can relate materials' properties to cell potential.

The basis for the model is the additive nature of the ideal potential (E_{oc}) and the non-idealities or losses caused by phenomena such as mass transfer limitations, electric resistance, and reaction kinetics. These losses are also known as overpotentials, denoted with η. Eq. 7.11 shows the model basis. This will also be referred to as the device-level model.

$$E(i) = E_{oc} - \eta_{act} - \eta_{con} - \eta_{ohm}. \tag{7.11}$$

Hybrid nature of this model lies on the fact that each of the overpotentials, as well as the ideal potential, is calculated using either a knowledge-based model, a fully data-driven model, or a combination of both. The details for each of the overpotentials is given in Appendix C. For this discussion, we will focus on the ohmic overpotential (Eq. 7.12), which carries information about the electric properties of the membrane materials.

$$\eta_{ohm} = i\left(\frac{\delta_{mem}}{\kappa_{mem}} + \frac{\delta_{io}}{\kappa_{io}}\right). \tag{7.12}$$

In Eq. 7.12, κ_{mem} and κ_{io} are the ionic conductivities of the membrane and the ionomer binder (see Appendix C), which are predicted using support vector regression (SVR) models trained on materials' characterization data.

Combination of materials' characterization data with operation and design data results in flexible hybrid modeling framework. Figure 7.3 shows a schematic view of the modeling procedure. Scientific knowledge is introduced in this case in the form of the model basis – the additive nature of the potential losses – and the incorporation of physics' laws such as the Nernst equation for the ideal potential and the Butler-Volmer equation for the cell kinetics.

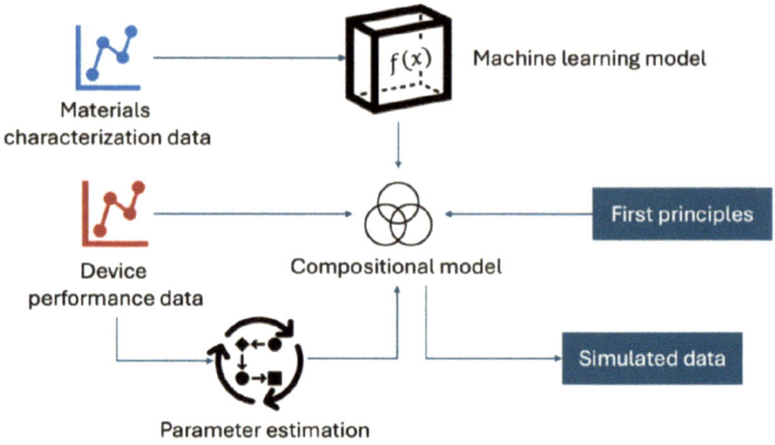

FIGURE 7.3 Bridging the gap between materials data and device data.

Data available for the materials' characterization included around 130 samples of conductivity measurements for different polymer electrolyte membrane formulations with different ionic exchange capacities (IEC) at different operation temperatures. IEC is assumed to be a generalizable descriptor of the membrane chemistry, meaning that it could be translated from one material to another. Meanwhile, temperature is one of the operational variables included in the cell potential model. By introducing a model for conductivity that is a function of a material descriptor and an operational variable, we can establish a relationship between cell performance and materials.

SVR was selected as model architecture because of the small size of the dataset and the expected high nonlinearity of the relationship between IEC, temperature, and conductivity. Here, SVR was implemented through the *sklearn* python package [17] with a radial basis function kernel, a kernel coefficient $\gamma = 5$, a regularization parameter $C = 100$, and $\varepsilon = 0.1$. Figure 7.4 conveys that the support vector machine models for both the polymer electrolyte membrane and the ionomer binder (characterized as a thin film) give an accurate prediction for the ionic conductivity as a function of material IEC (for non-acid imbibed values) and temperature.

Figure 7.5 presents the polarization curve from the device-level predictions and the polarization data (note: data is from beginning of life for the fuel cell). Figure 7.5 also shows the individual overpotential terms as a function of current density from the HT-PEMFC model. These overpotential terms govern the polarization of the HT-PEMFC.

7.9.2 Model Interpretation – Sensitivity Analysis

Sobol' GSA was used to identify which parameters in the knowledge-based and semi-empirical components of the model were the most influential on the variability of the model (see Figure 7.6) (for parameters definition refer to Appendix C). These results provide information about the physical significance of the parameters

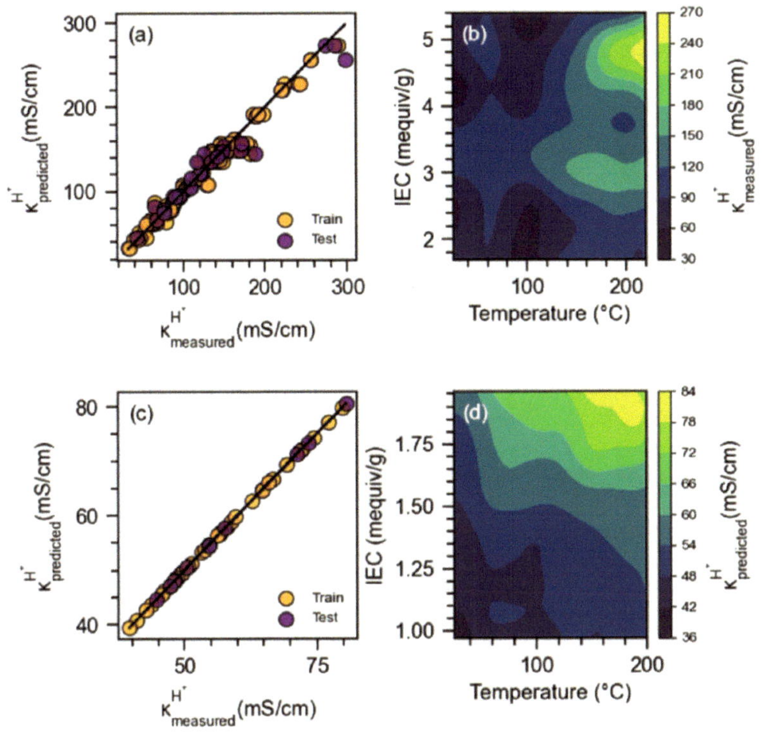

FIGURE 7.4 Model performance and predictions for ionic conductivity as a function of temperature and IEC using a support vector regression with a radial basis function kernel. (a–b) Polymer electrolyte membrane and (c–d) thin ionomer films that are similar in thickness to the ionomer binder in electrode layers.

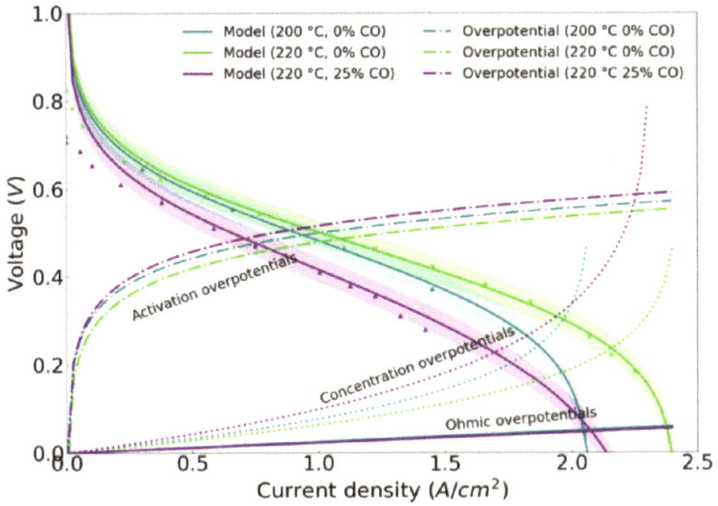

FIGURE 7.5 Device level model predictions.

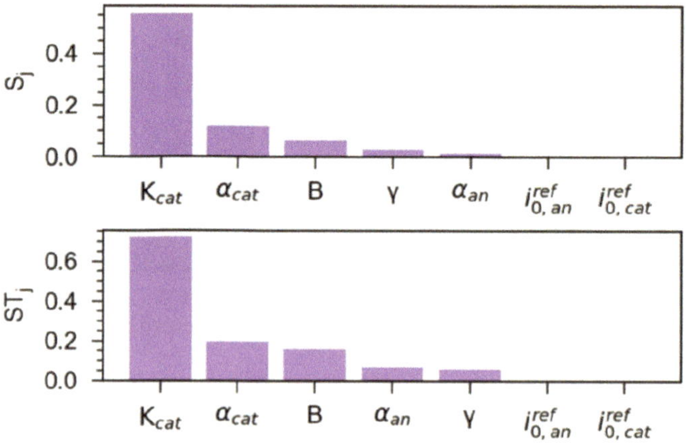

FIGURE 7.6 Sobol' sensitivity analysis for the model parameters.

involved in the model. The charge transfer coefficient for both the cathode and the anode, as well as the empirical parameter for the acid uptake in the concentration overpotential, was shown to be the most influential on polarization behavior in the low current density regime.

A sensitivity analysis was also performed on the model inputs. Figure 7.7 shows the ranking of importance for the inputs to the device level model. The variables identified as most influential by the sensitivity analysis were temperature, T, ion exchange capacity of the ionomer binder IEC_{io}, and the stoichiometric coefficient of oxygen, S_{O_2}. Out of these, T and S_{O_2} are expected to be highly influential for the system from physical intuition, as they influence reaction kinetics. The influence of IEC_{io}, however, is surprising as it reveals that it is the ionomer binder conductivity

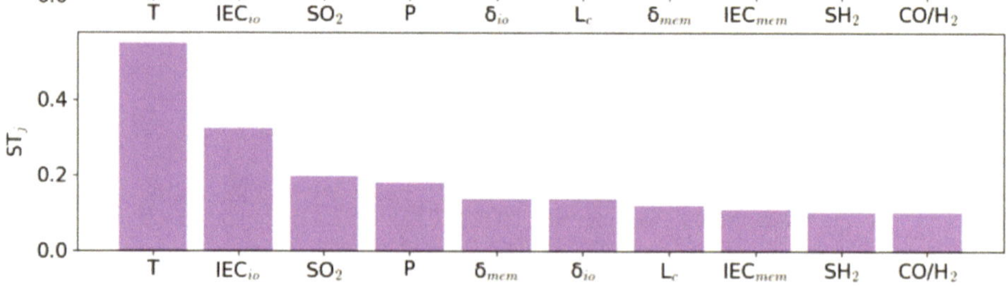

FIGURE 7.7 Sobol' sensitivity analysis for the model inputs.

and not the membrane conductivity which is the limiting factor in the system. This showcases the high performance of the polymer electrolyte membrane material under study and points future research directions to improve the performance of the ionomer binder.

7.9.3 Design Space Exploration Using Unsupervised Learning

With an experimentally substantiated semi-empirical HT-PEMFC model for ion-pair polymer electrolytes, it was possible to explore different configurations using a stochastic generator. To produce such configurations, the input variables were allowed to take different values around the base case in a uniform distribution within a given interval (e.g., ± base value × 40%). For example, a given configuration would include a lower catalyst loading, L_c, and higher temperature and pressure while another configuration would have the same catalyst loading but with a smaller membrane thickness, δ_{mem}. Gaussian noise was introduced into the model calculations to account for the variability of the model predictions.

Hierarchical density-based spatial clustering of applications with noise (HDBSCAN) clustering preceded by uniform manifold approximation and projection (UMAP) for dimension reduction was selected to find patterns (i.e., clusters) in the data associated with high power density values and to identify the more contributing variables to the clustering. We implemented HDBSCAN with 35 *min_samples*, and UMAP with 10 *n_neighors*, 0.1 *min_dist*, and 3 *n_components*. The resulting clusters were further analyzed using the SGS algorithm described in Chapter 7. The variables identified with the largest contributions to the separation between the classes are the same as those with the higher first-order sensitivity indices from the Sobol' GSA. Furthermore, a set of candidate combinations for the improvement of the fuel cell performance measured as peak power density was found from the cluster analysis.

Figure 7.8 shows the visualization for the clustering analysis using HDBSCAN with UMAP as preprocessing. Three major regions are seen in the cluster visualization: a region of low power density (far left in Figure 7.8), a high power density region (far right in Figure 7.8), and a transition region in between.

Although the combination of UMAP with HDBSCAN preserves the relationship of the high-dimensional space in the low-dimensional embedding, the physical interpretation of the low-dimensional variables Umap1, Umap2, and Umap3 is not readily accessible. Therefore, we applied an SGS analysis (see Figure 7.9) in which the variables S_{O_2}, T, and IEC_{io} were identified as the most influential in the formation of the clusters.

The variables designated by the clustering and SGS analyses as the more influential over the peak power density are the same as those with the higher first-order index in the modified Sobol' GSA – which serves as cross validation for the UMAP-HDBSCAN clustering and SGS analysis approach. With this, not only are a set of candidate configurations (values for design and operating variables) attained but it also provides insights as to which variables should be prioritized for enhancing HT-PEM performance and efficiency.

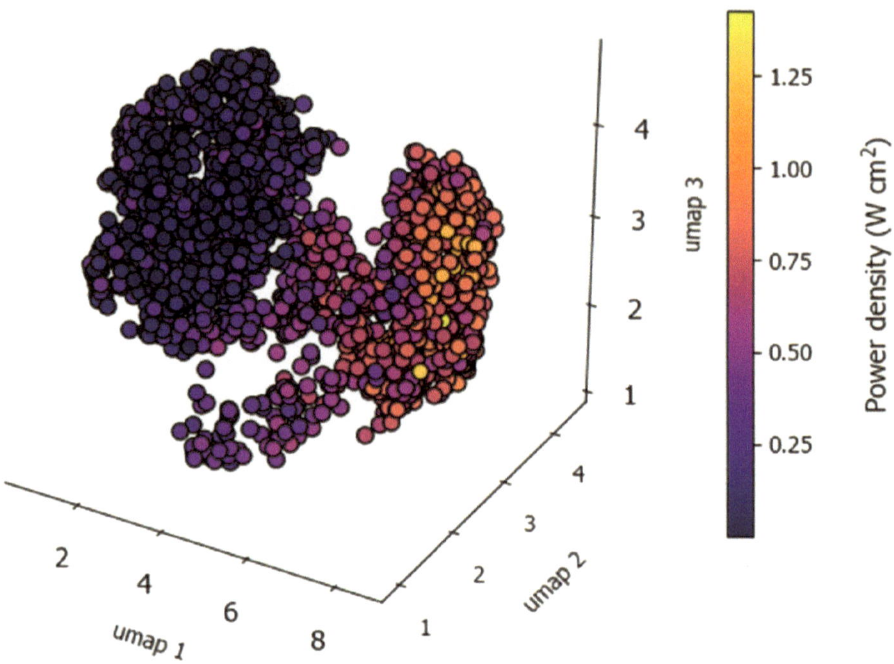

FIGURE 7.8 Visualization of the design space via UMAP and HDBSCAN.

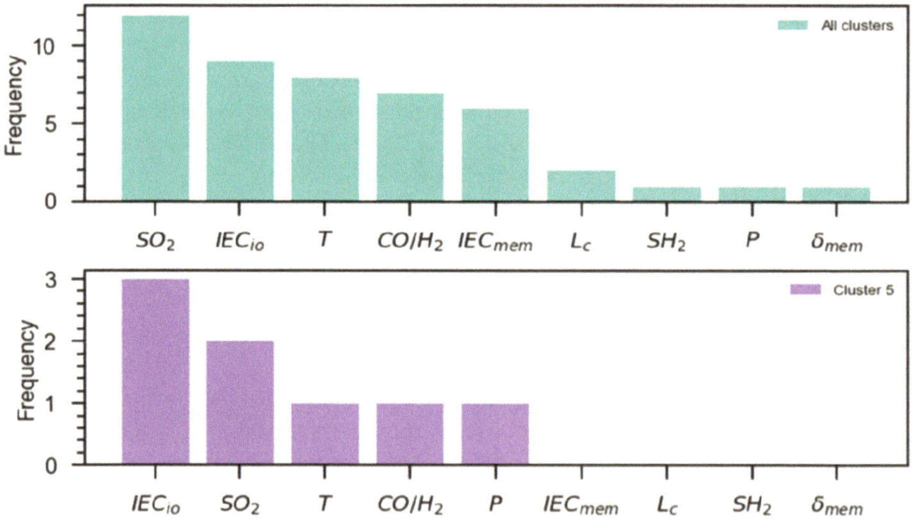

FIGURE 7.9 Influence of the input variables obtained from subspace greedy search applied on the HDBSCAN-UMAP clustering.

7.10 CONCLUSIONS

Modeling is deeply embedded in Chemical Engineering practice. The advent of new modeling techniques such as those within the realm of ML only expands the set of tools available for modeling chemical processes and phenomena. In this context, this chapter highlighted two major modeling approaches, namely knowledge-based modeling and data-driven modeling, and then delved into how combinations of these approaches can be used to yield cost-effective and accurate results via surrogate modeling and hybrid modeling. Moreover, the notion of parameter estimation from the perspective of search and optimization was discussed and illustrated with the Jaya algorithm. These ideas provide a solid foundation for the understanding of modeling approaches and, importantly, the possibilities for the integration of knowledge-based models that contain information about the physical nature of the phenomena and data-driven models that make the modeling task tractable. This chapter also showcased the importance of sensitivity analysis using the Sobol' method as a representative example. An example of hybrid modeling and its use to guide the design of a fuel cell was presented to illustrate how these techniques can be implemented in practice.

REFERENCES

1. McBride K., Sundmacher K. (2019), Overview of surrogate modeling in chemical process engineering. Chemie Ingenieur Technik, 91(3), pp. 228–239, doi: 10.1002/cite.201800091
2. Levenspiel O. (2002), Modeling in chemical engineering. Chemical Engineering Science, 57(22), pp. 4691–4696, doi: 10.1016/S0009-2509(02)00280-4
3. Falkenhainer B., Forbus K. D. (1991), Compositional modeling: Finding the right model for the job. Artificial Intelligence, 51(1), pp. 95–143, doi: 10.1016/0004-3702(91)90109-W
4. Sansana J., Joswiak, M.N., Castillo, I., Wang, Z., Rendall, R., Chiang, L.H., Reis, M.S. (2021), Recent trends on hybrid modeling for industry 4.0. Computers & Chemical Engineering, 151, p. 107365, doi: 10.1016/j.compchemeng.2021.107365
5. Bradley W., Kim, J., Kilwein, Z., Blakely, L., Eydenberg, M., Jalvin, J., Laird, C., Boukouvala, F. (2022), Perspectives on the integration between first-principles and data-driven modeling. Computers & Chemical Engineering, 166, p. 107898, doi: 10.1016/j.compchemeng.2022.107898
6. Nocedal J., Wright S. J. (2006), Numerical Optimization. 2nd ed. Springer.
7. Taha H. A. (2006), Operations Research: An Introduction. 8th ed. Prentice-Hall, Inc.
8. Rao R. V. (2016), Jaya: A simple and new optimization algorithm for solving constrained and unconstrained optimization problems. International Journal of Industrial Engineering Computations, 7, pp. 19–34.
9. Sobol I. M. (2001), Global sensitivity indices for nonlinear mathematical models and their Monte Carlo estimates. Mathematics and Computers in Simulation, 55(1), pp. 271–280, doi: 10.1016/S0378-4754(00)00270-6
10. Cosenza A., Mannina G., Vanrolleghem P. A., Neumann M. B. (2014), Variance-based sensitivity analysis for wastewater treatment plant modelling. Science of the Total Environment, 470–471, pp. 1068–1077, doi: 10.1016/j.scitotenv.2013.10.069
11. Saltelly A. (2002), Making best use of model evaluations to compute sensitivity indices. Computer Physics Communications, 145(2), pp. 280–297, doi: 10.1016/S0010-4655(02)00280-1

12. Qiong-Li W., Paul-Henry C., Amélie M. (2012), An efficient computational method for global sensitivity analysis and its application to tree growth modelling. Reliability Engineering & System Safety, 107, pp. 35–43, doi: 10.1016/j.ress.2011.07.001
13. Salas S. D., Brandão A. L. T., Soares J. B. P., Romagnoli J. A. (2029), Data-driven estimation of significant kinetic parameters applied to the synthesis of polyolefins. Processes, 7(5), p. 309.
14. Zhu W., Sun W., Romagnoli J. A. (2018), Adaptive k-nearest-neighbor method for process monitoring. Industrial & Engineering Chemistry Research, 57(7), pp. 2574–2586, doi: 10.1021/acs.iecr.7b03771
15. Lundberg S. M., Lee S. (2017), A unified approach to interpreting model predictions. In Guyon I., Luxburg U. V., Bengio S., Wallach H., Fergus R., Vishwanathan S., Garnett R. (Eds.), Advances in Neural Information Processing Systems 30, pp. 4765–4774. Curran Associates, Inc. http://papers.nips.cc/paper/7062-a-unified-approach-to-interpreting-model-predictions.pdf
16. Briceno-Mena L. A., Venugopalan G., Romagnoli J. A., Arges C. G. (2021), Machine learning for guiding high-temperature PEM fuel cells with greater power density. Patterns, p. 100187, doi: 10.1016/j.patter.2020.100187
17. Pedregosa F., Varoquaux G. V., Gramfort A., Michel V., Thirion B. (2011), Scikit-learn: Machine learning in python. Journal of Machine Learning Research, 12, pp. 2825–2830.

8 Supervised Learning Case Studies

8.1 CASE STUDY 1: MONITORING AN INDUSTRIAL PYROLYSIS REACTOR

8.1.1 Data-Driven Process Monitoring

Up to this point, we have discussed supervised and unsupervised learning applications separately, and for the case of supervised learning, a lot of the focus has been on regression models. In the case studies used in this chapter, we illustrate how both supervised and unsupervised learning can be coupled together to tackle process monitoring challenges casted into classification problems.

Chemical process monitoring can be separated into several tasks outlined in the flowchart in Figure 8.1. The four main tasks that compose process monitoring are fault detection (also known as *abnormal event detection* or AED), fault identification, fault diagnosis, and process recovery. Machine learning methods and ideas can be applied for each of these tasks, adding to the toolbox of process monitoring strategies.

Each process monitoring task uses data initially drawn from plant sensor measurements stored in the process's distributed control systems (DCS). Fault detection,

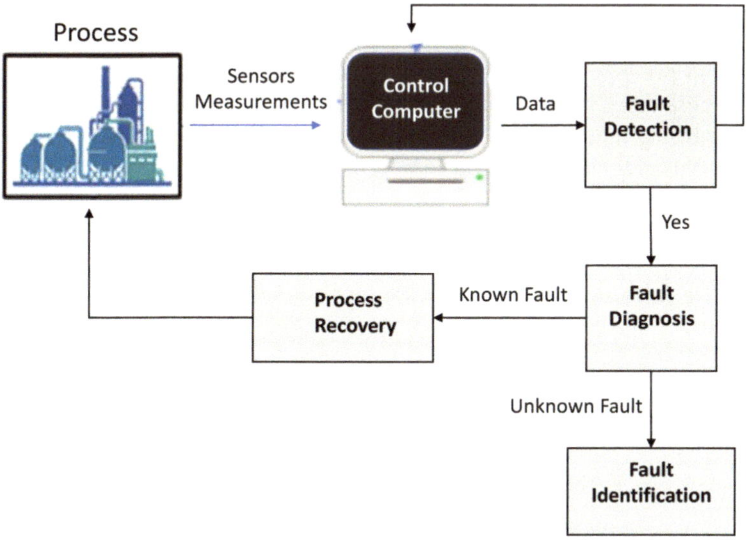

FIGURE 8.1 Industrial process monitoring workflow.

DOI: 10.1201/9781003455905-10

diagnosis, and identification all involve some amount of mathematical calculation using this digitized sensor measurement data. The data are first processed by a fault detection algorithm which often functions similar to a statistical test. Similar to the way that a z-test or Student's t-test might generate a test statistic and compare it to some threshold value, the fault detection task usually involves the generation of a test statistic against a bound representing the boundary between normal and faulty operations.

When our test statistic exceeds this bound, we say a fault has occurred, and while the statistic is below the bound, we say operations are normal. From the machine learning perspective, this step can be thought of as *binary classification problem* in which a new state of the plant can be identified as normal or abnormal. When an abnormal state is found, the next step is fault diagnosis. If data from previously experienced, expected faults is available, ideally a pattern recognition fault diagnosis algorithm can be trained to classify the data as corresponding to the known fault and enable operators to proceed immediately to the process recovery step to address the conditions creating the fault status. This scenario would be a multiclass classification problem, where we not only care about finding an abnormal state but also *which* abnormal state. If there is no data from previously available faults or the observed abnormal data does not fit into any known fault models, the workflow proceeds to fault identification. Fault identification uses the model of normal operations to assist operators in isolating the sensor measurements responsible for faulty observations. Techniques for sensitivity analysis and model interpretation are useful for fault identification. Without fault identification, operators may see a faulty plant status but not act if the sensors producing the faulty statistic are not obvious. Fault identification can guide operators' investigation into the root causes of the fault event by focusing attention to a certain area of the plant more relevant to the diagnosis of the fault.

The purpose of implementing a process monitoring AED system is not to automate the management of abnormal events but to assist in incorporating operators into the loop. Any approach to process monitoring must provide useful insight in quickly assessing an abnormal situation and offer a user-friendly way for operators to view and process large amounts of information. Additionally, they must accomplish all these tasks in a scalable, user-friendly manner without requiring expensive, time-consuming maintenance from application engineers or experts in data mining. Table 8.1 illustrates some general requirements for a fault detection scheme.

TABLE 8.1
General Requirements for a Fault Detection Scheme

	Nature of Process	Requirement of Fault Detection
(1)	Faulty data is rare in real operation. Commonly, only normal data is provided in training.	The binary classifier is able to determine boundaries on normal data.
(2)	Process could be either linear or nonlinear.	The classifier should be robust to nonlinearity of the process.
(3)	Normal process drifting could be caused by tube coking, deformation, and equipment aging.	The classifier can trace this process behaviors without misclassifying them.
(4)	Real-time monitoring.	High computational speed.

8.1.2 Background

Principal component analysis (PCA)-based methods for fault detection are well established in both industry and research [1–5]. These monitoring approaches aim to capture variables exhibiting the highest variance from normal data, facilitating effective process monitoring. The establishment of control limits for the normal region involves the computation of Hotelling's T^2 and squared prediction error (SPE) statistics. Any sample surpassing these control limits is flagged as a fault. Despite its utility, the PCA model presents notable limitations. Primarily, being a linear dimensionality reduction technique, it may lack efficacy in predicting control limits for nonlinear datasets. Furthermore, the thresholds for PCA T^2 are dependent on the training data provided, and if the training set does not represent the entire normal region, the model may produce a high false alarm rate. Moreover, a fixed PCA model cannot track normal process drifting, which is prevalent in most industrial processes due to catalyst deactivation, equipment aging, and tube coking [6].

To address the previously mentioned issues, two adaptation methods have been incorporated into the fixed PCA model: recursive PCA (RPCA) [7] and moving window PCA (MWPCA) [8, 9]. While these strategies effectively mitigate false alarms arising from normal process variations, their implementation in real-world processes poses challenges. RPCA continuously updates process data recursively, resulting in a substantial expansion of the training database and a notable increase in computational load. In response to RPCA's data size challenge, MWPCA was developed by discarding older samples; however, its robustness in a real process context may be insufficient. The model's vulnerability arises when confronted with repetitive samples in a steady state before a drift occurs, potentially causing irreversible tightening of control limits and a deterioration in model quality. Furthermore, it is important to note that both RPCA and MWPCA are grounded in a linear PCA model.

The *k*-nearest neighbor algorithm, commonly referred to as *k*-NN, stands out as a widely used method for classifying unknown objects [10]. When confronted with an unlabeled object, the *k*-NN algorithm assigns it to the class exhibiting the highest similarity. This approach finds extensive application in diverse fields such as pattern recognition, data mining, and outlier detection. Despite its prevalence, limited research has delved into the utilization of *k*-NN for process monitoring and fault detection. He and Wang [11, 12] pioneered a *k*-NN-based method for identifying faults in semiconductor manufacturing processes. In He's methodology, the model exclusively incorporates normal data, designating a faulty sample as an outlier positioned distantly from the normal samples. The squared nearest neighbor distance threshold is derived from a non-central chi-square distribution, acting as the boundary delineating the normal region. However, He's approach lacks an adaptation strategy.

In this section, we delve into the application of the adaptive *k*-NN (Ak-NN) approach discussed in Chapter 6, which is built upon the *k*-NN algorithm. The primary objective of this approach is to tackle various challenges, such as nonlinearity, insufficient training data, computational complexity, and time-varying behaviors [13].

8.1.3 Pyrolysis Reactor Monitoring Using A*k*-NN

The monitoring of the pyrolysis process is crucial to ensure safe operation, but it poses challenges due to the complex conditions involved, such as high temperatures, tube coking, and process drifting. Typically, chemical plants rely on DCS for process monitoring. These systems display sensor measurements and enable operators to observe plant operations. In many DCS systems, measurements are visualized on individual charts or process diagrams. Faulty conditions are determined by the violation of predefined thresholds for each measurement.

This study used approximately 70,000 data records, representing 90 days of process operation. By coloring the data according to their time index (Figure 8.2), it is possible to observe slow drifting over time. During the process, faulty data points appear as sparse points around the dense clusters. At the end of the process, a dark red cluster indicates a step change due to changes in steam flows. Without prior data, this step change would be classified as a faulty condition. The goal is to track normal process drifting while simultaneously detecting abnormal conditions.

As pointed out previously, in our study, the sorted *k*-NN distance curve is used to characterize the distribution of normal samples. The last "valley" point on the curve is used as the distance threshold (see Figure 8.3), which is estimated from the derivative of the curve. The "valley" distance can naturally catch the central dense regions and avoid the interference from noise.

In Figure 8.4, the A*k*-NN method is shown tracing the pyrolysis reactor process for a selection of values for the updating factor, p, and the number of neighbors, k. The first 1000 data points (after the startup condition) were used to train the model for normal conditions, while the rest were utilized for adaptive monitoring. The normal drifting region is effectively traced, and faulty samples that deviate notably from the normal region are accurately detected. Additionally, the plot highlights green values in regions where model updating occurs due to process drift.

A zoomed-in region of the first part of the process cycle, showcasing the startup condition, is depicted in Figure 8.5. The data in this area are labeled as faults since it is out of the norm. The chart indicates normal operation once the process reaches its regular state.

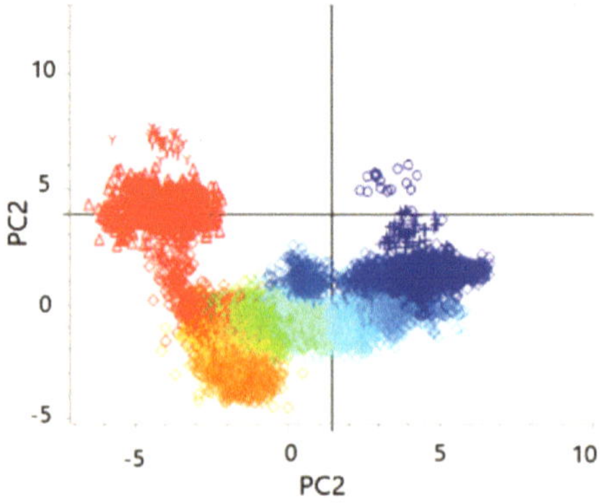

FIGURE 8.2 Pyrolysis reactor data projected by PCA.

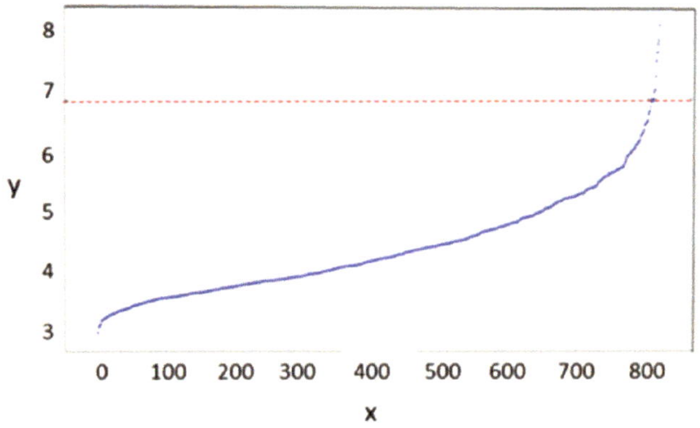

FIGURE 8.3 Distance threshold determination for pyrolysis process.

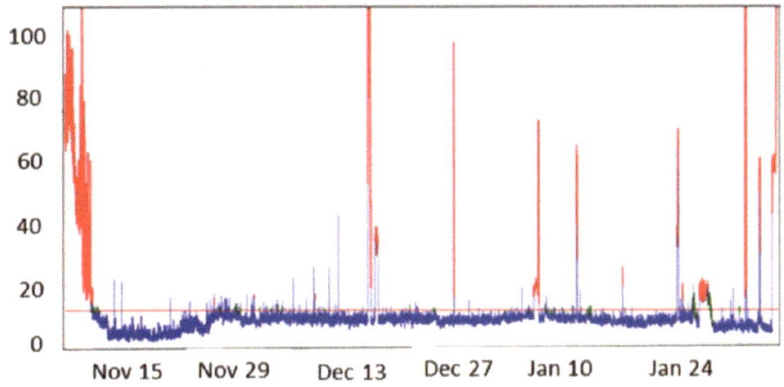

FIGURE 8.4 Fault detection result for pyrolysis reactor over whole operating cycle ($p = 0.75$ and $k = 30$).

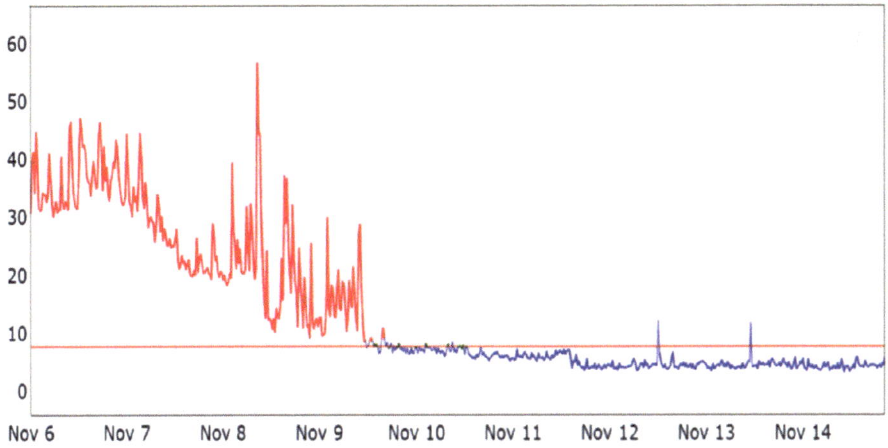

FIGURE 8.5 Fault detection result for pyrolysis reactor. Expanded view of first portion of cycle (startup).

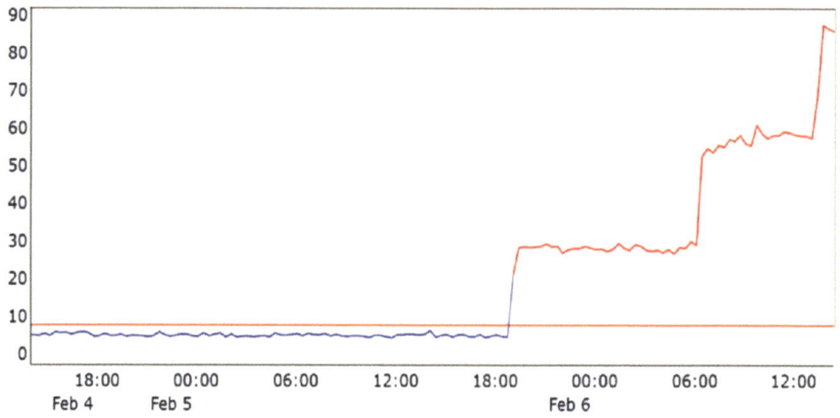

FIGURE 8.6 Fault detection result for pyrolysis reactor. Expanded view of last portion of cycle.

The zoomed-in region of the last part of the unit cycle, where the detection of the faulty condition is clearly shown, is presented in Figure 8.6. In Figure 8.7, a zoomed-in section of the cycle is illustrated, where green values in the plot indicate regions of model updating due to changing conditions in the furnace. Lastly, Figure 8.8 depicts the monitoring system that utilizes a fixed-model PCA approach. However, the fixed PCA model fails to predict the normal drifting in the real pyrolysis process. In other words, after the normal condition, the entire normal drifting is classified as faulty.

Figures 8.9–8.12 demonstrate the effect of parameter p on performance across a range of values. The value of parameter p is closely tied to the process itself, as it is intended to describe the degree of process variation. In the case of the pyrolysis process, the process variation arises from natural drifting, which is gradual and consistent. As a result, a smaller p value results in a monitoring system that is highly sensitive to faults, as shown in the figures. Conversely, increasing p from 0.25 to 0.75 substantially reduces the occurrence of faulty conditions, as indicated by the decrease in red spots. Figure 8.9 also highlights a more radical approach, allowing for model updating even when the data is dissimilar. This means that the model will

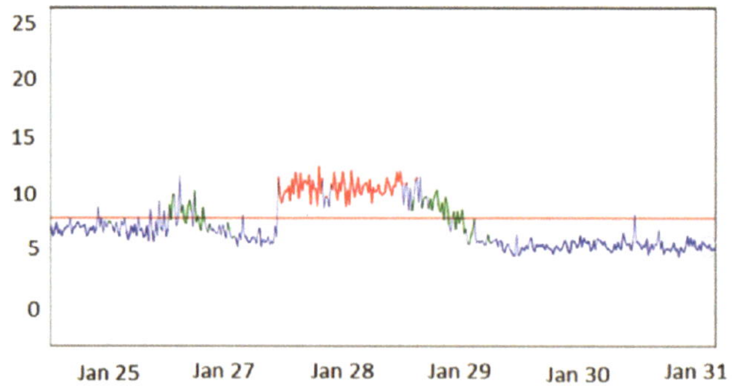

FIGURE 8.7 Fault detection result for pyrolysis reactor. Model updating.

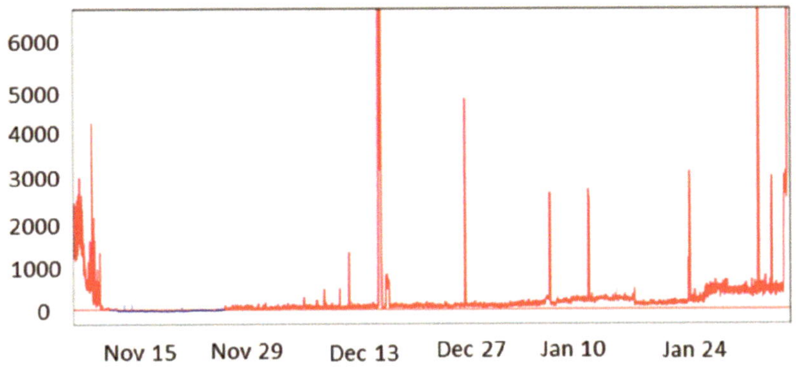

FIGURE 8.8 Fault detection result for pyrolysis reactor over whole operating cycle with PCA.

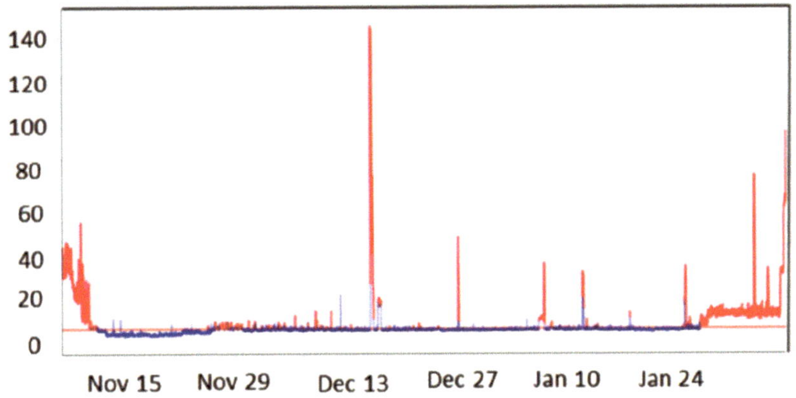

FIGURE 8.9 Fault detection result for pyrolysis reactor over whole operating cycle $p = 0.25$.

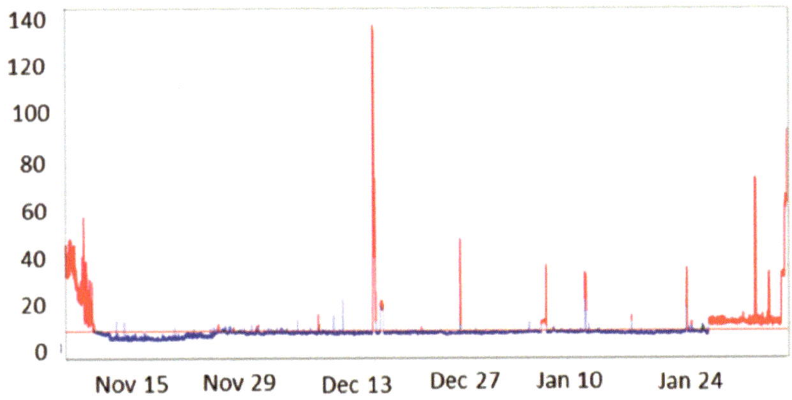

FIGURE 8.10 Fault detection result for pyrolysis reactor over whole operating cycle $p = 0.50$.

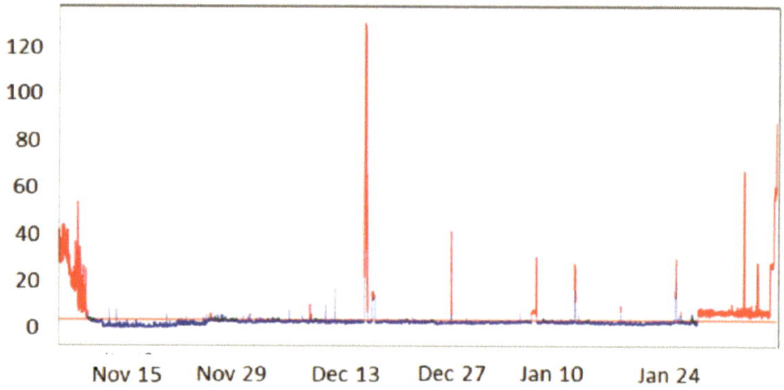

FIGURE 8.11 Fault detection result for pyrolysis reactor over whole operating cycle $p = 0.75$.

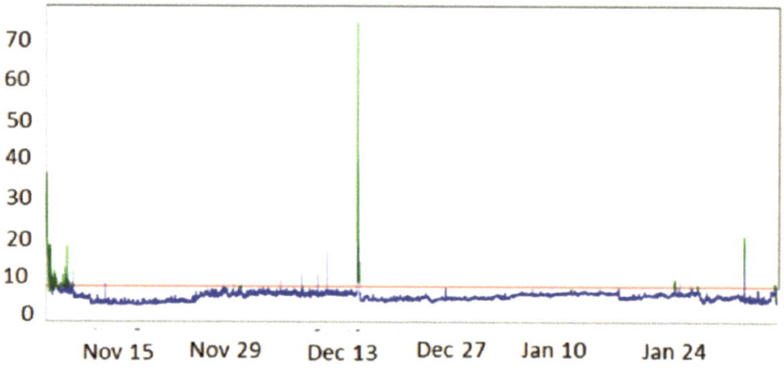

FIGURE 8.12 Fault detection result for pyrolysis reactor over whole operating cycle: forced by not using the threshold for adaptation.

be updated even under faulty conditions. In this case, both the value of the parameter p and the number of neighbors, k, are adjusted.

In Chapter 4, we discussed the use of the subspace greedy search approach to identify the combination of variables that contribute the most to a detected faulty condition. Figure 8.13 shows the results of this approach for the fault detected in Figure 8.4 (at the end of the reactor cycle). The figure displays the contributing variables and the time evolutions of the two most significant contributors, which are the stem flowrates of coils 1 and 3. The sharp and drastic changes observed in these two flowrates (around 6000 datapoints) are responsible for the process anomaly.

The monitoring architecture enables visualization of the algorithm's evolution over the unit full cycle using Pairwise Controlled Manifold Approximation and Projection (PaCMAP) projection. Figures 8.14–8.16 show the PaCMAP projection on the 2D space. Figure 8.14 shows the data projected in the plane and colored by time, allowing the appreciation of the process evolution through the different clusters (operating conditions). Figure 8.15 displays the Ak-NN method tracing the pyrolysis reactor process for $p = 0.75$ and $k = 30$. Initially, the process starts at normal conditions and most points correspond to normal situations (green). Then,

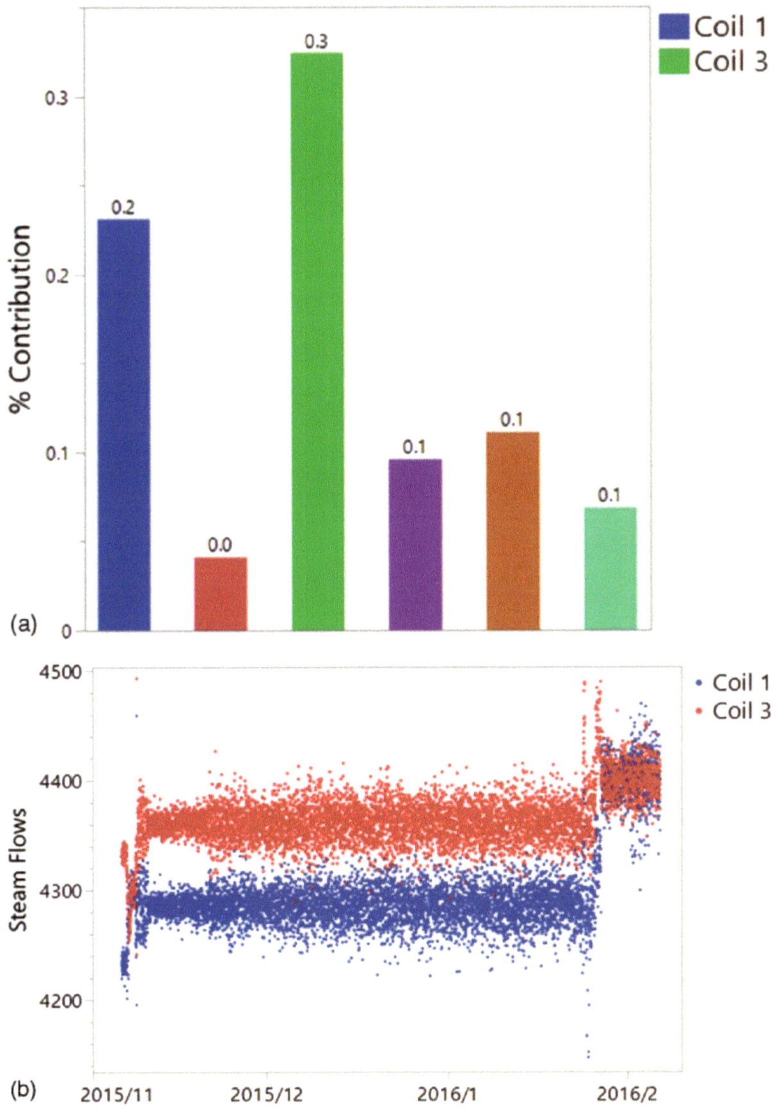

FIGURE 8.13 Fault identification for pyrolysis reactor: (a) SGS variable contribution plot and (b) time evolution of most contributing variables.

the process shifts to different operating conditions where the adaptive monitoring operates (green and blue) until the end of the cycle, where the faulty condition is detected (red). The normal drifting region is effectively traced, and faulty samples that deviate notably from the normal region are accurately detected. Figure 8.16 illustrates the monitoring results using PCA. In this case, all states, except for the starting normal condition where the PCA model was trained, are categorized as faulty (red).

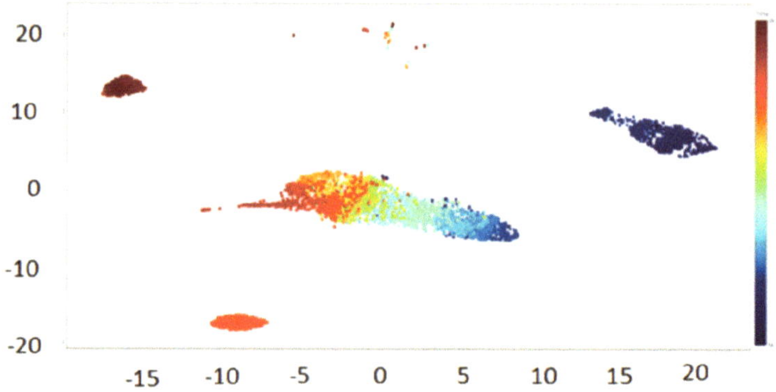

FIGURE 8.14 Data project into 2D colored based on time.

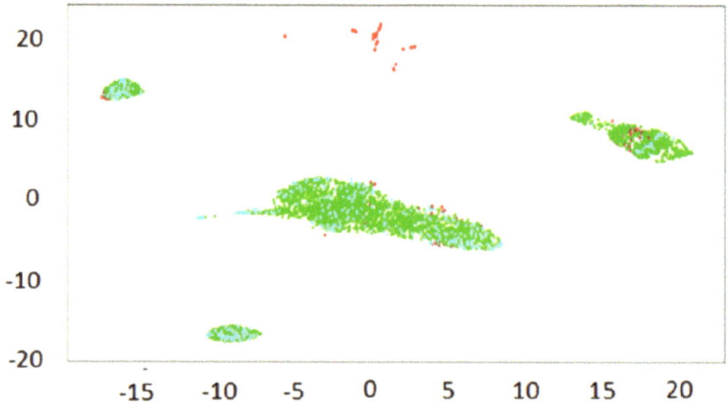

FIGURE 8.15 Monitoring results for Ak-NN monitoring scheme with $p = 0.75$ and $k = 30$: red dots – faulty conditions, green dots – normal.

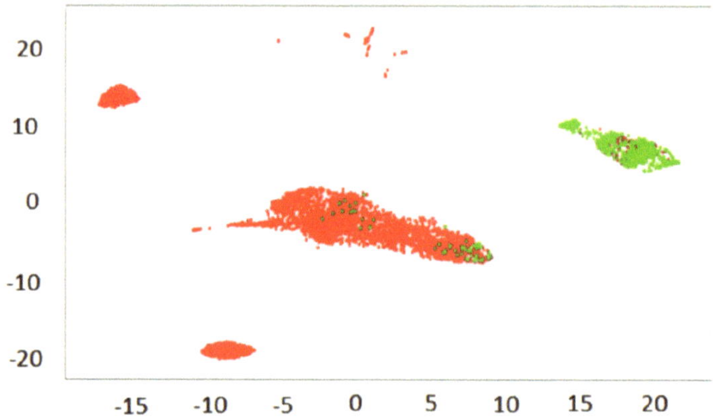

FIGURE 8.16 Monitoring results for PCA monitoring scheme: red dots – faulty and green dots – normal.

8.1.4 Online Monitoring Tool (OMT)

After creating an offline model that includes a clustering result for a set of training data and a visualization model, the next step is to feed this model to the OMT for supervised learning. The OMT features a data processing pipeline specifically designed for use in a rapidly changing online environment, making it ideal for chemical plant data.

To develop the OMT for a real chemical plant environment, the calculation engine, where algorithms are run using Python code, was separated from the front-end Graphical User Interface (GUI), which was created using Ignition (see Figures 8.17 and 8.18),

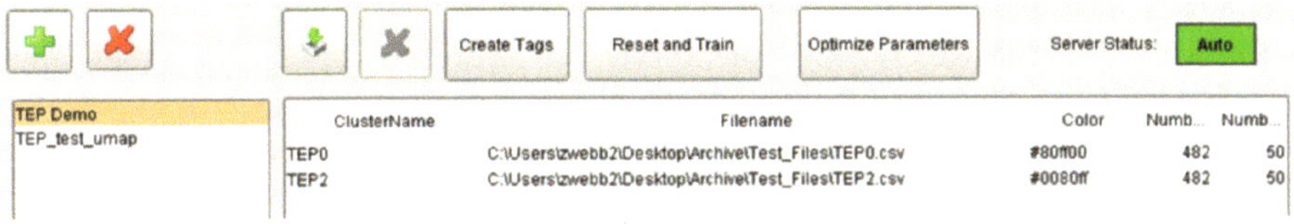

FIGURE 8.17 OMT implementation into Ignition.

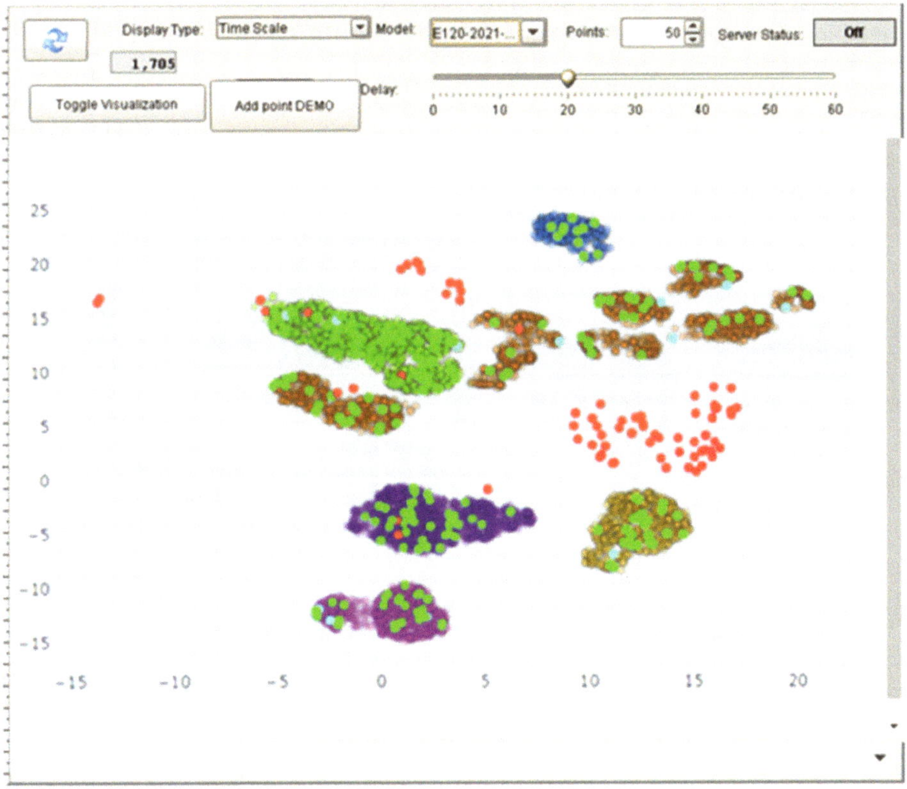

FIGURE 8.18 OMT graphical display.

a popular Supervisory Control and Data Acquisition (SCADA) software by Induction Automation. This separation of the front and back ends of the application serves three purposes. First, it maintains security at the plant, preventing end-users from accessing arbitrary Python code. Second, some algorithms require a server-level machine to run due to their computational complexity. Finally, as the OMT is intended to be scaled to many processes and multiple users, separating the server/client architecture streamlines authentication, version control, and database management.

Ignition is an excellent software solution for managing the GUI of the OMT. It has a special ability to connect various displays, handle Open Platform Communications (OPC) server management, and offer customizable user interaction through buttons and selection boxes. The software generates reports and alarms as outputs. The back-end application that runs the calculations is hosted on a Flask web server (Figure 8.19). The functions of the server are accessed through http calls in the Ignition server. The flexibility of this architecture allows a plant to run multiple instances of the web server for different areas and manage various instances of the Ignition client/server programs for extensive organizational options.

Extensive testing and troubleshooting are crucial for a monitoring workflow, such as the one outlined in OMT. At the time of the publication of this book, the real plant installation is still in the prototype phase. To facilitate efficient testing, a connection to a plant simulation would be highly beneficial. The OMT architecture is highly flexible, allowing for database connectivity to various types of data. Consequently, connecting to a simulation environment would enable us to perform several testing types. With a simulation, testing engineers can monitor a process while the controller adjusts for a process disturbance or a simulated process fault. Additionally, a simulation connection would make it much easier to test multiple process types than in an actual plant environment.

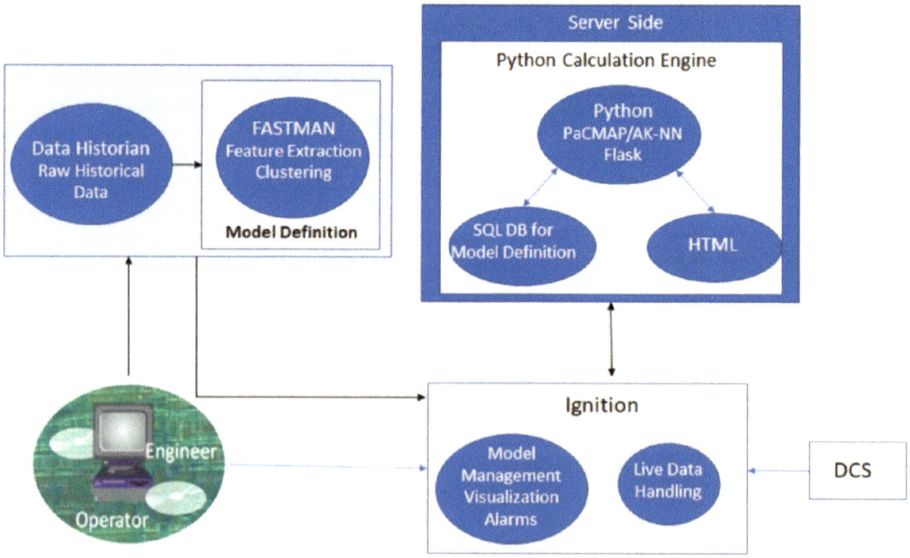

FIGURE 8.19 OMT architecture.

8.2 CASE STUDY 2: HIGH-TEMPERATURE PEM FUEL CELL: SURROGATE MODELING AND TRANSFER LEARNING

8.2.1 Introduction

The experimental high-temperature polymer electrolyte membrane (PEM) fuel cell, previously introduced in Section 7.9, serves again as an excellent candidate for illustrating the formulation and implementation of the concepts of surrogate modeling and transfer learning (TL). More details about this implementation can be found elsewhere [14]. The overarching strategy for Few-Shot Learning (FSL) in the high-temperature PEM fuel cell (HT-PEMFC) implementation is depicted in Figure 8.20.

In this approach, a knowledge-based explicit equation model (EEM) for a HT-PEMFC, as detailed in Appendix C with empirical models as described in Section 7.9, is utilized as the source domain. The target domain consists of an experimental dataset. The implementation involves a fully connected neural network that is optimized using a genetic algorithm-based Automated Machine Learning (AutoML) framework.

8.2.2 Data Preparation

For the general training stage, simulated data were generated using the knowledge-based model reported in Appendix C. The model, hereafter referred to as equation-based model or EEM, incorporates the influence of 11 input variables (hydrogen stoichiometric ratio, S_{H_2}; oxygen stoichiometric ratio, S_{O_2}; temperature, T; pressure, P; membrane ion exchange capacity, IEC_{mem}; ionomer binder ion exchange capacity, IEC_{io}; membrane thickness, δ_{mem}; ionomer binder thickness, δ_{io}; carbon monoxide to hydrogen ration in feed stream, CO/H_2; and cathode and anode platinum loading, $L_{c_{cat}}$ and $L_{c_{an}}$) as well as current density (i) to predict the voltage of the fuel cell.

To obtain a balanced dataset, the values for the input variables were structured following a full three-level factorial experimental design using 11 variables (3^{11}). It

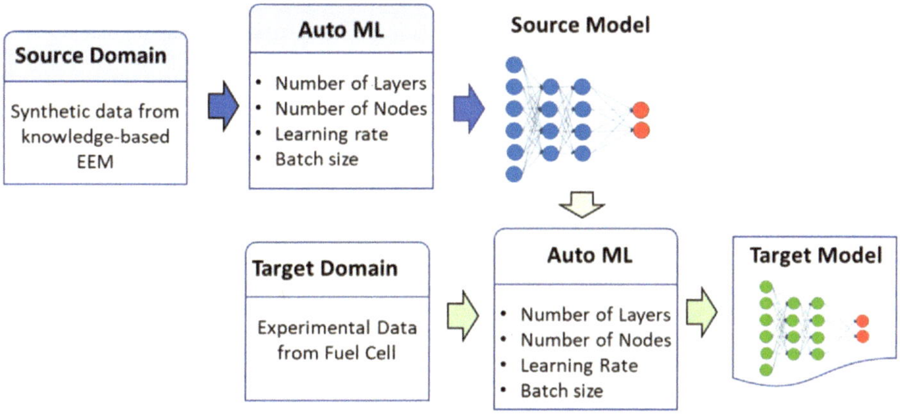

FIGURE 8.20 Transfer learning-based modeling strategy.

is important to note that all HT-PEMFC models and data in this contribution only consider pure oxygen as the oxidant. For each set of inputs, 5 points in the polarization curves were generated. The resulting datasets were arranged to generate a total of 887,735 input vectors of size 12 (the 11 input variables plus the current density) and the corresponding labels of size 1. For the target domain, experimental data reported by Venugopalan et al. [15] was used. This dataset contains a total of 28 samples, and all inputs and outputs are the same as those in the source domain, with the same ranges.

Out of the 28 samples in the target domain dataset, 25 were used for training/validation and 5 were left out for testing. This testing set was never available for any of the models and was used only to assess their performance against experimental data. For both the source and target models, the training/validation data was split according to a five-fold cross validation scheme for hyperparameter optimization step and a 70:30 (training and validation, respectively) for the final models reported. Data was scaled using over min-max scaling in all cases.

8.2.3 Few-Shot Learning and Automated Machine Learning Framework

Broadly, FSL strategies are based upon the notion of preserving information obtained through a first stage of training and reusing it in some fashion to develop a target model. FSL was implemented through regularization where the target model has the same architecture as the source model, and the learned parameters (weights) are adjusted during a new training stage, but some restriction is applied to prevent the model from overfitting the new data [16]. In artificial neural networks (ANNs), regularization takes the form of restricting how much the weights of the model can be updated. By setting different restrictions for each layer in the ANN, influence of the new data over the model can be controlled. The problem of obtaining the machine-learning-based surrogate model (MLBSM) can then be divided in two stages:

i. Find a good source model.
ii. Find the best way to perform the task-specific training.

The first stage involves finding the proper architecture and training conditions (number of hidden layers [L], number of nodes [N], the learning rate [lr], and the batch size [n]) for the given source domain. The second stage requires finding a proper set of learning rates and batch size for a given target domain.

Here, the selection of L, N, lr, and n is defined as a Mixed Integer Nonlinear Programming problem and solved using an evolutionary algorithm [17], namely the Non-dominated Sorting Genetic Algorithm (NSGA-II) [18, 19]. For tuning the hyperparameters of the neural network at source training, an optimization problem is formulated:

$$\textbf{Optimization problem 1} \quad \begin{cases} \min\limits_{x} & f_m(x) = E^{s}_{validation}, \\ \text{Subject to} & x \in R^4, \\ & x_i \in \mathbb{Z},\ i = 1, 2, 3 \end{cases},$$

where x_1, x_2, and x_3 are L, N, and n, respectively, and x_4 is the learning rate. $f_m(x) = E^s_{validation}$ is the mean value for the validation error in the five-fold cross validation.

To obtain a model for a given source domain, a new neural network is built for each individual using the set of hyperparameters x (in addition to a fixed input layer $12:N$ and two fixed output layers N: $floor\left[\frac{N}{2}\right]:1$), trained using the mini-batch gradient descent algorithm was used via the Adam optimizer [20], and then cross validated.

For the FSL stage, a similar optimization problem can be formulated to obtain the appropriate learning rates:

$$\textbf{Optimization problem 2} \quad \begin{cases} \min\limits_{x} & f_m(x) = E^s_{validation}, \\ \text{Subject to} & x \in R^4, \\ & x_i \in \mathbb{Z},\ i = 4 \end{cases}.$$

In this case, x_1, x_2, and x_3 are lr_{input}, $lr_{general}$, and lr_{task}, respectively, and x_4 is the batch size and $f_m(x) = E^s_{validation}$ is the mean value for the validation error in the five-fold cross validation for a given target domain. The learning rates (lr_{input}, $lr_{general}$, and lr_{task}) correspond to the different sections of the ANN, namely the input layer (lr_{input}), the last two layers (closer to the output) (lr_{task}), and the bulk of the hidden layers ($lr_{general}$).

8.2.4 Handling Extrapolation

Performance of data-driven models when predicting a point of extrapolation (POE) is hard to assess without exhaustive experimental validation. Unlike knowledge-based models, where insights about extrapolation capabilities can be drawn from the nature of the equations involved in the model, data-driven models are opaque, and therefore the quality of the predictions for POEs cannot be known a priori. This issue is aggravated as the dimensionality of the predictor space and the level of correlation among predictors increase [21]. As a result, a new prediction might be a POE even when each of the predictors is within the range of the training data. During derivative-free optimization (DFO), a great number of combinations in the predictor space are to be considered as candidate solutions, which can easily lead to POEs. Hence, it is of interest to account for POE when doing optimization using data-driven models. Here, the POE issue is addressed by detecting if the new sample to be predicted lies out of the training data distribution. Intuitively, if a point is located close enough to a set of training examples, it is not a POE. The out-of-distribution detection (OODD) was implemented using the k-NN distance, following a similar approach to that described by Sun and collaborators [22].

Figure 8.21 shows an illustration of the OODD procedure. A k-NN distance threshold is calculated using the synthetic data used for training the source model. For every new point to be estimated, the k-NN distance is evaluated and compared to the threshold; if the k-NN distance is greater than the threshold, the point is

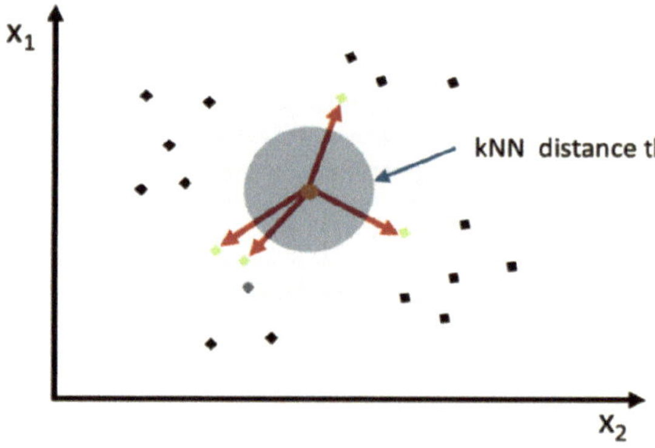

FIGURE 8.21 Example of the out-of-distribution detection method using k-nearest neighbors' distance. For a given number of nearest neighbors, a point is considered out-of-distribution if the computed k-NN distance is larger than a set threshold.

designated as a POE. As it will be explained below, this provides useful information to guide the optimization.

8.2.5 Dimensionality Reduction Using PaCMAP

A 2D visualization of the feature space was obtained via PaCMAP [23]. PaCMAP was adopted because of its capacity to preserve the local and global structure of the high-dimensional space in the low-dimensional space projection. The hyperparameters of the algorithm are furthermore very intuitive and can be used to smoothly transition between a focus on local to global structure.

8.2.6 Model Interpretation

To further understand how the model comes to a prediction, Shapley additive explanation (SHAP) was used. SHAP considers each predictor as an individual in a cooperative game setting and finds its individual contribution to the overall output of the model. From this, the relative importance of each input variable can be evaluated. SHAP analysis was performed on the final MLBSM with 5000 samples.

8.2.7 Derivative-Free Optimization Using MLBSMs

Once the surrogate model is obtained, the next step is to use it to inform both the design of new materials for the fuel cell and its operating conditions. This task can be defined as a multi-objective optimization problem as follows:

$$\textbf{Optimization problem 3} \quad \begin{cases} \min\limits_{x} & f_m(x), \\ \text{Subject to} & x \in R^{10}, \\ & l_i < x_i < u_i. \end{cases}$$

Here $f_1(x)$ is the negative of the peak power density of the fuel cell estimated using surrogate model, $f_2(x)$ is the k-NN distance between a candidate solution and the samples in the training space, and l_i and u_i are the lower and upper bounds of each decision variable (inputs in the MLBSM) set by their physical realizability. By maximizing the power density while minimizing the k-NN distance, the optimization is guided away from POEs and toward more trustworthy solutions. Because knowledge about the derivatives of the objective function is unavailable, the problem is well suited for DFO [24]. Furthermore, DFO benefits from the ability of ANN to perform well in high-dimensional spaces [25], opening the opportunity to exploit the flexibility of both MLBSMs to represent complex systems and DFO to perform optimization using black-box models. As with optimization problems 1 and 2, NSGA-III was implemented.

8.2.8 Results

8.2.8.1 Source Model Training

From **Optimization Problem 1**, it was found that a neural network with three hidden layers, each with 36 nodes (12:36:36:36:18:1), a batch size $n = 331$, and a learning rate $lr = 9.94 \times 10^{-5}$ was the best architecture. For the five-fold cross validation source model training using the optimized neural network, the mean relative root mean squared error (rRMSE) values for training, validation, and testing were 2.56%, 2.44%, and 29.02%, respectively. The simulated data was obtained from an explicit equations model reported in Appendix C. Table 8.2 shows the accuracies and their variability for the source model and the target model. The high validation error for the source model stems from the simplicity of the original explicit equations model used to generate the simulations for training.

8.2.8.2 Target Model Training

When the source domain is a simulation, that is, synthetic data from a source model, and the target domain is real experimental data for the same system, TL can improve the source model. As shown in Table 8.2, although the optimized neural network reached very low errors for the training and validation data, the error for the testing data is still high. This is consistent with the fact that at pretraining the ANN has not been exposed to real experimental data and the source domain is known to have inferior performance at very low current densities. For the task-specific

TABLE 8.2
Five-Fold Cross Validation Results for Model Improvement

	Source Model			Target Model		
Dataset	Training	Validation	Testing	Training	Validation	Testing
Mean	2.56	2.44	29.02	2.19	5.56	7.63
Σ	0.85	1.47	1.34	0.74	1.40	2.34

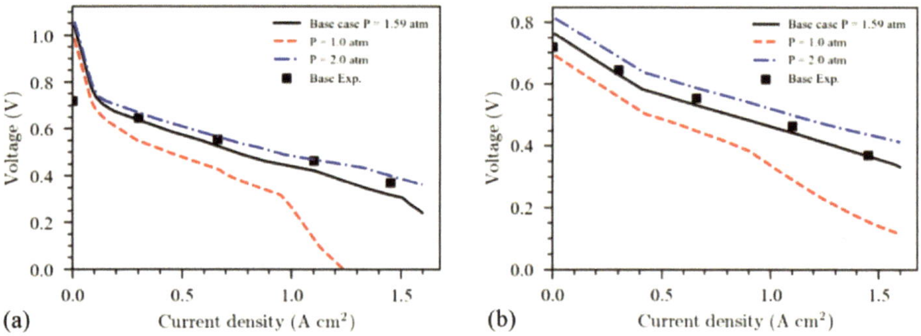

FIGURE 8.22 Effect of pressure as predicted by the source model (a) and the target model (b). The base case corresponds to 200°C.

training, the optimized learning rates and batch size from **Optimization Problem 2** ($lr_{input} = 1.99 \times 10^{-8}$, $lr_{general} = 8.80 \times 10^{-6}$, $lr_{task} = 0.00097$, $n = 7$) were used. Testing data corresponds to the 200°C polarization curve. As can be observed, the validation error decreases by a factor of 3 for the target model. These results demonstrate the applicability of TL to improve existing models using small datasets, thus reducing the burden of generating additional experimental data to arrive at a more accurate model.

As the EEM models contain physical knowledge (e.g., Butler-Volmer kinetics for informing the activation overpotential and Ohm's Law for determining the ohmic overpotential from membrane conductivity and ionomer binder conductivity in the electrode), TL enables the introduction of this information into the MLBSM. An illustrative example of this is shown in Figure 8.22. Here, the change in the polarization curve with changes in pressure is represented for both the source model (Figure 8.22a) and the target model (Figure 8.22b). Although the target domain does not contain information about the effect of pressure over the system (all examples correspond to the same pressure), the target model preserves some of this knowledge from the pretraining stage, which was introduced via the Nernst equation and physics-informed expressions for the exchange and limiting current density values. These results pave the way for attaining more accurate models in less time and is anticipated to inform new material properties and device operating parameters for improved fuel cell performance (e.g., current density at 0.7 V). Furthermore, the MLBSM can assist with future optimization activities such as pathways to reduce platinum group metal (PGM) loadings in the membrane-electrode assembly (MEA) while minimizing device performance losses such as power density.

8.2.8.3 Search Space and Extrapolation

To get a better understanding of the distribution of the synthetic data used to train the source model (pretraining data), a 2D projection (sample of 5000 points) was obtained using PaCMAP ($n_{neighbors} = 10$, $MN_{ratio} = 5$, $FP_{ratio} = 20$, and $lr = 1.0$). As shown in Figure 8.23a, pretraining data is evenly distributed across the space

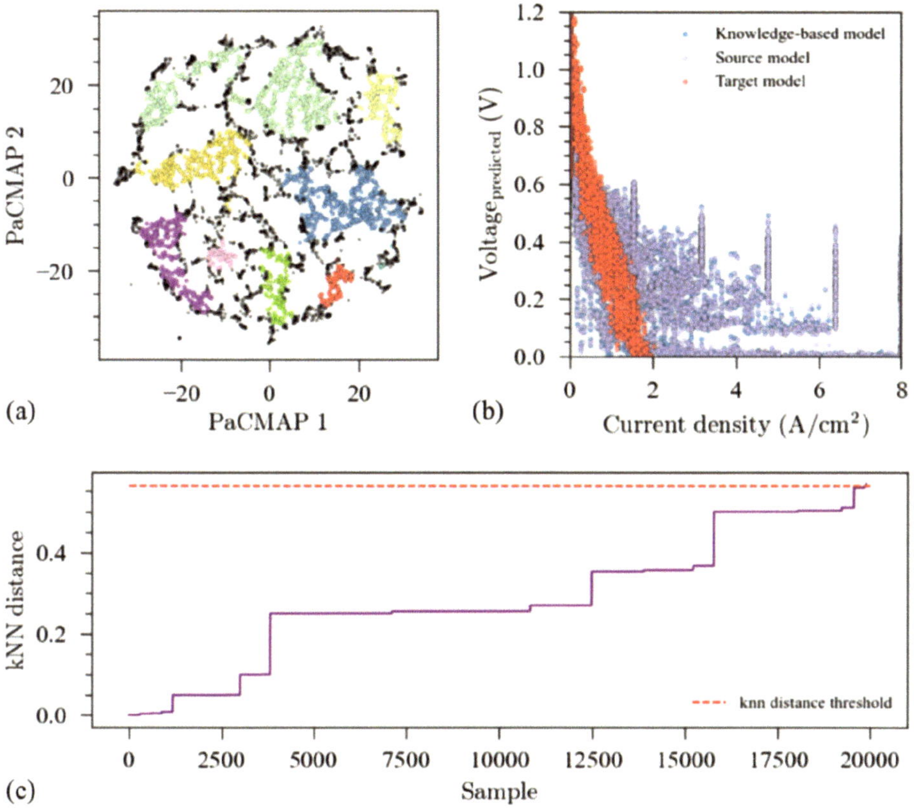

FIGURE 8.23 (a) PaCMAP projection ($n_{neighbors} = 10$, $MN_{ratio} = 5$, $FP_{ratio} = 20$, $lr = 1.0$) for a sample of pretraining data ($n = 5000$). (b) Predictions for the pretraining data from the knowledge-based model, the source model (surrogate model before transfer learning), and the target model (surrogate model after transfer learning). (c) k-nearest neighbor distances for the pretraining data and distance threshold found using the knee method.

defined by the factorial design used to generate it. However, empty space can be found within the range of the reduced dimensions. In terms of extrapolation, points located in these empty regions, as well as points outside of the range, could be designated as points of extrapolation applying the k-NN distance threshold criteria. Figure 8.23b shows the predictions for the pretraining data as estimated by the original knowledge-based model (hybrid), the source surrogate model (source, before TL), and the final MLBSM (target, after TL). As expected, the source model captures all the information available in the knowledge-based model. Predictions from the target model, on the other hand, seem to be more tightly constrained. Moreover, the target model does a better job at capturing the limiting current behavior expected for the fuel cell, evidenced by the sharp drop in the cell voltage at high current densities. From the pretraining data, a k-NN distance threshold was obtained using the knee method proposed by Satopaa et al. [26] (see Figure 8.23c).

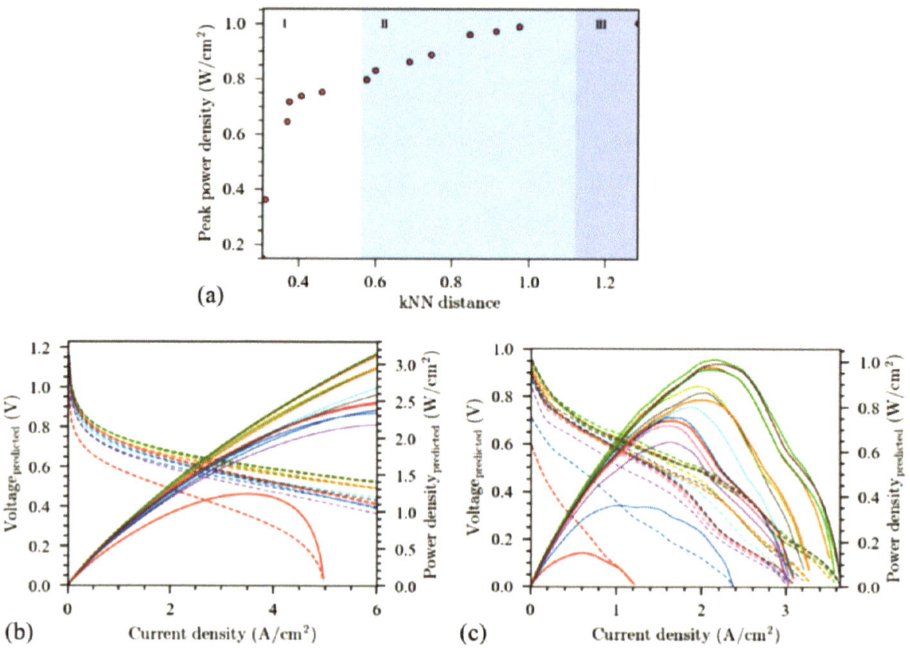

FIGURE 8.24 (a) Pareto front for Optimization Problem 3 (roman numerals indicate the level of extrapolation: I : $kNN_{distance} < threshold$, II : $threshold < kNN_{distance} < 2 \times threshold$, and III : $kNN_{distance} > 2 \times threshold$), and predicted polarization curves for the pareto front solutions from (b) knowledge-based and (c) machine learning-based surrogate model.

8.2.8.4 Derivative-Free Optimization

Solutions in the pareto front for optimization problem 3 (see Figure 8.24a) show that the best performance for the HT-PEM fuel cell is achieved at a POE with ~1.0 W/cm^2, while the highest peak ***power*** density achieved without extrapolation is ~0.75 W/cm^2. However, beyond a $kNN_{distance}$ around twice the threshold, performance does not improve considerably. These observations suggest that while model extrapolation is needed for optimization, the level of extrapolation could be monitored and managed. This provides useful information in terms of deciding which candidate solutions to pursue experimentally if less confidence is conceded to POEs. It is worth noting that the MLBSM produces more conservative estimations for the fuel cell performance than those from the knowledge-based model. Figure 8.24b and 8.24c shows the predictions for the solutions in the pareto front as estimated by the knowledge-based model and the target model, respectively. In particular, the behavior observed on the predictions for the pretraining dataset, in which the MLBSM does a better job capturing the voltage losses at higher current densities, becomes more important in the search for plausible solutions. This could suggest that after the TL step, the surrogate model becomes more realistic and, in some fashion, aware of some of the physical limitations of the system.

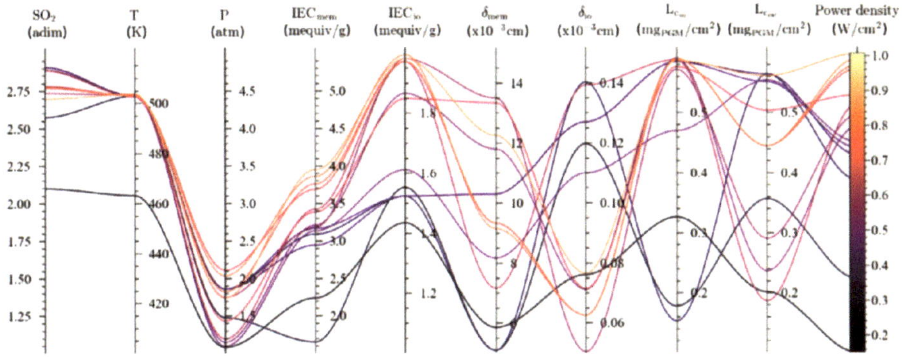

FIGURE 8.25 Values of the design and operation variables of the HT-PEM fuel cell for the solutions in the pareto front. Range of axes represents the bounds for each decision variable in Optimization problem 3. Color of the line represents the peak power density for each configuration (brighter colors represent higher values).

Further inspection of the solutions found by the optimization shows trends that agree with experimental intuition (Figure 8.25). The stoichiometric ratio for oxygen and temperature appear as crucial conditions to drive up the performance of the fuel cell, as most solutions fall within a narrow interval close to the upper bound for each decision variable. At the same time, the ranges allowed for pressure and ion exchange capacity of the membrane are not fully exploited by the optimizer, although a direct proportionality is identified. On one hand, this recognizes that the membrane chemistry, and in particular the ionic conductivity, for this system is not the limiting factor in the cell performance. Despite this current limitation, the results show that the optimizer avoids the naive approach of increasing the pressure to its upper bound. This is consistent with the physical intuition that there are additional hindering mechanisms, such as mass transfer of the reactant to the electrocatalyst surface through the ionomer binder, and that pressure alone could not overcome.

The absence of clear trends in the catalyst loading could suggest the lower values (less catalyst availability) could be compensated by improving other materials design and operation variables. This analysis was complemented with a SHAP analysis (Figure 8.26) to determine the relative importance of the input variables on the model predictions. Again, oxygen stoichiometric ratio and temperature, along with ion exchange capacity of the ionomer (IEC_{io}), which relates to ionomer binder conductivity, and cell pressure, appear with the highest relative importance. Out of these, IEC_{io} is a design variable that can be improved through the chemistry of the ionomer binder (e.g., blending perfluorosulfonic acid materials with phosphonic acid ionomer materials) [27]. It is worth noting that comparable results about the importance of the input variables were also identified for the knowledge-based model previously reported in Section 7.9. Overall, results from the optimization can be used to inform new developments in materials and operating conditions for the HT-PEM fuel cell.

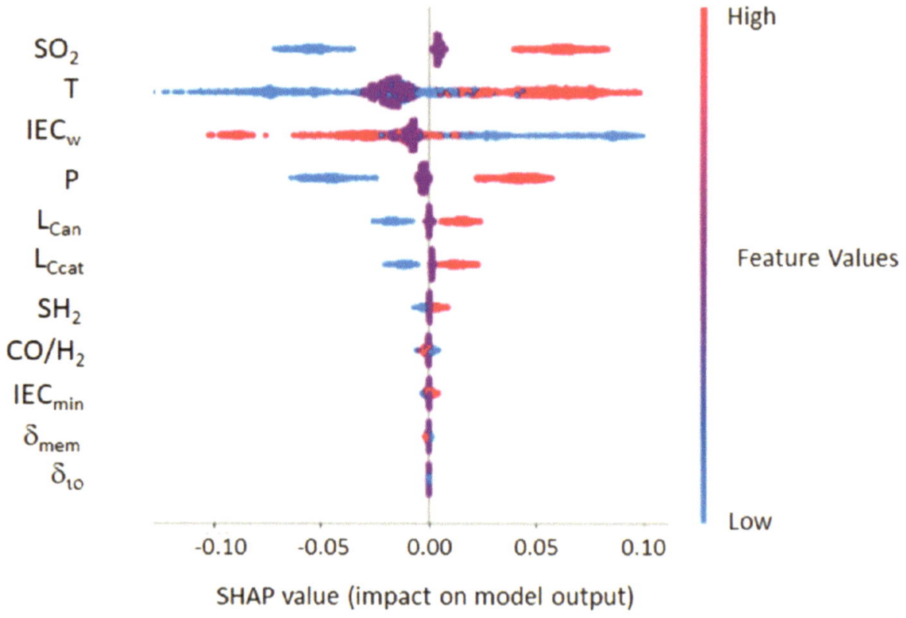

FIGURE 8.26 SHAP analysis (sample size $N = 500$) on the machine-learning-based surrogate model after TL.

8.3 CONCLUSIONS

This chapter explores two case studies demonstrating the application of supervised learning in Chemical Engineering. The first case study focuses on process monitoring in an industrial pyrolysis reactor, employing the A*k*-NN approach. Traditional methods like PCA are shown to be limited in addressing challenges such as nonlinearity and insufficient training data. A*k*-NN proves effective in detecting abnormal conditions and tracking normal process drifting, offering a robust solution for fault detection. Additionally, the development of an OMT enhances practical applicability by providing a user-friendly interface for real-time monitoring and decision-making.

The second case study explores the application of surrogate modeling and TL techniques to enhance modeling accuracy and optimization capabilities of HT-PEMFCs. Integrating knowledge-based EEMs with MLBSMs improves model accuracy, especially with small datasets, by incorporating physical insights. FSL strategies and AutoML frameworks enable efficient training and optimization of MLBSMs, adapting effectively to new target domains and yielding significant predictive performance improvements.

Moreover, this chapter addresses challenges related to extrapolation and dimensionality reduction in HT-PEMFC modeling, employing innovative techniques such as OODD and PaCMAP projection. These methods help identify points of extrapolation and provide insights into the distribution of training data, enhancing the reliability of model predictions. Application of DFO techniques to multi-objective optimization problems illustrates the utility of MLBSMs in guiding the design and operation of HT-PEMFCs, offering insights into optimal material properties and operating conditions while minimizing the risk of extrapolation.

Overall, the case studies underscore the potential of supervised learning techniques in Chemical Engineering applications. From process monitoring in industrial reactors to enhancing the modeling and optimization of HT-PEMFCs, the integration of machine learning with domain knowledge proves instrumental in addressing challenges, improving accuracy, and accelerating advancements in the field. These approaches not only offer practical solutions but also pave the way for discovering novel materials and design strategies for improved performance in various engineering domains.

REFERENCES

1. Ge Z., Song Z. (2013), Distributed PCA model for plant-wide process monitoring. Industrial & Engineering Chemistry Research, 52(5), pp. 1947–1957.
2. Li S., Wen J. (2014), A model-based fault detection and diagnostic methodology based on PCA method and wavelet transform. Energy Build, 68, pp. 63–71.
3. Hu Z., Chen Z., Gui W., Jiang B. (2014), Adaptive PCA based fault diagnosis scheme in imperial smelting process. ISA Transactions, 53(5), pp. 1446–1455.
4. Jiang Q., Yan X., Huang B. (2016), Performance-driven distributed PCA process monitoring based on fault-relevant variable selection and Bayesian inference. IEEE Transactions on Industrial Electronics, 63(1), pp. 377–386.
5. Zhang H., Palazoglu A., Zhang X., Zhang W., Zhao Z., Sun W., Liu S. (2014), Prediction of surface ozone exceedance days using PCA with a non-parametric T2 control limit. Chemometrics and Intelligent Laboratory Systems, 133, pp. 42–48.
6. Gallagher N. B., Wise B. M., Butler S. W., White D. D., Barna G. G. (1997), Development and benchmarking of multivariate statistical process control tools for a semiconductor etch process: Improving robustness through model updating. IFAC Proceedings, 30(9), pp. 79–84.
7. Li W., Yue H. H., Valle-Cervantes S., Qin S. J. (2000), Recursive PCA for adaptive process monitoring. Journal of Process Control, 10(5), pp. 471–486.
8. Jeng J.-C. (2010), Adaptive process monitoring using efficient recursive PCA and moving window PCA algorithms. Journal of the Taiwan Institute of Chemical Engineers, 41(4), pp. 475–481.
9. Wang X., Kruger U., Irwin G. W. (2005), Process monitoring approach using fast moving window PCA. Industrial & Engineering Chemistry Research, 44(15), pp. 5691–5702.
10. Duda R. O., Hart P. E., Stork D. G. (2000). Pattern Classification. 2nd ed. Wiley-Interscience.
11. He Q. P., Wang J. (2010), Large-scale semiconductor process fault detection using a fast pattern recognition-based method. IEEE Transactions on Semiconductor Manufacturing, 23(2), pp. 194–200.
12. He Q. P., Wang J. (2007), Fault detection using the k-nearest neighbor rule for semiconductor manufacturing processes. IEEE Transactions on Semiconductor Manufacturing, 20(4), pp. 345–354.
13. Zhu W., Sun W., Romagnoli J. (2018), Adaptive k-nearest-neighbor method for process monitoring. Industrial & Engineering Chemistry Research, 57(7), pp. 2574–2586.
14. Briceno-Mena L. A., Arges C. G., Romagnoli J. A. (2023), Machine learning-based surrogate models and transfer learning for derivative free optimization of HT-PEM fuel cells. Computers and Chemical Engineering, 171, p. 108159.

15. Venugopalan G., Chang K., Nijoka J. D., Livingston S., Geise G. M., Arges C. G. (2020), Stable and highly conductive polycation-polybenzimidazole membrane blends for intermediate temperature polymer electrolyte membrane fuel cells. ACS Applied Energy Materials, 3(1), pp. 573–585.
16. Yang Q., Zhang Y., Dai W., Pan S. J. (2020), Transfer Learning. Cambridge University Press.
17. Vishwakarma G., Haghighatlari M., Hachmann J. (2019), Towards autonomous machine learning in chemistry via evolutionary algorithms. ChemRxiv, doi: 10.26434/chemrxiv.9782387
18. Deb K., Pratap A., Agarwal S., Meyarivan T. (2002), A fast and elitist multiobjective genetic algorithm: NSGA-II. IEEE Transactions on Evolutionary Computation, 6(2), pp. 182–197.
19. Blank J., Deb K. (2020), Pymoo: Multi-objective optimization in python. IEEE Access, 8, pp. 89497–89509.
20. Kingma D. P., Ba J. (2015), Adam: A method for stochastic optimization. Proceedings of the 3rd International Conference on Learning Representations (ICLR 2015).
21. Hooker G. (2004), Diagnosing extrapolation: Tree-based density estimation. Proceedings of the tenth ACM SIGKDD International Conference on Knowledge Discovery and Data Mining, pp. 569–574. ACM Digital Library.
22. Sun Y., Ming Y., Zhu X., Li Y. (2022), Out-of-distribution detection with deep nearest neighbors. arXiv:2204.06507v3. https://doi.org/10.48550/arXiv.2204.06507
23. Wang Y., Huang H., Rudin C., Shaposhnik Y. (2021), Understanding how dimension reduction tools work: An empirical approach to deciphering t-SNE, UMAP, TriMap, and PaCMAP for data visualization. Journal of Machine Learning Research, 22(1), 1–73, Article 201.
24. Rios L. M., Sahinidis N. V. (2013), Derivative-free optimization: A review of algorithms and comparison of software implementations. Journal of Global Optimization, 56(3), pp. 1247–1293.
25. Bhosekar A., Ierapetritou M. (2018), Space mapping-based derivative-free optimization framework for supply chain optimization. Computer Aided Chemical Engineering, 44, pp. 985–990.
26. Satopaa V., Albrecht J., Irwin D., Raghavan B. (2011), Finding a 'Kneedle' in a Haystack: Detecting knee points in system behavior. 31st International Conference on Distributed Computing Systems Workshops, pp. 166–171. IEEE Computer Society.
27. Lim K. H., Lee A. S., Atanasov V., Kerres J. A. (2022), Protonated phosphonic acid electrodes for high power heavy-duty vehicle fuel cells. Nature Energy, 7, pp. 248–259.

9 Deep Learning

9.1 INTRODUCTION

In Chapter 6, we discussed the issue of feature selection and its influence on the model's ability to capture the underlying information in the data. An implicit assumption in that discussion was the fact that we have a *set of features to select from* that are suitable for the task at hand. But what happens when we do not have such features, and they must be extracted from the data first? The manual approach would be to use domain knowledge to compute and organize input features in such a way that they are useful for training. However, this approach will easily prove unfeasible if the number of features to be found is too large or, as it is the case in many applications, we just simply do not know which are the features that we need. Also, and perhaps more consequential, having to manually extract features hinders the model's applicability as data would need to be heavily preprocessed before it can be used to make predictions or retrain the model.

Deep learning (DL) is a subset of machine learning methods in which the model is also provided with the means to extract features from data. To achieve this, DL implements a wide variety of strategies and operations that allow for the efficient processing of input data in various formats. Since the feature extraction step is now part of the model, it can also be trained automatically resulting in powerful and accurate models.

Data types like images, sound records, graphs, text, and time series, for which the structure of the data, that is, the relative position between data points, is important, are especially suitable for DL methods. In this chapter, we will discuss a selection of methods and strategies deployed in DL models, with an emphasis on image and time series analysis.

9.2 CONVOLUTIONAL NEURAL NETWORKS (CNN)

Convolutional neural networks (CNNs) are a class of artificial neural networks (ANNs) that were designed to be effective at processing images. Rather than taking the complete input vector (or matrix) of pixels like in traditional ANNs, CNNs focus on one portion of the input at a time and then feed the corresponding outputs of each region into the next layer. This seemingly simple idea unlocked a vast array of advances in ML. The origin of the CNN can be traced back to Fukushima's "neocognitron" approach [1] in 1980. Based on a similar idea, Rumelhart et al. [2] and LeCun et al. [3] composed an error-oriented backpropagation convolutional network for character recognition. However, the high computational cost for its implementation limited is proliferation until new technologies such as graphical processing units (GPUs) became widely available.

DOI: 10.1201/9781003455905-11

In 2012, CNN models made breakthroughs in different applications specially in image recognition problems. In the annual ImageNet Large Scale Visual Recognition Challenge (ILSVRC) [4] that requires participants to correctly classify the contents of images, a CNN model lowered the error rate to 16% error, while the runner up achieved 25%. In the following years, improved CNN architectures such as VGG [5] and ResNet [6] improved the records consecutively, which led to the boom of the industry-wide artificial intelligence that we see now. Different applications have been proposed in recent years, including skin cancer detection [7], crack detection for many construction tasks [8], and photo post-processing for many smartphone cameras [9].

Before further discussion of CNN models, it is important to understand the different operations or building blocks that are typically used in a CNN.

9.2.1 Convolution Operation

In the context of ML, convolution is the application of a filter to an input that results in an activation. Repeated application of the same filter to an input results in a map of activations called a feature map (Figure 9.1).

A filter applied on an $M \times N$ input matrix is typically defined in terms of size, $m \times n$, where $m < M$ and $n < N$, and a given *stride* that determines how much it moves before being applied to another portion of the input. In most cases, the activation is the result of the dot product between the filter and a portion of the input of the same size as the filter. In the example illustrated in Figure 9.1, the size of the filter is 3×3 and the stride is 3 which means that there is no overlap between the first element in the feature map and the second. The filter size and stride are hyperparameters to be tuned.

One useful analogy is to think of the filter in a CNN model as a metal detector that can scan a piece of land for the presence/absence of metal. The metal detector can be dragged over the land slowly, and every time it detects metal, it makes a sound. Similarly, filters in a convolution operation are used to search for specific

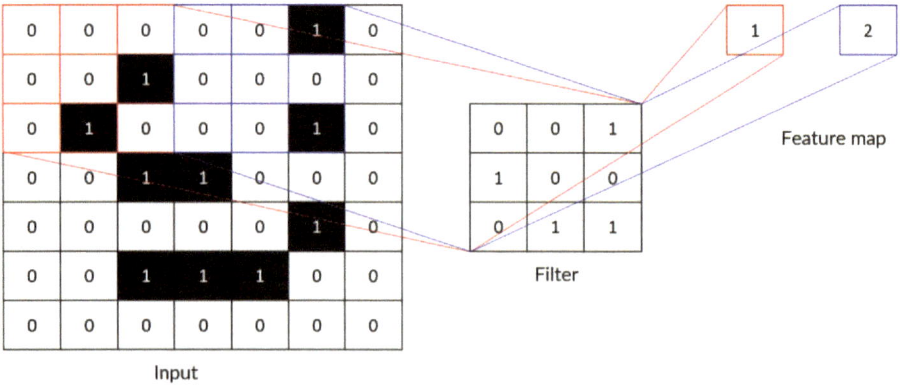

FIGURE 9.1 Schematic representation of a convolution operation.

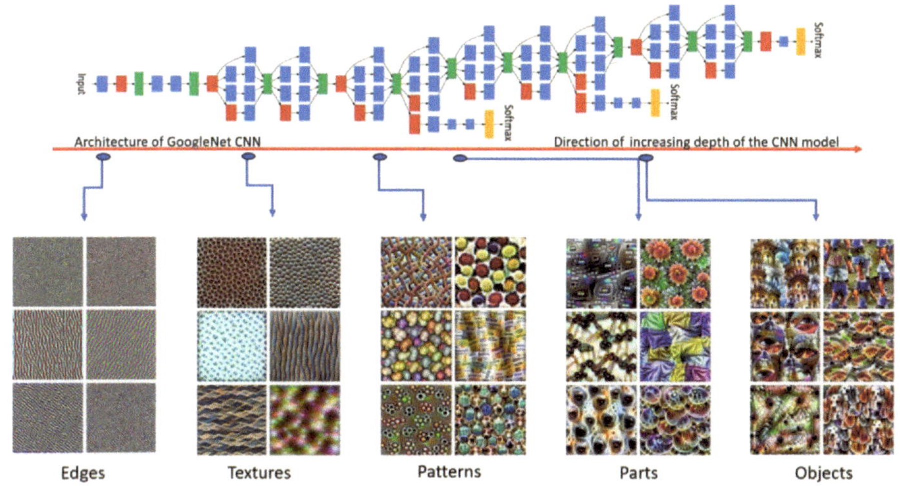

FIGURE 9.2 Increasingly complex features detected by filters in a convolutional neural network architecture. (Adapted from [10].)

items in the images' real estate. Every time the filter spots the item it is looking for, it makes a sound (activation). What each filter is looking for is not manually defined by an expert but automatically learned from the data in a supervised manner using the backpropagation algorithm. Researchers have found that filters are commonly looking for things like edges of objects, textures, patterns, and parts. Some of these are shown in Figure 9.2.

Commonly used together with convolution, *padding* involves the addition of zeros around the edges of the input matrix (see Figure 9.1). Padding affects the activation that is passed to the next layer by allowing the filter to extract information on the edge of the input object. The amount of padding is typically one position beyond the edge of the input, but it can also be a hyperparameter to be tuned.

9.2.2 Receptive Field

Given a filter size and stride, it is evident that some of the input entries will never fall in the same convolution operation and relationships between these inputs will not be considered in the model. This means that these relationships are outside of the *receptive field* of the current convolution operation. A narrow receptive field could limit the model's ability to capture useful underlying patterns in the data, which is a disadvantage. To address this, multiple filters can be stacked, meaning that a new filter is applied on the feature map produced by the previous filter. By doing this, variables that are far apart in the original input matrix can now be considered together, while still preserving information about their original position.

In the schematic shown in Figure 9.3, the third layer includes information from the entire input to the first layer even though a filter of size 3×3 with a stride of 1 was used.

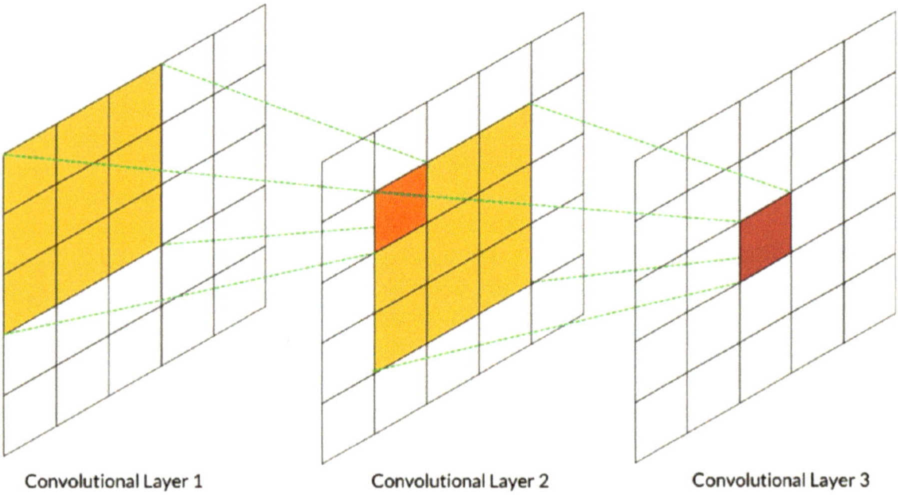

FIGURE 9.3 Receptive field with multiple convolutional layers.

9.2.3 Max-Pooling Operation

The goal of pooling is to reduce the size of the feature map so that we have fewer parameters in the network. An illustration of this is given in Figure 9.4. From every 2×2 block of the input feature map, we select the maximum pixel value and thus obtain a *pooled feature map*. The idea is to retain only the important features from each region. It should be noticed that both convolution operation and specially the pooling operation reduce the size of the input, which often helps with the computational efficiency of the model.

Pooling also serves as a regularization tool. By keeping the largest values only, we effectively pass only the strongest signals to the next layer, which in turn prevents the model from picking up irrelevant patterns in the data, such as noise. Typically, pooling layers are placed between convolutional layers, and their placement, as well

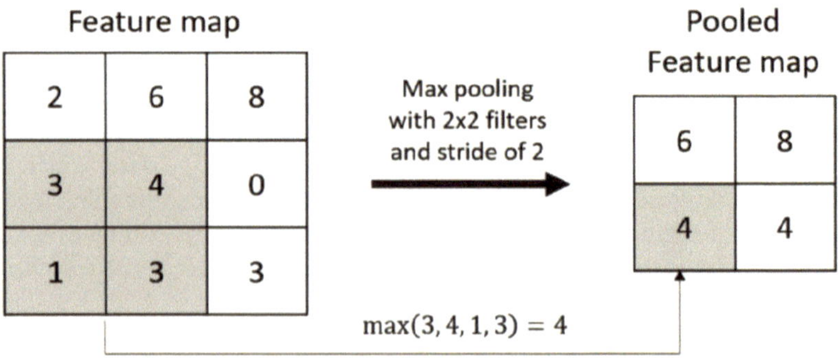

FIGURE 9.4 Schematic representation of a pooling operation.

as the size of the filter and the stride, is to be tuned and designed for each specific application.

9.2.4 Down Sampling and Up Sampling

In the context of CNNs applied to image analysis, down sampling means converting a high-resolution image to a lower-resolution image. In doing so, the original information in, say, a 4×4 image is now represented in a 2×2 image after pooling. When applying the convolution operation again, the filters in the next layer will have a larger receptive field. The result is that as we go deeper into the network, the size of the image reduces while the receptive field increases.

Up sampling an image means converting the low-resolution image after convolution and pooling to a high-resolution image. In most state-of-the-art networks, transposed convolution is the preferred choice for up sampling an image. Transposed convolution (sometimes also called as deconvolution) is a technique to perform up sampling of an image with learnable parameters.

Figure 9.5 shows the various operations performed by a CNN while processing an image. After repeating the convolution and pooling operations in stacked layers, the feature maps are commonly flattened, that is, converted to a 1D vector, and fed into a fully connected (FC) neural network section with a SoftMax activation function for final class prediction. For the predicted probability of the j-th class in total K categories, the SoftMax function is defined as follows:

$$P(y = j \mid x) = \frac{\exp\left(x^T w_j\right)}{\sum_{k=1}^{K} \exp\left(x^T w_k\right)}, \tag{9.1}$$

where x denotes the input vector into the FC layer and w denotes the weights for the FC layer.

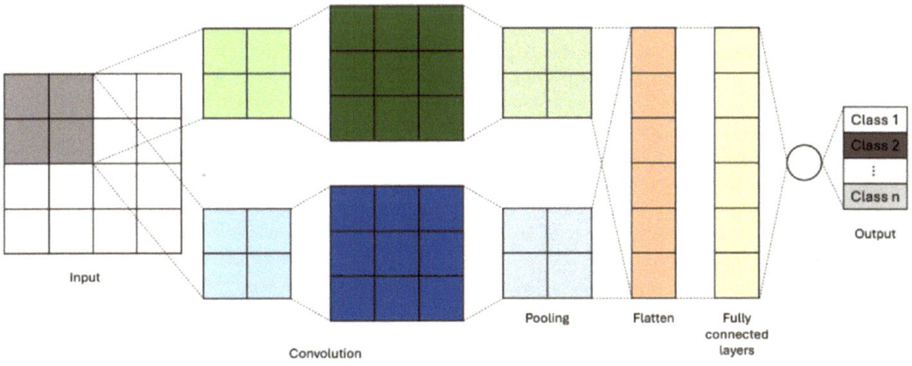

FIGURE 9.5 Operations performed by a CNN model on an image while making a prediction.

9.3 COMPUTER VISION

One of the most salient applications of CNN models is computer vision (CV). CV is the field of computer science that focuses on creating digital systems that can process, analyze, and make sense of visual data (images) in the same way that humans do. The concept of CV is based on teaching computers to process an image at a pixel level and understand it. It is a multidisciplinary field that could broadly be called a subfield of artificial intelligence and machine learning, which may involve the use of specialized methods and make use of general learning algorithms.

The goal of CV is to understand the content of digital images. There are various levels of granularity in which the computers can gain an understanding of images. For each of these levels, there is a problem defined in the CV domain (Figure 9.6):

Object Classification: This is the task of recognizing the primary object in the image. It is typically framed as a classification problem in which the algorithm has to choose the appropriate category the image falls into from a predefined list of categories. In image classification, we assume that there is only one (and not multiple) objects in the image.

Object Localization: This is the task of locating the position of the object within the image. It is typically implemented using a bounding box which can be defined by some numerical parameters (coordinates of the bounding box) with respect to the image boundary.

Object Detection: This task is a mix of object classification and localization in that we want to both locate the object within the image and recognize the category of the object.

Image Segmentation: Image segmentation is the process of partitioning a digital image into multiple segments (sets of pixels, also known as image objects). The goal of segmentation is to simplify and/or change the representation of an image into something that is more meaningful and easier to analyze.

Image segmentation is typically used to locate objects and boundaries in images. More precisely, image segmentation is the process of assigning a label to every pixel in an image such that pixels with the same label share certain characteristics.

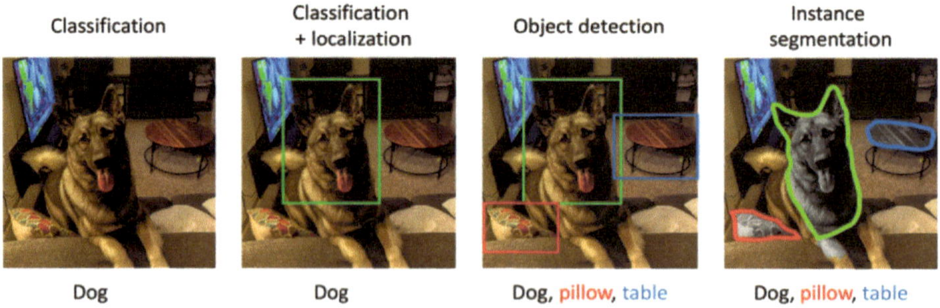

FIGURE 9.6 Computer vision domain.

The result of image segmentation is a set of segments that collectively cover the entire image or a set of contours extracted from the image. Each of the pixels in a region are similar with respect to some characteristic or computed property, such as color, intensity, or texture.

With respect to segmentation, there are two broad categories of objects. The first category, which is sometimes called "things", involves objects that are countable such as people, animals, and toys. The second category (called "stuff") includes objects that are uncountable and/or amorphous such as sky, water bodies, and roads. These two categories are not mutually exclusive since, for example, countable objects like multiple people can be clumped together into one single "people" class instead of considering them as separate individual people. We can broadly divide image segmentation techniques into two types based on the kinds of objects they detect.

- ***Semantic segmentation*** is the problem of identifying all the "stuff" like objects that make up an image. This is done by assigning to every pixel in the image its own object category. For example, all cats in an image can be grouped together as one object and every non-cat pixel in the image can be classified as the background.
- ***Instance segmentation*** is defined as the task of detecting the precise boundaries of the countable objects in an image. This task is not framed as a per-pixel classification problem and is generally considered distinct from semantic segmentation.

In addition to the above, there are many other interesting tasks in the field of CV such as image captioning, depth estimation, pose estimation, image generation, style transfer, 3D image understanding, super resolution, image coloring, and video understanding.

Historically, these problems in CV were thought to require manually crafted features to improve performance. But a CNN model learns relevant features like edges, textures, and patterns [11] automatically based on supervised feedback using the backpropagation algorithm. As the complexity of the learning task increases, so does the depth (number of layers and filters) of the model needed to achieve it. However, with increasing network depth, accuracy gets saturated and then degrades rapidly, which constrains further improvement from stacking more layers [6].

To deal with the issues related to CNNs' depth, a number of strategies and network architectures have been proposed. Here, we introduce some of these approaches as an illustration of the capabilities of CNNs for CV.

9.3.1 VGG16

VGG16 is a model proposed by Simonyan and Zisserman from the University of Oxford in the paper "Very Deep Convolutional Networks for Large-Scale Image Recognition" [5]. The model achieves 92.7% top-five test accuracy in ImageNet, which is a dataset of over 14 million images belonging to 1000 classes. Figure 9.7 illustrates the architecture of VGG16. As the name suggests, the VGG16 model is 16 layers deep and is composed of the building blocks described earlier.

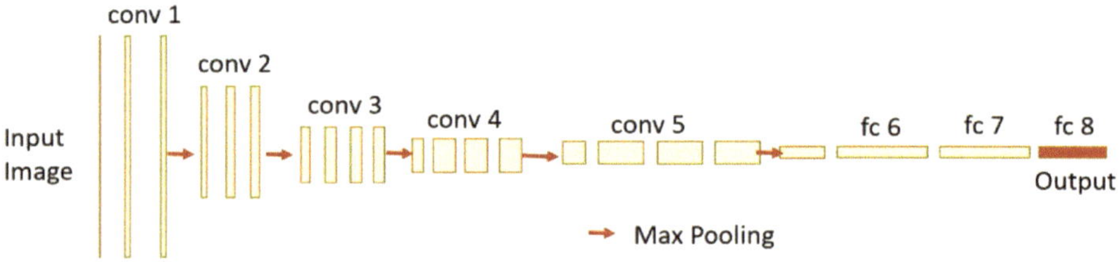

FIGURE 9.7 VGG16 architecture [5].

The input to the first convolutional layer, cov1, is of fixed size 224 × 224 RGB image. The image is passed through a stack of convolutional layers, where the filters were used with a very small receptive field: 3 × 3 (which is the smallest size to capture the notion of left/right, up/down, and center). In one of the configurations, it also utilizes 1 × 1 convolution filter, which can be seen as a linear transformation of the input channels (followed by nonlinearity). The convolution stride is fixed to one pixel; the spatial padding of it is such that the spatial resolution is preserved after convolution, that is, the padding is one pixel for 3 × 3 convolutional layers. Spatial pooling is carried out by five max-pooling layers, which follow some of the convolutional layers (not all the convolutional layers are followed by max-pooling). Max-pooling is performed over a 2 × 2-pixel window, with stride 2.

Three FC layers follow a stack of convolutional layers (which has a different depth in different architectures): the first two have 4096 channels each, and the third performs 1000-way classification and thus contains 1000 channels (one for each class). The final layer is the soft-max layer. The configuration of the FC layers is the same in all networks.

9.3.2 ResNet

With the success of VGG models, researchers started stacking more and more convolutional layers onto CNN models to improve their performance. While the performance of these models initially improved with depth, they soon hit a wall. Beyond a certain number of convolutional layers, the performance of the CNN models started to flatten and subsequently deteriorate with further increase in depth. He et al. [6] discovered that the gradients in these very deep CNN models were not effectively propagated back to the shallower layers of the model.

To solve this problem, they proposed a residual learning framework that introduces the "gradient highways" into the CNN, named ResNet. Figure 9.8 illustrates the comparison between a plain CNN and one with a ResNet block, where a highway connection performs the identity mapping of the input and element-wise addition to the output of the residual block. Each building block consists of multiple batch normalization (BN) layers [12], bi-linear activation layers, and convolutional layers. Such architecture can be expressed as follows:

$$y = F(x,\{Wi\}) + x \quad \textit{if} \text{ same dimension,} \tag{9.2}$$

$$y = F(x,\{Wi\}) + Ws\ x \quad \textit{else}, \tag{9.3}$$

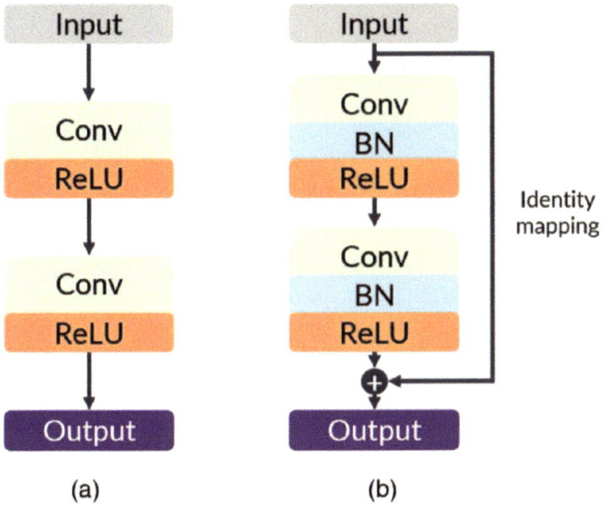

FIGURE 9.8 Demonstration of building blocks of deep convolutional neural network. (a) Plain convolutional structure. (b) ResNet blocks with identity mapping.

TABLE 9.1
Comparison of Different Models in ImageNet Benchmark

Model	Parameters	Top-1 Acc	Top-5 Acc
VGG16 [5]	138M	0.715	0.901
VGG19 [5]	143M	0.727	0.910
ResNet50 [6]	25M	0.759	0.929

where F denotes the residual functions and x denotes the input to the residual block. If the dimension of the x and $F(x,\{Wi\})$ are not equal, linear projection is often used to regulate the dimension before the addition of the two terms. Benefiting from the residual architecture, the ResNet outperforms its plain network competitors such as VGG [5], while dramatically reducing the number of parameters that need to be learned. The comparison is summarized in Table 9.1.

9.3.3 U-Net

Fully convolutional networks (FCN) [13] is an approach that converts an image classification model into an image segmentation model, in order to make pixel-level predictions possible. CNNs designed for image classification (like VGG) compress each image into a fixed-length vector, and this vector is classified by a FC layer. While this is appropriate for image classification where the goal is to classify the whole image into a category, this vector compression approach loses the pixel-level information necessary to classify every single pixel in the image. FCN models

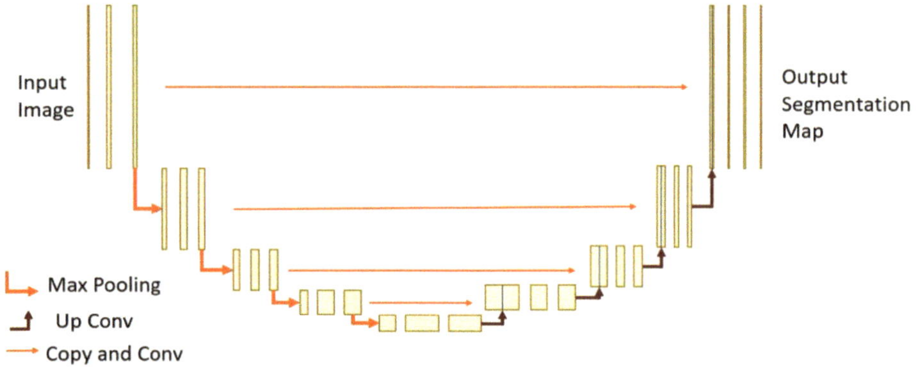

FIGURE 9.9 Architecture of the U-Net [14].

recover the granularity lost in compression by introducing some additional convolution operations. These operations, called deconvolution or up sampling, resize the compressed vector back into the size of the original image. FCN starts from a pretrained VGG network with removal of the final flatten and SoftMax layers in the VGG classification model. Using intermediate feature maps from different layers of the network, predictions are made in each map by 1×1 convolution filters. Then, the predictions in different scales are up sampled by bi-linear interpolation where the corresponding parameters are learned in training. Hence, up sampled coarse predictions are concatenated with finer predictions to improve the pixel-level details in the final prediction.

U-Net [14] is an alternative segmentation network developed from the original FCN. Unlike the FCN that produces predictions directly from up sampling feature maps, U-Net utilizes extra feature channels for decoding to allow the network to propagate high-resolution context information. U-Net is designed in an encoder-decoder architecture, where the encoding path is a typical convolutional network that consists of a number of convolutional layers with rectified linear unit (ReLU) activations. In the decoding path, extra convolution operations are applied in intermediate feature layers in different stages of the network, which is then combined with feature maps through a sequence of up sampling and concatenations with high-resolution features from the encoding path. The architecture of the U-Net is shown in Figure 9.9. Such design gives a u-shaped architecture, which shows a remarkable performance in cases with limited amount of available data.

9.3.4 Mask R-CNN

Data scientists and researchers at Facebook AI Research (FAIR) pioneered a DL architecture, called Mask Region-based Convolutional Neural Network (Mask R-CNN) [15], that can create a pixel-wise mask for each object in an image. Mask R-CNN is an extension of the popular Faster R-CNN [16] object detection architecture by adding a branch to the already existing Faster R-CNN outputs.

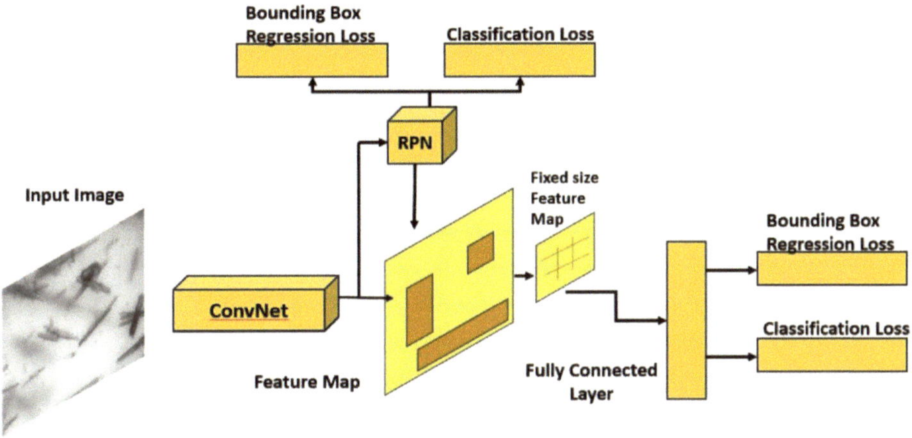

FIGURE 9.10 Faster R-CNN is a single, unified network for object detection.

Faster R-CNN is a region-based CNN that returns bounding boxes for each object and its class label with a confidence score. Faster R-CNN works in two stages (Figure 9.10):

Stage1: The first stage consists of two networks: a backbone network (e.g., ResNet or VGG) that extracts features and a region proposal network (RPN). The image is passed sequentially through these two networks to generate a set of region proposals. Region proposals are regions in the image or feature maps that are likely to enclose an object. You can think of a region proposal stage as an initial filter to narrow down the regions of interest in the image.

Stage2: In the second stage, the network predicts bounding box refinements and object class for each of the proposed regions obtained in stage1. Since objects in images come in different sizes and shapes, each proposed region can be of arbitrary size. But FC layers in the networks always require fixed size vector to make predictions. The size of these proposed regions is therefore normalized by using either region of interest (RoI) pool (which is very similar to MaxPooling) or RoIAlign method.

Mask R-CNN adds a third branch to the Faster R-CNN architecture in order to generate pixel-level masks for the objects detected by the first two branches As shown in Figure 9.11, the Mask R-CNN consists of a RPN that generates proposals of object locations which are then sent to the second stage of the network for further refinement. The second stage consists of a box sub-network to output the coordinate locations of the objects, a class sub-network to classify the objects into one of several categories and in addition, and a mask sub-network to generate binary masks for each object location, thereby allowing the model to make pixel-level predictions of objects.

Despite its large structure, which packs over 44M parameters, the Mask R-CNN can process up to six frames per second (GPU). However, on a CPU, the inference

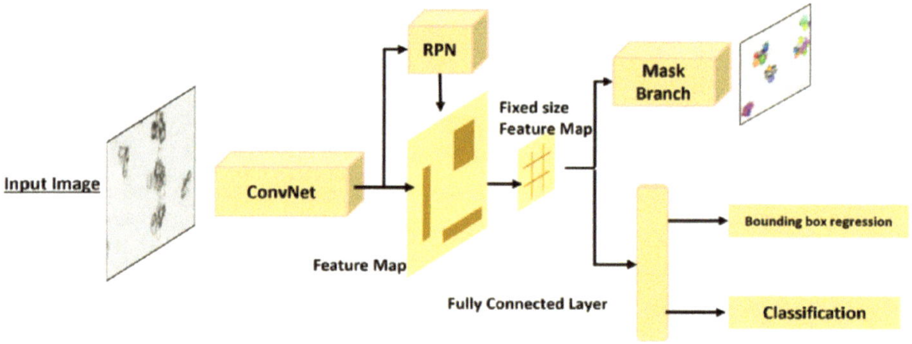

FIGURE 9.11 Architecture of the Mask R-CNN.

speed lags and can take up to 13 seconds for a single frame. One of the reasons for this lag is the two-part structure of the Mask R-CNN which is designed this way to handle the large foreground-background imbalance present in images.

9.3.5 RetinaNet

Lin et al. [17] discovered that a custom loss function could remove the need for an RPN and reduce the size of the CNN model (Figure 9.12). In their paper, they developed a focal loss function that assigned weights to examples based on their difficulty. Smaller weights were assigned to the easier background examples while the more difficult foreground examples were assigned larger weights. By including this loss function in a single-stage detector (RetinaNet), they saw large gains in inference speeds without a drop in the accuracy.

9.3.6 Modified Method

Despite its improvements, a significant drawback of the RetinaNet is its inability to generate pixel-level masks which is essential to characterize for example crystal particles in a crystallization process (Case Study 2 in Chapter 10). To address this problem, a one-stage segmentation model that uses the RetinaNet framework is

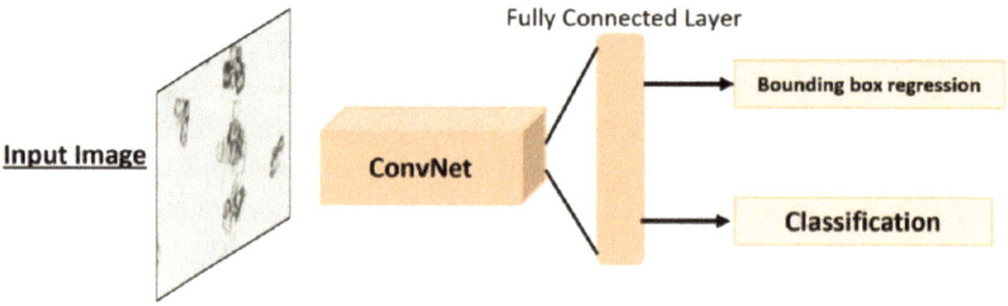

FIGURE 9.12 Architecture of the RetinaNet.

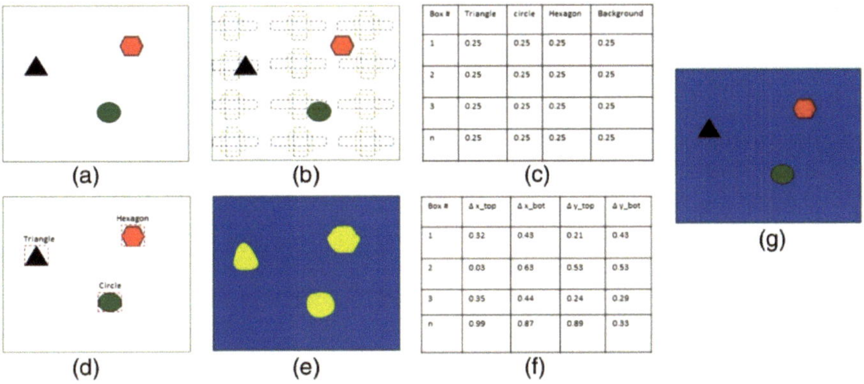

FIGURE 9.13 Shows a toy image (a) and the outputs of the sub-networks. (b) Anchor boxes overlaying the image. (c) Output of the class sub-net. (d) Visual representation of the outputs. (e) Output of the segmentation sub-net. (f) Output of the class sub-net. (g) Final segmented output.

introduced to detect objects but has an additional segmentation head to generate a semantic map of the image. In this model, the task of detecting contours is divided into three sub-tasks, each of which is crucial to the overall performance of the model.

The first sub-task, named bounding box detection, is necessary to localize or identify the locations of objects in the image. One way to do this is to train a convolutional operator to scan the image and directly predict the bounding box coordinates for every object in the image [18]. But in practice, this approach does not work well. A more rewarding approach is to pack the image with predefined bounding boxes called anchor boxes at closely spaced locations as shown in Figure 9.13b. In the figure, four anchor boxes of similar size but different aspect ratios are spread across the image whereas in the actual model, nine anchor boxes are used. This dense usage of boxes helps to ensure that every nook and corner of the image is thoroughly scanned for objects. The convolutional operator moves across the image in patches, and at every patch, it predicts a 36D vector (nine anchors boxes × four coordinates) that adjusts the position and shape of the nine anchor boxes in that patch (Figure 9.13f). Since the operator only predicts coordinates and not bounding box scores, it is difficult to distinguish the boxes that enclose objects from those in the background. To do this, a second sub-task is introduced for class detection.

The class detection sub-task is similar to bounding box detection where a convolutional operator moves across the image in patches and outputs a vector for each patch. But while the earlier operator outputs coordinate deltas, the convolutional operator here predicts a $(9n + 9)$-D vector (nine anchor boxes × $[n + 1]$ classes) (Figure 9.13c). Each element of the vector contains the probability that the bounding box encloses an object belonging to one of the n categories or the background. The probabilities are used to filter out the background boxes while the remaining top scoring boxes are retained. At the end of these two sub-tasks, the locations of the objects in the image and their class labels are determined. However, for some problems, it is insufficient to merely localize the object, and instead, precise pixel-level coordinates are necessary. This can be seen in Figure 9.13d where a triangle is enclosed by a bounding box, but the exact contours of the triangle are unavailable. To solve this problem, a third sub-task is introduced that segments the image into foreground and background.

In the segmentation sub-task, the convolutional operator again moves across the image in patches, and in each patch, it predicts a 2D vector. The elements of the vector represent the foreground and background probabilities of the corresponding patch. The output of this sub-task is a binary map with the same spatial size as the original image as shown in Figure 9.13e. While it may seem that this sub-task can individually solve the crystal contour characterization problem, the binary output of this task cannot differentiate between different objects or their classes. As a result, the outputs of the three sub-tasks are combined to obtain pixel-level contours of individual objects shown in Figure 9.13g.

At the outset, the image is passed through 50 convolutional layers, called the ResNet-50 stage, the weights of which are pre-trained on the ImageNet dataset. This dataset has class labels for over a million images but lacks more detailed labels needed for object detection. Pretraining the weights on this large corpus of data is necessary to learn low-level information such as textures, edges, and patterns. Intermediate outputs or activation maps from this stage, each with a different spatial dimension, are extracted for further processing. The activation maps with large spatial dimensions have low-level information that are not adequate for object detection whereas the activation maps with more detailed information have very low resolution. To resolve this problem, a feature pyramid network (FPN) is constructed that combines the information from the smaller maps with the spatial resolution of the earlier maps. The construction of the feature pyramid is shown in Figure 9.14.

Once the FPN is constructed, the three sub-tasks are handled by three distinct pathways, each one designed to solve the corresponding task. All three pathways build on top of the FPN while still operating in parallel. The pathways for the class sub-task and the bounding box sub-task are constructed in a manner similar to the original paper with one minor difference. While the paper uses the feature pyramid levels P_3–P_7, we make use of levels P_2 and P_3 for the segmentation sub-task and retain the levels P_3–P_7 for the class and box sub-tasks. This is because the segmentation sub-task requires the higher resolution P_2, P_3 maps to make accurate pixel-level predictions, and since the pathways share a common FPN, a bit of speed is sacrificed for the sake of accuracy.

The architecture of the segmentation pathway is shown in Figure 9.14. The activation maps in the FPN are passed through two convolution blocks that include a 3×3 convolution, BN, ReLU activation, and 2× up sampling using bi-linear interpolation. Once the activation maps have been resized to the spatial size of P_2, the resulting activations are concatenated and passed through an additional convolution block before being finally up sampled to the size of the original image. The difference between the predicted map and the target map is measured using a per-pixel SoftMax loss function. While the segmentation pathway can be used to predict class-wise maps, limiting the output to class-agnostic masks ensures that the work is divided equally between the three sub-tasks.

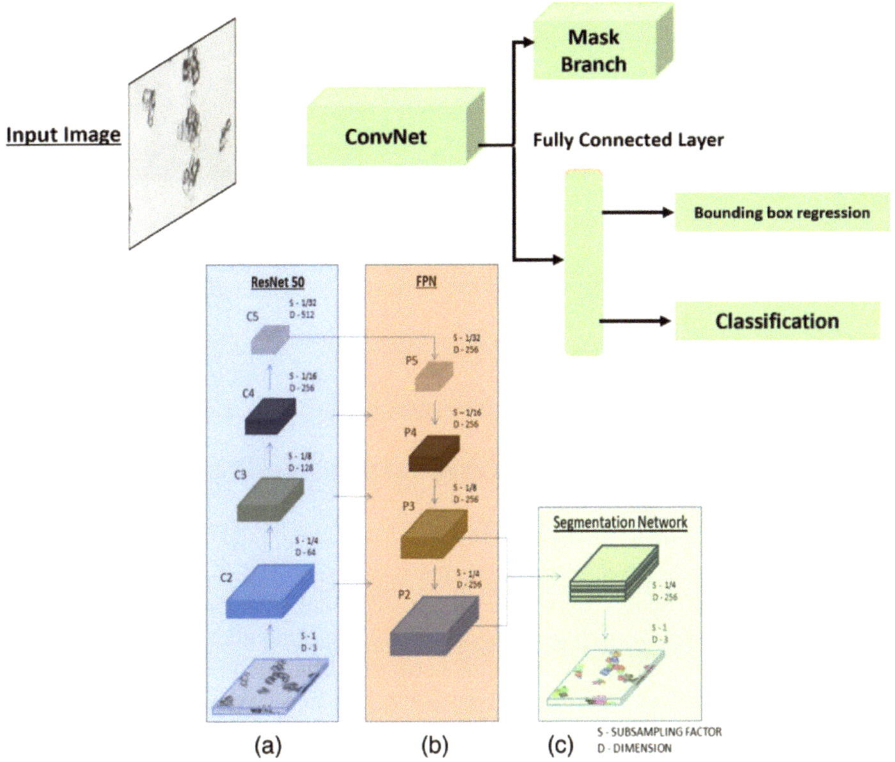

FIGURE 9.14 Architecture of the ResNet-50 backbone (a), feature pyramid network (FPN) (b), and the segmentation sub-network (c).

9.4 RECURRENT NEURAL NETWORKS (RNN)

Relationships among the entries in the input matrix may carry valuable information and be helpful to better capture the underlying nature of the system being modeled. In image analysis, we not only care about the value in each pixel but also *where* in the input matrix it is located. A key advantage provided by CNNs is precisely the ability to handle these relationships effectively. But, what about relationships between samples? If we do not assume that each sample is independent, and we can distribute these samples over time or space, then we have sequence data. Examples of sequence data are text, audio signals, and time series. In all these cases, we are interested in predicting what will happen next, that is, what is the next word in a speech, the next measure in music piece, or the next trend in the stock market. For data with such properties, FC and even CNNs often fail. In this section, we present one more type of neural network that has proven to be highly effective in these applications.

Recurrent neural networks (RNNs) are artificial neural networks with a recurrent structure applied for time series or sequential data. RNNs use outputs of network units from the previous time point t as input to the next unit at time point $t+1$. An illustration of a recurrent network and the unfolded structure is shown in Figure 9.15.

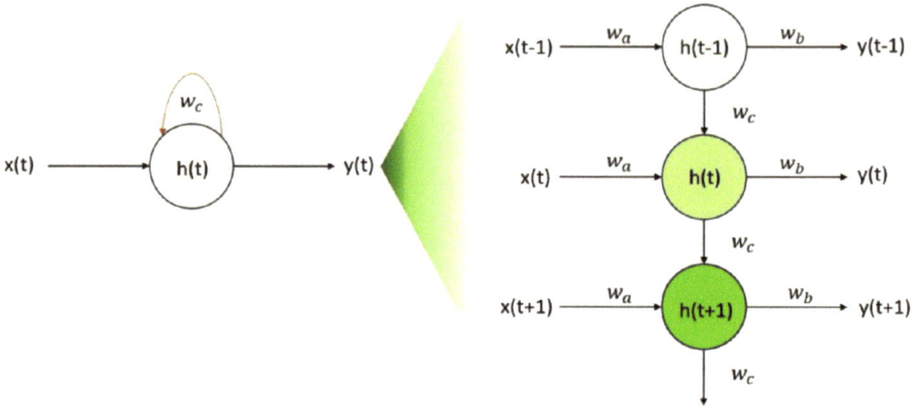

FIGURE 9.15 Schematic diagram of an RNN structure with time series input.

In the figure, the left-most item shows the recurrent connections as the arc. It is "unfolded" in time to produce the appearance of layers.

The logic behind an RNN is to consider the sequence of the input. For example, for us to predict the next word in a sentence, we need to remember what word appeared in the previous time step. These neural networks are called recurrent because this step is carried out for every input. As these neural networks consider the previous value during predicting, it acts like a memory storage unit which stores it for a short period of time.

The diagram in Figure 9.15 represents a three-layer RNN which is unrolled to understand the inner iterations. Looking at each step:

- $x(t)$ is the input at time step t. $x(t-1)$ will be, for example, the previous word in the sentence or the sequence.
- $h(t)$ will be the hidden state at time step t. The output of this state will be nonlinear and considered with the help of an activation function.
- $y(t)$ will be our output at the time step t. It is a value which has the highest probability considered using an activation function.

The RNN in Figure 9.15 has the same evaluation at each step considering the weights w_a, w_b, and w_c, but the inputs differ at each time step making the process fast and simpler. It *remembers* only the previous value and not the values before, thus acting like a memory.

9.4.1 Long-Short-Term-Memory Network (LSTM)

Although RNNs have the right framework to model temporal data, they suffer from some inadequacies. NNs learn by backpropagating the error to modify the network weights. If the network is too large or the number of time steps is too long, the distance over which the error must be transmitted becomes infeasible and the network

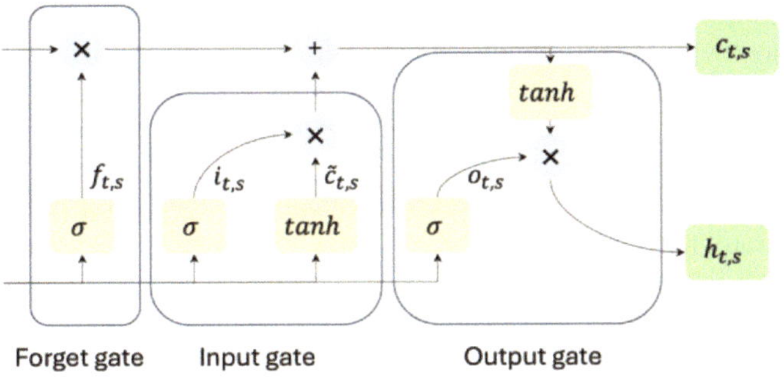

FIGURE 9.16 Schematic diagram of an LSTM structure with time series input.

stops learning. This problem is referred to in literature as the vanishing gradient problem. RNNs have trouble with short-term memory, that is, if a sequence is long enough, they have a hard time carrying information from earlier time steps to later ones. Consequently, if you are trying to process a paragraph of text to predict the next block of text, RNNs may leave out important information from the beginning. These issues gave rise to the idea of LSTM [19] which is a special kind of RNNs, capable of learning long-term dependencies.

The addition of the LSTM feature to the original RNN structure allows the model to predict the output based on long-distance features. Different from the traditional RNN, LSTM introduces three gates including input gate, forget gate, and output gate and a cell memory state (see Figure 9.16) that allows the network to "memorize" important information on long-distance features and "forget" information that are less important. The *cell memory state* represents the memory of the LSTM, and it undergoes changes via forgetting of old memory (forget gate) and addition of new memory (input gate). The *forget gate* decides the contents of the cell state that are relevant to the problem and discards everything else. The *input gate* ensures that relevant information in the input is stored in the cell state. Finally, t*he output gate* adds the relevant input to the cell state and passes it on to the next time step. This architecture ensures that the gradients are propagated back into the earlier time steps in the network.

The Gated Recurrent Unit (GRU) [20] is another modified RNN cell that is similar to the LSTM in many ways. It differs from the LSTM in that it has two gates in place of three. The update gate and the reset gate preserve long-term information in the data. While the reset gate determines the information that is important to hold over the long term, the update gate ensures that this information is added to the hidden state. This architecture ensures that the GRU has fewer parameters to train than an LSTM. It saves memory and take less time to train. However, in small sequences this typically does not make much of a difference.

9.5 ATTENTION AND TRANSFORMERS

Predicting events over time, whether in the context of stock market behavior or natural language processing, has proven to be highly profitable, and it's perhaps one the most sought out applications of ML. Because of this, RNNs have raised a great amount of interest and investment. Today, some of the larger ML models publicly available and with more ubiquitous presence across users include some form of RNN. A major step in the development of these models is that of *attention* in the context of ML. Building upon the notion of LSTM, attention allows the computer to derive relationships between blocks of data at any position in the input sequence, thus prioritizing how useful a piece of information is. This not only addresses the vanishing gradient problem but also accelerates the training of these models.

Unlike RNNs, in which elements of the input sequence are passed to the learning agent one at the time, in transformers, all elements are passed simultaneously. An attention mechanism then prioritizes the relative importance of the elements in the sequence with respect to each other. Figure 9.17 shows a simplified schematic of the transformer architecture introduced by Google in 2017 [21]. Inputs are fed to the model and then translated into vectors with an embedding. Then, a positional encoder captures distances between elements of the input sequence, thus providing a notion of context. The embedded vector with the positional information is then passed through a multi-head attention block that learns the relative importance of each element in the input sequence to the rest. The output of the attention block passes through as feedforward neural network to make easier to understand by the next step. Like the inputs, the outputs are embedded, encoded to provide context, and pass through an attention block. Then, the attention vectors, that is, the outputs of the attention blocks, for both the inputs and outputs are combined in a single attention block. This operation seeks to learn relationships between elements of input sequence and elements of the output sequence. The last attention vectors are then passed through a feedforward network and some activation layers to report the probability of an element in the output sequence to come up next.

Although DL architectures such as transformer are leading to qualitative changes in fields like language processing, their applications for engineering problems are still limited. In the last chapter of this book, we will discuss the outlook of new technologies and some visions of future developments.

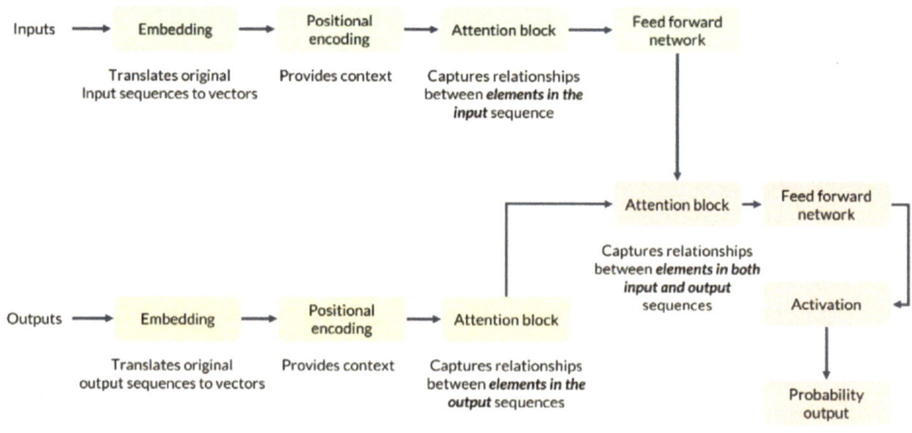

FIGURE 9.17 Simplified schematic of a transformer architecture.

9.6 GRAPH NEURAL NETWORKS

Graph neural networks (GNNs), introduced by Scarselli et al. [22], have emerged as a powerful paradigm for learning from graph-structured data, offering a versatile framework for tasks ranging from social network analysis to drug discovery. In machine learning, the ability to represent and reason about graph-structured data is crucial due to the ubiquity of such data in various domains including social networks, molecular structures, and recommendation systems. Unlike traditional data formats like images or text, graphs capture complex relationships and dependencies among entities, making them indispensable for modeling relational data.

At its core, the intuition behind GNNs lies in the notion of message passing among nodes within a graph. The fundamental idea is to iteratively update node representations by aggregating information from neighboring nodes, allowing the network to capture local and global graph structure. Mathematically, this process can be formalized as follows:

$$h_v^{(l+1)} = \sigma\left(\sum_{u \in \mathcal{N}(v)} f\left(h_u^{(l)}, h_v^{(l)}, e_{uv}\right)\right). \tag{9.4}$$

Here, $h_v^{(l)}$ represents the representation of node v at layer l, $\mathcal{N}(v)$ denotes the neighbors of node v, f is a differentiable function that aggregates information from neighboring nodes, and e_{uv} represents the edge features between nodes u and v. The process continues for multiple layers, allowing the network to capture increasingly abstract and high-level representations of the graph.

Training GNNs involves optimizing a suitable objective function such as node classification or graph-level prediction tasks. The process typically involves backpropagation through the graph structure, where gradients are propagated through the message passing mechanism. Stochastic gradient descent (SGD) variants or more sophisticated optimization algorithms like Adam are commonly employed to update the network parameters. Additionally, techniques such as dropout or BN can be utilized to improve generalization and training stability.

When using GNNs, several key hyperparameters must be carefully considered to ensure optimal performance. These include the number of layers in the network, the dimensionality of node embeddings, the choice of aggregation function, learning rate, and regularization strength. The number of layers determines the depth of the network and its capacity to capture complex relationships within the graph. The dimensionality of node embeddings controls the expressiveness of the learned representations, while the choice of aggregation function influences the way information is propagated through the graph structure. Fine-tuning these hyperparameters is crucial to achieving good performance and generalization across different graph datasets and tasks.

9.7 CONCLUSIONS

In this chapter, we focused on DL with special attention different strategies for efficient data representation including CNNs, RNNs, and GNNs.

We discussed the main ideas behind CNNs, a pivotal development in artificial neural networks designed specifically for processing images. Tracing back to the groundwork laid by Fukushima's "neocognitron" in 1980, CNNs evolved through notable contributions by Rumelhart et al. and LeCun et al. The breakthrough moment for CNNs arrived in 2012 during the ILSVRC, where a CNN model drastically reduced the error rate in image classification, setting off a cascade of advancements in various applications from skin cancer detection to smartphone camera post-processing.

CV, a multidisciplinary field at the intersection of artificial intelligence and machine learning, aims to imbue computers with the ability to comprehend visual information akin to human perception. At its core, CV involves tasks ranging from object classification and localization to image segmentation, each with its unique challenges and methodologies. Within image segmentation, semantic segmentation assigns each pixel in the image to a specific object category, while instance segmentation delineates precise boundaries for countable objects, enabling pixel-level predictions.

RNNs offer a potent framework for modeling sequential data, capturing dependencies over time in tasks like language processing and time series prediction. However, their limitations, such as the vanishing gradient problem, spurred the development of specialized architectures like LSTM and GRU, which excel at retaining and updating information over long sequences. Furthermore, the advent of attention mechanisms and transformer architectures has revolutionized sequence modeling by enabling simultaneous processing of all elements and learning their relative importance, marking a significant shift in machine learning paradigms. Looking forward, the integration of RNNs, attention mechanisms, and transformers promises to drive unprecedented innovations in natural language understanding, time series analysis, and beyond, propelling us toward a new era of intelligent systems capable of comprehensively modeling complex temporal phenomena.

GNNs have emerged as powerful tools for learning from graph-structured data in various domains. They operate by iteratively updating node representations based on neighboring node information, capturing both local and global graph structures. Training GNNs involves optimizing objective functions using techniques like backpropagation, SGD, and regularization. Careful tuning of hyperparameters such as layer depth, node dimensionality, and aggregation function is crucial for achieving optimal performance across different datasets and tasks.

REFERENCES

1. Fukushima K. (1980), Neocognitron: A self-organizing neural network model for a mechanism of pattern recognition unaffected by shift in position. Biological Cybernetics, 36(4), pp. 193–202, doi: 10.1007/BF00344251
2. Rumelhart D. E., Hinton G. E., Williams R. J. (1986), Learning representations by back-propagating errors. Nature, 323(6088), pp. 533–536, doi: 10.1038/323533a0
3. Lecun Y., Bottou L., Bengio Y., Haffner P. (1998), Gradient-based learning applied to document recognition. Proceedings of the IEEE, 86(11), pp. 2278–2324, doi: 10.1109/5.726791
4. Russakovsky O., Deng J., Su H., Krause J., Satheesh S., Ma S., Huang Z., Karpathy A., Khosla A., Bernstein M., Berg A. C., Fei-Feiet L. (2015), ImageNet large scale visual recognition challenge. International Journal of Computer Vision, 115(3), pp. 211–252, doi: 10.1007/s11263-015-0816-y

5. Simonyan K., Zisserman A. (2014), Very deep convolutional networks for large-scale image recognition. CoRR, 1409(1556), p. 2014.
6. He K., Zhang X., Ren S., Sun J. (2016), Deep residual learning for image recognition. IEEE Conference on Computer Vision and Pattern Recognition (CVPR), pp. 770–778, doi: 10.1109/CVPR.2016.90
7. Brinker T. J., Hekler A., Utikal J. S., Grabe N., Schadendorf D., Klode J., Berking C., Steeb T., Enk A. H., von Kalle C. (2018), Skin cancer classification using convolutional neural networks systematic review. Journal of Medical Internet Research, 20(10), p. e11936.
8. Ali R., Chuah J. H., Talip M. S. A., Mokhtar N., Shoaib M. A. (2022), Structural crack detection using deep convolutional neural networks. Automation in Construction, 133, p. 103989, doi: 10.1016/j.autcon.2021.103989
9. Morikawa C., Kobayashi M., Satoh M., Kuroda Y. (2021), Image and video processing on mobile devices: A survey. The Visual Computer, 37(12), pp. 2931–2949, doi: 10.1007/s00371-021-02200-8
10. Olah C., Mordvintsev A., Schubert L. (2017), Feature visualization: How neural networks build up their understanding of images. Distill, doi: 10.23915/distill.00007
11. Goodfellow I., Bengio Y., Courville A. (2016), Deep Learning (Adaptive Computation and Machine Learning). The MIT Press (in Eng).
12. Ioffe S., Szegedy C. (2015), Batch normalization: accelerating deep network training by reducing internal covariate shift. Proceedings of the 32nd International Conference on International Conference on Machine Learning, 37, doi: 10.5555/3045118.3045167
13. Long J., Shelhamer E., Darrell T. (2015), Fully convolutional networks for semantic segmentation. IEEE Conference on Computer Vision and Pattern Recognition (CVPR), 7–12, pp. 3431–3440, doi: 10.1109/CVPR.2015.7298965
14. Ronneberger O., Fischer P., Brox T. (2015), U-Net: Convolutional networks for biomedical image segmentation, ArXiv, vol. abs/1505.04597. https://doi.org/10.48550/arXiv.1505.04597
15. He K., Gkioxari G., Dollár P., Girshick R. (2017), Mask R-CNN. IEEE International Conference on Computer Vision (ICCV), 22–29, pp. 2980–2988, doi: 10.1109/ICCV.2017.322
16. Ren S., He K., Girshick R., Sun J. (2015), Faster R-CNN: towards real-time object detection with region proposal networks. Proceedings of the 28th International Conference on Neural Information Processing Systems, 1, pp. 91–99, doi: 10.5555/2969239.2969250
17. Lin T. Y., Goyal P., Girshick R., He K., Dollár P. (2020), Focal loss for dense object detection. IEEE Transactions on Pattern Analysis and Machine Intelligence, 42(2), pp. 318–327, doi: 10.1109/TPAMI.2018.2858826
18. Redmon J., Divvala S., Girshick R., Farhadi A. (2016), You only look once: Unified, real-time object detection. IEEE Conference on Computer Vision and Pattern Recognition (CVPR), 27–30, pp. 779–788, doi: 10.1109/CVPR.2016.91
19. Hochreiter S., Schmidhuber J. (1997), Long short-term memory. Neural Computation, 9(8), pp. 1735–1780, doi: 10.1162/neco.1997.9.8.1735
20. Chung J., Gülçehre C., Cho K., Bengio Y. (2014), Empirical evaluation of gated recurrent neural networks on sequence modeling, ArXiv, vol. abs/1412.3555. doi: 10.48550/arXiv.1412.3555
21. Vaswani A., Shazeer Parmar N., Uszkoreit J., Jones L., Gomez A. N., Kaiser L., Polosukhin I. (2017), Attention is all you need. Proceedings of the 31st International Conference on Neural Information Processing Systems, 6000–6010, https://dl.acm.org/doi/10.5555/3295222.3295349
22. Scarselli F., Gori M., Tsoi A. C., Hagenbuchner M., Monfardini G. (2009), The graph neural network model. IEEE Transactions on Neural Networks, 20(1), pp. 61–80.

10 Deep Learning Case Studies

10.1 CASE STUDY 1: THERMAL IMAGE PROCESSING IN A PYROLYSIS FURNACE

10.1.1 Introduction

In Chapter 5 we emphasized the critical role of monitoring pyrolysis reactor conditions to ensure safe and efficient manufacturing processes. The inherent challenge arises from the elevated temperatures (exceeding $800°C$) in the cracking regions, preventing the implementation of conventional sensors to observe the reactor's internal operating conditions. A viable solution lies in the utilization of infrared thermography, offering practical insights into temperature-related information. This method proves instrumental in identifying operational anomalies like overheating, tube coking, and tube deformation.

To automate the monitoring process, it becomes essential to extract pertinent features from infrared images. Subsequently, automatic monitoring systems can be implemented to support plant operators in making well-informed decisions regarding reactor operation. The approach involves utilizing pixel-wise temperature signals derived from thermographic cameras strategically positioned throughout the furnace [1], as illustrated in Figure 10.1.

10.1.2 Data Preparation

The measured temperature matrix serves as the basis for temperature monitoring and generating RGB-colored infrared photos (see Figure 10.2) for tube shape monitoring and visual inspection. Although there are approximately 5000 photos taken by the fixed-position cameras, their diversity is limited. Therefore, to enhance the training set's diversity and reduce training memory usage, we sample image pieces from the original photos. The image sampling process involves

- splitting each photo into nine pieces of equal dimensions (240×426),
- sampling two 240×240 pieces with a stride of 186 from each split piece, and
- generating a new piece by flipping the piece from step 2.

By following these steps, we can generate 36 image pieces with a resolution of 240×240 from each photo. We sample four photos taken from normal operating conditions for each camera from the raw dataset and label tube regions and irrelevant backgrounds in each photo. This process generates a total of 864 image pieces for training.

DOI: 10.1201/9781003455905-12

(a) (b)

FIGURE 10.1 Onsite furnace with infrared camera: (a) single camera and (b) six cameras.

FIGURE 10.2 A sample of generated infrared photo from the temperature matrix.

10.1.3 Segmentation Architecture

To effectively monitor the operating conditions of infrared cameras, it is essential to identify and recognize the pyrolysis tube areas in the photos. Therefore, the first step of the monitoring framework involves accurately segmenting tube regions from the bulk background. While training a deep network with millions of parameters typically requires a large amount of training data, transfer learning is a common solution to reduce the data size needed for deep learning. In this work, a pre-trained 50-layer ResNet was chosen for image segmentation due to its accuracy and size [1]. The architecture of ResNet-50 was implemented by He et al. [2] and only has two pooling layers for down sampling, with the remaining down sampling operations achieved by setting the stride size of convolution filters to two.

To convert a classification model into a segmentation model, the implementation proposed by Shelhamer et al. [3] was used in this approach, replacing fully connected layers and the final SoftMax layer with convolutional layers. In the decoding stage, the approach proposed by Ronneberger et al. [4] was considered, introducing extra feature channels for decoding to allow the network to propagate high-resolution context information. The U-Net decoding architecture was found to provide better segmentation performance for this specific application.

The architecture of the tube segmentation network is demonstrated in Figure 10.3. To integrate U-Net decoding into a ResNet-50 network, as described in Chapter 9, feature maps with different down sampling factors ($f = [2, 4, 8, 16, 32]$) are combined into decoding layers for prediction. In this study, transposed convolution is employed for up sampling intermediate feature maps. Finally, the last layer of the decoding network utilizes SoftMax to predict the final binary classes (tube and background). Thus, the proposed segmentation network is referred to as Res50-UNet.

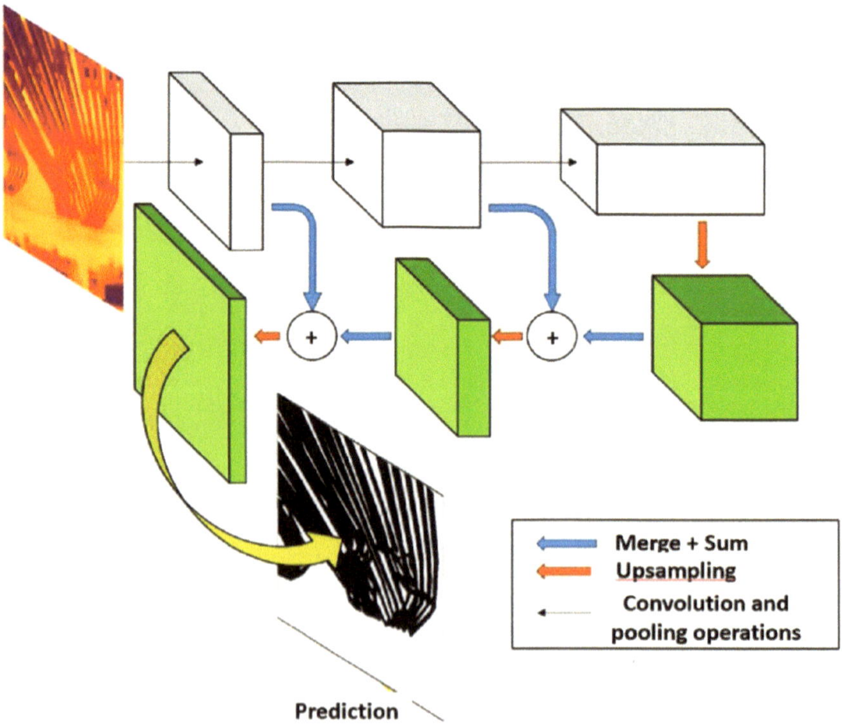

FIGURE 10.3 Schematic of the segmentation network.

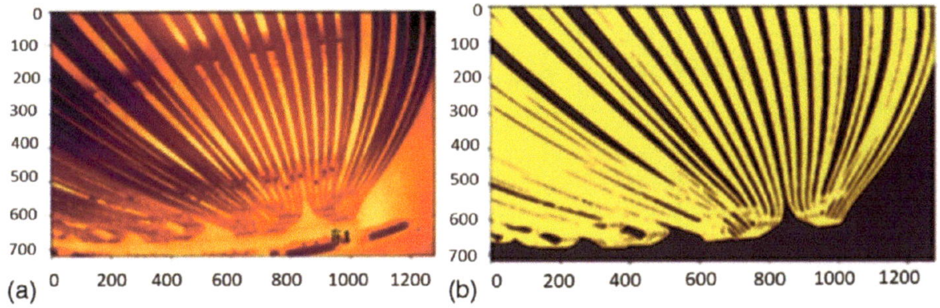

FIGURE 10.4 Segmentation results for the photos from one camera. (a) Original image and (b) image after segmentation.

Using the proposed method, a pre-trained 50-layer ResNet served as the base model to segment the tube region from raw photos. The model was built using TensorFlow [5] in a Python 3.6 environment, and each epoch took roughly 36 seconds to train with a batch size of 20 on a GPU. Various architectures were tested to select the optimal combination of feature layers with different f values, ranging from the highest down sampled level of $f = 32$ to the lowest level of $f = 2$. During training, it was observed that gradually adding lower layers with finer context information into higher layers with more global information can improve the model's performance [1]. However, beyond a certain number, additional layers failed to enhance the accuracy any further. In contrast, including extra layers increased the number of parameters required.

Figure 10.4 presents the visualization of the segmentation results achieved by the proposed Res50-UNet, which demonstrates excellent performance. Although flame nozzles exhibit similar temperature patterns to the pyrolysis tubes, they can still be identified. Additionally, post-processing techniques were explored in this study. Specifically, the Conditional Random Fields (CRF) [6] was applied to refine the raw predictions generated by the Res50-UNet. This additional step led to improvements in terms of noise suppression and edge smoothing. However, it should be noted that this improvement comes at the cost of extra computational time, especially for high-resolution images.

10.1.4 Monitoring Framework

The image segmentation model has the capability to automate the monitoring of tube shape and temperature within the fired furnace. Input data for this study comprises the temperature matrix, which is measured from the infrared cameras within the furnace. Although the temperature matrix contains abundant information, precise monitoring requires further analysis and feature extraction from the raw matrix. To achieve the primary monitoring objective – shape and surface temperature monitoring – a monitoring framework is proposed as shown in Figure 10.5. The raw temperature matrix is provided by the infrared cameras, which are used to construct the infrared pictures. The information from the temperature matrix is first transformed into an RGB-colored image, where each pixel is colored based on its temperature value.

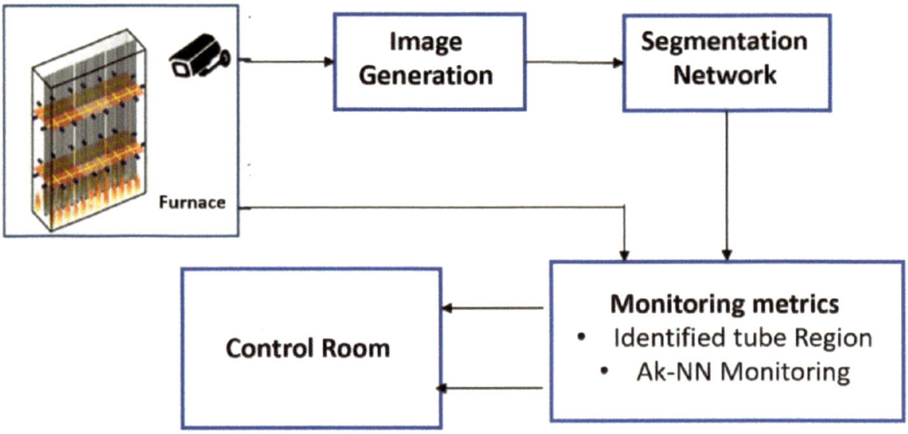

FIGURE 10.5 Monitoring framework.

The proposed segmentation network is then used to identify the tube areas. The segmentation results, along with temperature measurements, are processed to monitor temperature and shape changes. The extracted temperature and shape information is marked on the infrared photos, while the adaptive *k*-NN method introduced in Chapter 8 is used to automatically detect abnormal conditions from these variables.

To monitor shape changes during real-time operation, we start by assuming quasi-steady-state conditions where the tube's shape does not change significantly in short periods, typically within hours. This assumption is reasonable, considering that significant deformations usually take years to occur. With this assumption in mind, we use a moving window to store information about the tube region continuously from previous time steps (as shown in Figure 10.6).

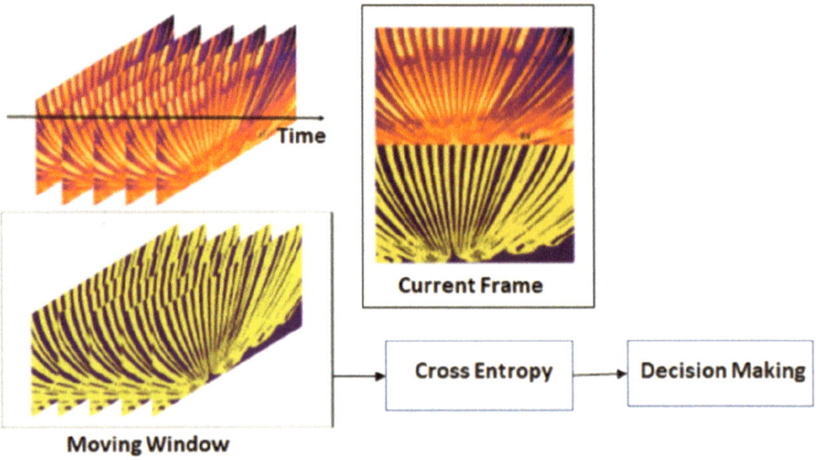

FIGURE 10.6 Moving window approach.

Hence, the shape-changing can be caught by the comparison between the moving window averaged prediction (Y_{mw}) and current prediction (Y_t) as follows:

$$\nabla A = \left|\overline{Y}_{mw} - Y_t\right|. \tag{10.1}$$

The difference defined in Eq. 10.1 can be marked on the infrared image to indicate the root cause of the shape-changing. Besides the visualization of photos, the sum of the squared error (SSE) is used to quantify the shape-changing in a single statistical metric:

$$SSE\left(\overline{Y}_{mw}^{ij} - Y_t^{ij}\right) = \sum_i \sum_j \left(\overline{Y}_{mw}^{ij} - Y_t^{ij}\right)^2. \tag{10.2}$$

In addition to shape monitoring, the tube surface temperature can be calculated by incorporating the original temperature matrix, T_{ref}. The temperature information corresponding to tube regions can be precisely located as follows:

$$T_{tube} = T_{ref} \odot \overline{Y}_{mw}, \tag{10.3}$$

where $\odot$ denotes component-wise multiplication. To improve the robustness against noises in the prediction, the binary moving window prediction, $\overline{Y}_{mw}$ is used to represent the tube region instead of the current tube region prediction, Y_t. With the calculated tube surface temperature, two key metrics for temperature monitoring can now be defined: the average temperature and maximum temperature. These two metrics can provide different scales of statistical information. The average temperature monitoring offers global information that can be used for decoking management and maintenance schedules. Meanwhile, the maximum temperature monitoring focuses on current operating conditions, such as regional overheating and coking, that could potentially affect tube endurance. Meanwhile, overheated regions can be marked by the infrared photos, helping process operators identify the abnormal locations clearly. Hence, the necessary adjustment can be implemented.

Finally, the calculated temperature and shape-changing information are monitored by multivariate statistical methods. Using available historical data, the control limits can be defined, and future measurements can be automatically analyzed. By considering the nature of the cracking process where coking is typical during the operation, the operating conditions can be slowly drifting. Recall from Chapter 8 that given several training samples, the A*k*-NN method determines a threshold for such operating conditions from the *k*-nearest neighbor curve of the given dataset [7]. The updating factor p is used in this method to update prototypes to characterize the process drifting.

10.1.5 Application of Monitoring Strategy

To validate the proposed model, the whole framework is tested over 150 continuous data records to simulate real plant conditions. For detection condition, a current prediction from the raw image is compared to predictions from previous several steps for both temperature and shape (Figure 10.7).

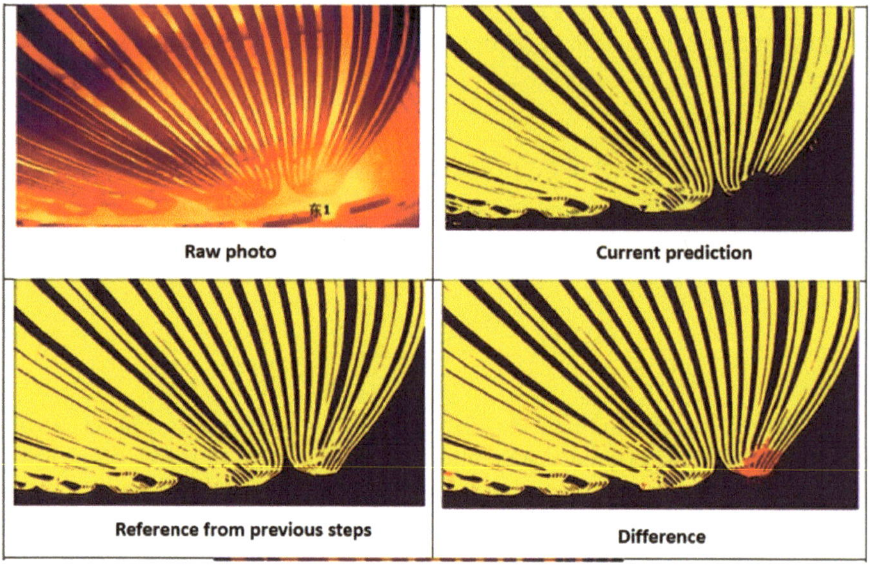

FIGURE 10.7 Current and previous steps comparison.

To implement the Ak-NN fault detection method, the first 50 data samples are used to train the Ak-NN model to determine a threshold for normal conditions, while the other 100 samples are used for testing. Figure 10.8 presents the calculated results in terms of SSE changes and monitoring results.

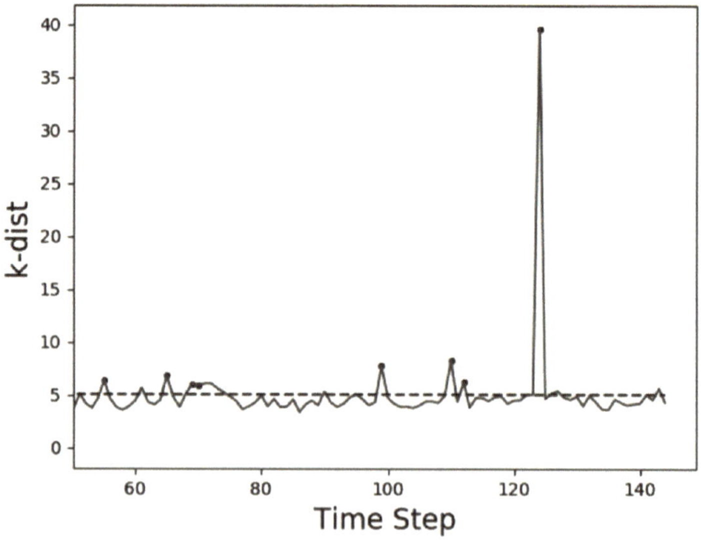

FIGURE 10.8 Quantitative monitoring results from the proposed framework from the Ak-NN method.

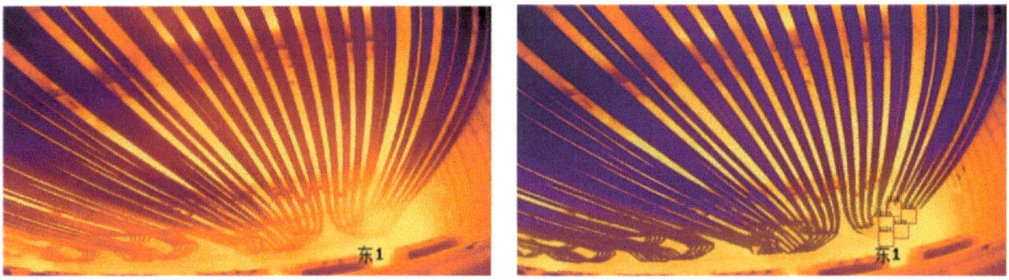

(a) (b)

FIGURE 10.9 Fault reasoning for time step 125. (a) Generated infrared image. (b) The output result for visual monitoring; area differences are highlighted, and high-temperature regions are marked.

From the statistical results, a large peak can be noticed at time step 125, which is expressed directly on the images themselves for a more intuitive visualization for the plant operator. The basic principle of image-based abnormal condition reasoning is to map temperature and shape changes on the image itself, where high-temperature regions and significant shape differences can be highlighted. In Figure 10.9, the analyzed result from one camera is demonstrated. Comparing with the normal condition in Figure 10.2, an overheated region in light yellow that overlaps some tubes can be observed in the bottom part.

The corresponding segmentation prediction in Figure 10.9b excludes all abnormal tube regions. According to Eq. 10.1, the abnormal regions can be simply identified by their differences with previous predictions in the moving window, which can be visualized on the image for precise abnormal location identification. Additionally, the high-temperature regions are marked on the image, which provides extra information to analyze the root cause of the problem. Images from the other five cameras are all analyzed in the same way. Subsequently, the abnormal condition detected at this time step was further validated by plant engineers. The abnormal condition was caused by improper flame control, temporarily sending the flame too high and wrapping around pyrolysis tubes. Despite using a very deep network, the computational time for segmenting each image is about 0.14 seconds with a single Graphics Processing Unit (GPU). The total processing time for one set of six images including result visualization is about 3 seconds, enabling online support for process operators in the control room (Figure 10.10).

10.2 CASE STUDY 2: CHARACTERIZING CSD FOR CRYSTALLIZATION OPERATIONS

10.2.1 Introduction

Crystals serve as fundamental building blocks, capable of growth through diverse methods that yield materials endowed with distinctive textural and physical attributes. The properties inherent in each crystal dictate the composition of the final product, with even slight variations – such as size, shape, and purity – exerting significant

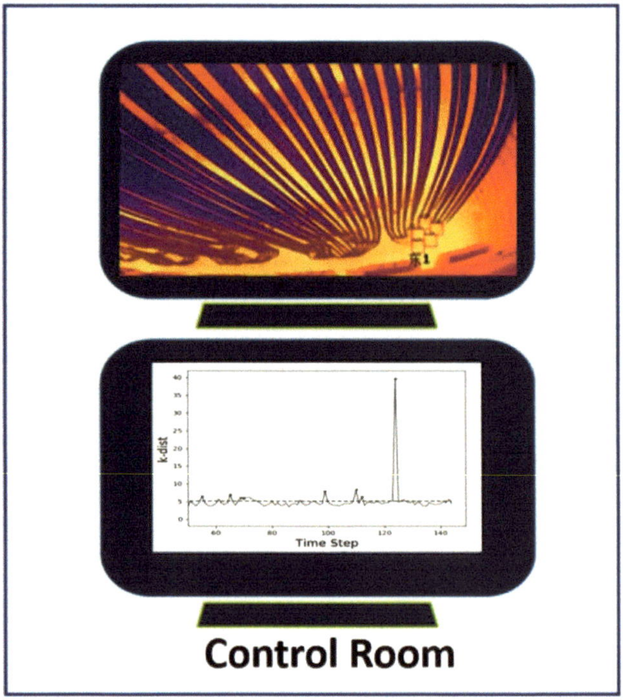

FIGURE 10.10 Results visualization in the control room.

influence on the resulting quality. For example, in drug manufacturing, the size of crystals plays a pivotal role in determining the efficacy of the active pharmaceutical ingredient (API) [8]. Crystals that exceed optimal dimensions may diminish the bioavailability of the drug, while those too small can become susceptible to oxidation. In the realm of the food industry, diverse crystal shapes contribute to an array of textures. Smooth, round crystals that effortlessly glide past each other create a creamy texture in ice cream, whereas popsicles necessitate jagged-edged crystals for a hard and brittle consistency [9]. Given the imperative for uniform crystal properties in the large-scale production of these items, the control of crystal growth emerges as a pivotal challenge extensively explored in the existing literature.

To assess the system's state and attain the desired condition, controllers gauge crystal properties and implement appropriate control measures. Commonly, these properties are determined through process measurements, including solute concentration, particle size distribution, and morphology. Various tools have been devised to acquire these measurements across different conditions [10–16]. However, their efficacy might be compromised in instances of excessively high solute concentrations or significant overlap among the crystals under examination. While these challenges are manageable in small-scale experiments, they become pervasive in industries emphasizing rapid and large-scale production. Consequently, the development of tools capable of effectively functioning across a broad spectrum of conditions persists as an ongoing research challenge without a definitive solution [18].

This application employs a machine-learning paradigm, wherein images and labels are fed to the algorithm, enabling it to uncover underlying patterns in the data [17]. This approach deviates from the traditional practice of identifying patterns in images and manually encoding them into pattern recognition algorithms. The development includes a segmentation model that extends the RetinaNet framework by incorporating a dedicated segmentation branch as discussed in Chapter 9. This addition becomes crucial for accurately gauging the size and shape of crystal particles, as we will demonstrate later. To validate the approach, a comparative analysis was conducted with the Mask Region-based Convolutional Neural Network (Mask R-CNN), a benchmark object detection model. The findings support the advocacy for utilizing a smaller and faster segmentation model in crystal characterization.

10.2.2 Experiment

The effectiveness of the segmentation network was tested on a crystallization experiment involving the growth of NaCl crystals from H_2O in the presence of an ethanol anti-solvent [18]. In anti-solvent crystallization, the solute to be crystallized is separated from its solvent by the addition of an anti-solvent in which the solute is only sparingly soluble. The anti-solvent lowers the solubility of the solute in the resulting solution and initiates nucleation. In the case of NaCl, it is highly soluble in water with a solubility of 36 g/100 g and weakly soluble in ethanol with a solubility of only 0.065 g/100 g. When ethanol is added to a saturated aqueous solution of NaCl, the solubility limit of NaCl in the mixed solution is lowered leading to supersaturation. The rate of addition of ethanol can be manipulated to control the growth of NaCl crystals.

In the experiment, 34 g of NaCl was dissolved in 100 g of deionized (DI) water to prepare a slightly under-saturated salt solution while the anti-solvent solution was prepared by mixing 950 g of ethanol in 50 g of DI water. The reaction was carried out in a glass crystallizer at various temperatures (10°C, 20°C, and 30°C) and anti-solvent flowrates (7, 15, and 30 mL/min). The reactor was equipped with an impeller to promote mixing so that uniform concentration could be assumed. A controlled stream of solvent was delivered to the reactor using a peristaltic pump. The temperature of the mixture was maintained at the desired value using a recirculating heating jacket with a resistance temperature sensor used as a monitoring tool. Samples of the reacting mixture were continuously removed from the crystallizer and pumped into a flow cell fastened below a video microscope. Images were recorded by the microscope at regular intervals after which the sample was redirected back into the reactor.

Since the images were captured during flow, motion blur influenced their quality, hindering the use of several images. The images were examined to isolate the clear pictures and the rest were discarded. Approximately 2000 clear images were captured during the experiments with each image having a maximum spatial size of 1280 pixels. The crystals in the images were manually labeled using the AMSCOPE software with the mean size and standard deviation of the crystals computed for each image. These measurements were used as the gold standard labels to evaluate the performance of the segmentation network. The overall setup of the experiment is shown in Figure 10.11.

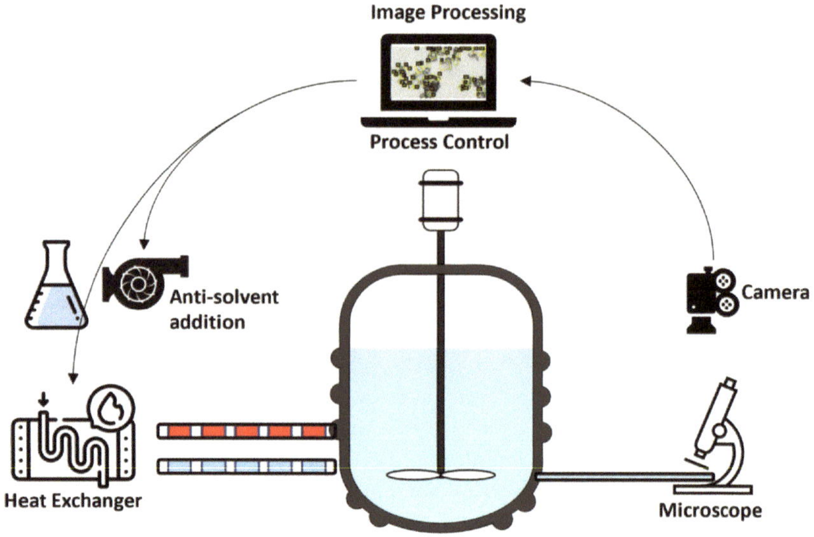

FIGURE 10.11 Overall setup of the experiment.

10.2.3 CNN for Crystal Characterization

A convolutional neural network (CNN) was employed to analyze images of crystals captured by the microscope. In the context of crystallization, CNNs can identify individual boundaries of adjoining crystals, a task that is easy to do visually but challenging to convert into code. Another advantage of CNNs is their ability to generalize learning to a variety of experimental conditions. Data augmentation techniques can be used to simulate common artifacts like blurred images and variations in brightness. This can teach the network to be more robust to such variations.

In this implementation, the raw images serve as inputs, while the annotated images (Figure 10.12) act as labels. This approach works particularly well for crystal images as it allows for the measurement of crystal particle size and differentiation between crystal clusters and individual instances, while also effectively handling instances where crystals overlap.

CNNs can take on a variety of architectures depending on the image analysis task at hand. For instance, in object detection, the objective is to characterize all objects present in an image. The most efficient architecture to accomplish this using CNNs is that in which the problem is divided into two sub-tasks. In the first sub-task, the CNN localizes every object in the image by predicting their coordinates. In the second sub-task, the CNN classifies the objects into one of several categories. Both of these sub-tasks share a common backbone to extract patterns and are solved in parallel, resulting in the RetinaNet model. RetinaNet is capable of characterizing objects in an image by localizing them and predicting their class, using a feature pyramid architecture to ensure the final layers learn high-level features. This simple architecture enables RetinaNet to process images for object detection at very high speeds.

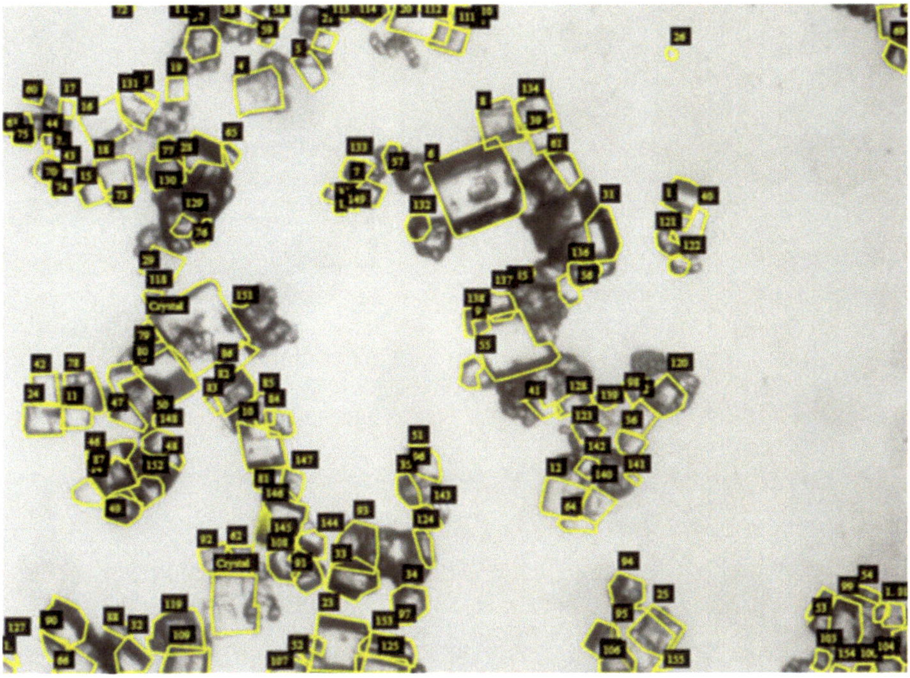

FIGURE 10.12 Manually labelled images.

In contrast, instance segmentation requires a CNN architecture that involves solving three sub-tasks. The first sub-task is to localize all the objects of interest in the image. The second is to classify the objects into categories, and the final sub-task is to generate segmentation maps for the objects. By combining the output of these three tasks, precise pixel-level maps for each object in the image can be generated. This architecture is used in this work to characterize the boundary of each crystal instance and estimate the size distribution of the crystals. Figure 10.13 illustrates the structure of the modified CNN, as discussed in Chapter 9.

10.2.4 Training and Inference

One hundred images randomly sampled from different experimental conditions were used to train the segmentation network with an additional 20 images used for testing. The images were resized to 1024 pixel-wide squares and batched into groups of two. Each image had a label associated with it containing information about the boundary coordinates of individual crystal instances, their class, and their binary masks. Crystal boundaries that were unclear due to lighting effects and excess overlap were ignored during labeling.

Since the quantity of labeled images available was insufficient to train the entire network from scratch, transfer learning was implemented. Pre-trained weights from the COCO dataset and ImageNet were used during the initialization phase. Only the

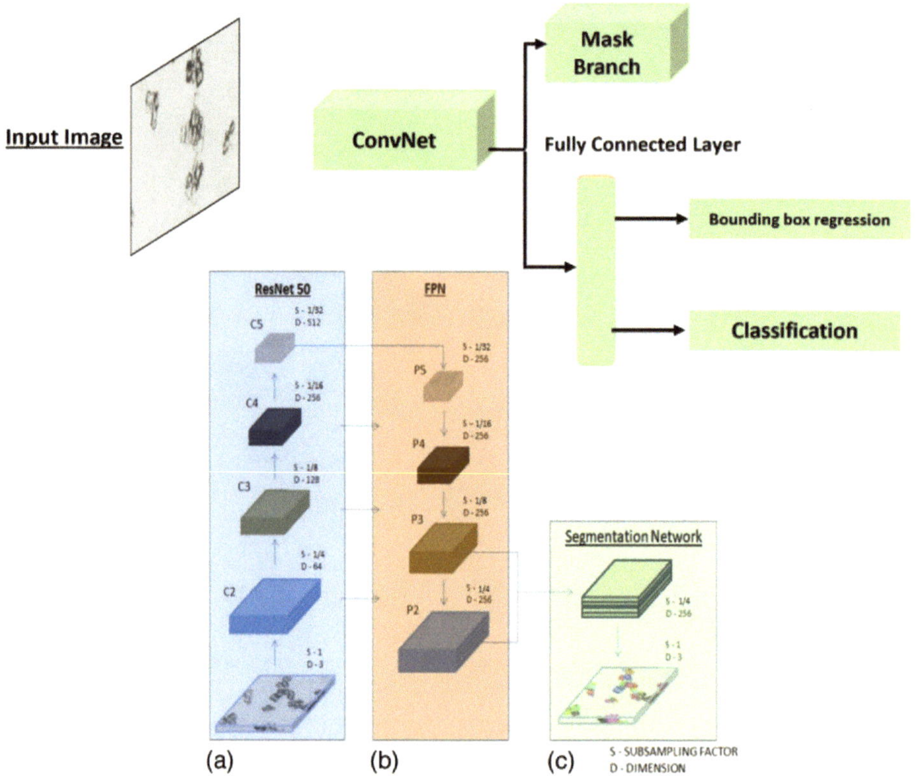

FIGURE 10.13 Convolution neural network architecture for the characterization of crystals. (a) ResNet backbone, (b) feature pyramid network (FPN), and (c) segmentation sub-network.

upper layers of the segmentation network – those closer to the output – were fine-tuned on the crystal images while the rest of the weights were frozen. This allowed the model to transfer the feature extraction learned from the initial training set to the type of images that we are interested in. A Nvidia GTX 1070 Ti GPU with 8 GB of memory was used to train the network for 3 hours.

The inference stage of the segmentation network is similar to the training stage. Instead of computing the errors made by each head, the outputs of the individual heads are assembled into a single pixel-level map with each crystal instance highlighted using a different color. This information-rich map is then used to compute the parameters of interest including mean size and standard deviation.

10.2.5 Results and Discussion

Figure 10.14 displays the results of the segmentation network. Before being fed into the model, the raw images obtained from the video microscope were padded and resized to 1024-pixel squares. The segmentation sub-network produced binary images with the same spatial size as the input images, which separated the crystals

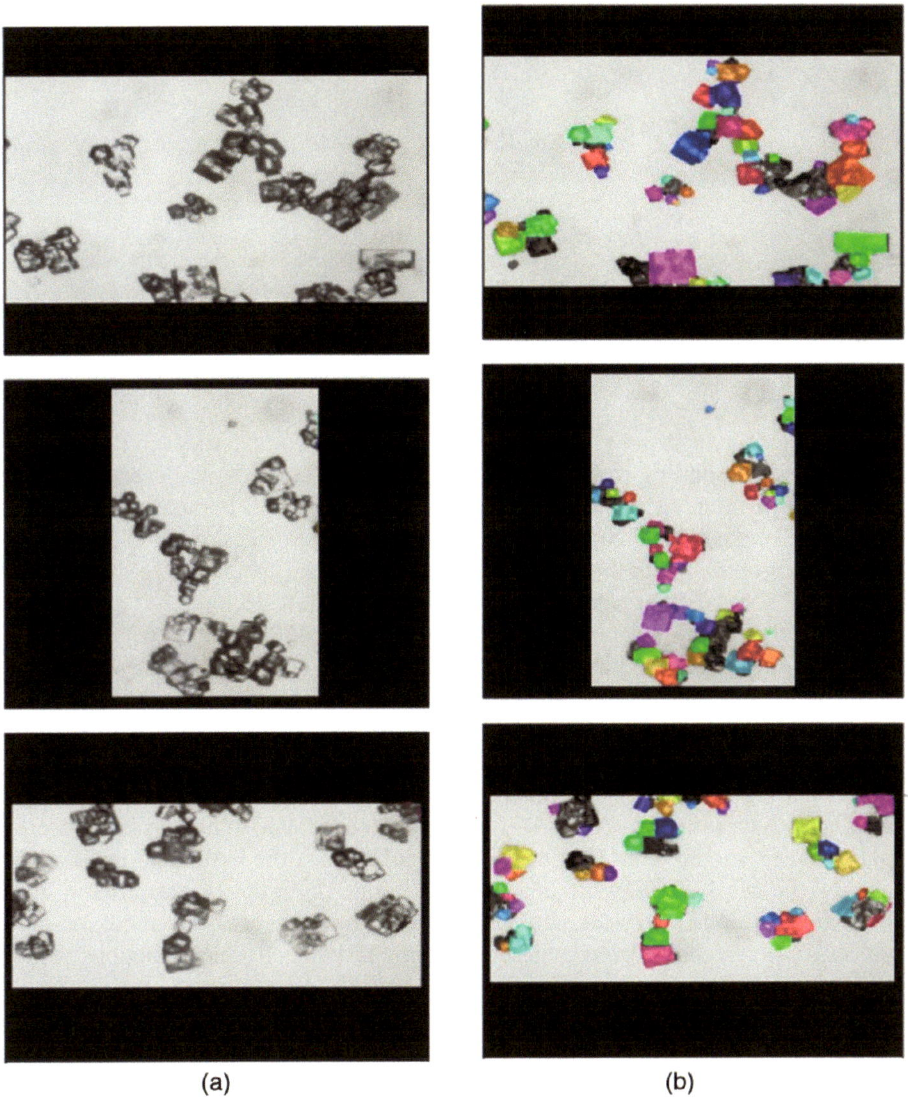

FIGURE 10.14 The results of our segmentation model on crystal images. From left to right, the columns represent (a) the raw image obtained from the video microscope and (b) the combined output of the network.

from the background. These binary images, combined with the outputs of the box sub-network and class sub-network, were utilized to generate pixel-level masks for each crystal instance, considering their irregular shapes. The size of each crystal instance was defined as the area of its 2D surface, and the area of each crystal instance was determined by summing the number of foreground pixels in the crystal mask. To convert the calculated pixel area to microns, a micron to pixel ratio

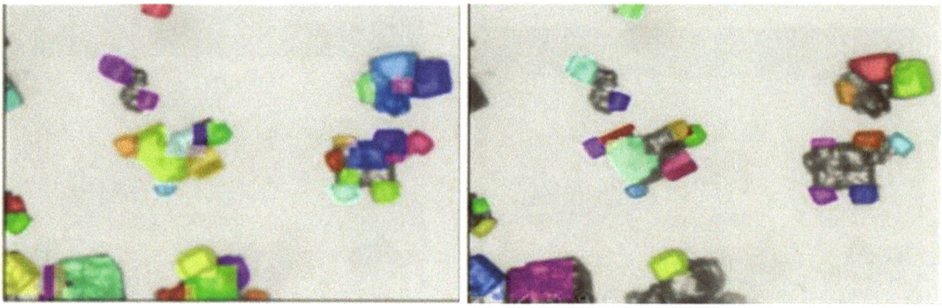

(a) (b)

FIGURE 10.15 Crystal detection outputs produced by: (a) New segmentation model and (b) Mask R-CNN.

of 0.775 corresponding to 25× zoom was employed since crystal size is generally expressed in microns. Mean size, standard deviation, and population distribution were then employed to characterize the overall crystal population. Finally, these values were compared with those obtained from manual labeling and the outputs of the Mask R-CNN and RetinaNet models, as shown in Figure 10.15.

While the model generally performs well, the Mask R-CNN outperforms it during the initial stages of some experimental conditions, particularly when crystals are small and agglomerated. The Mask R-CNN's two-stage construction yields superior results because masks are generated independently for each bounding box. In contrast, in a one-stage model, a single mask is initially generated for the entire image, and bounding boxes are subsequently used to extract masks for individual crystal instances. As a result, the Mask R-CNN is better equipped to handle small overlaps than the new model. However, under other conditions, the individual masks generated by the Mask R-CNN are not always accurate or complete, resulting in underestimates of the true crystal size. Under these circumstances, the modified model produces more reliable masks and consistently provides better predictions than the Mask R-CNN.

The performance of the CNN sensor was also validated by evaluating its performance on three different experimental conditions as shown in Figure 10.16. This was done by capturing images at the end of each experiment and estimating the mean size and standard deviation using the CNN. These results were compared with manual labels prepared by an expert labeler. The results illustrate that the sensor provides excellent estimates of the size distribution of crystals.

10.2.6 Comparative Analysis among CNN Models

The new model has fewer parameters than the Mask R-CNN resulting in faster processing speeds and smaller memory requirements. Table 10.1 shows the speed gains of our one-stage model over two-stage models like the Mask R-CNN. The experiments were run on a Nvidia GTX 1070 Ti GPU using TensorFlow [5] as the backend

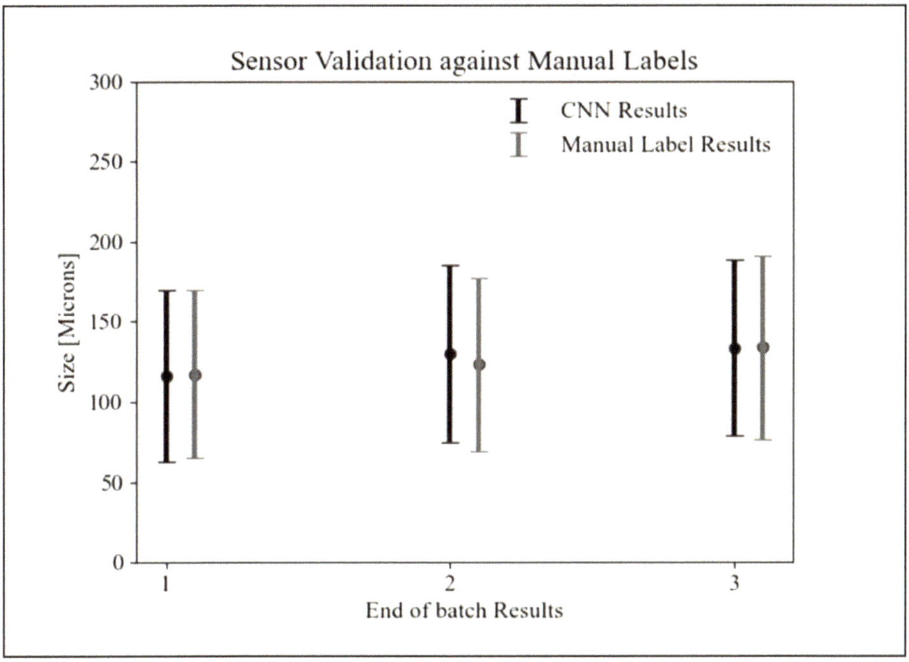

FIGURE 10.16 Validation for samples obtained at the end of the experiment for three different experimental conditions: 1) Temp: 10°C and Flow: 3 mL/min; 2) Temp: 20°C and Flow: 3 mL/min; 3) Temp: 30°C and Flow: 3 mL/min.

platform. The RetinaNet model, which has 10M fewer parameters than the Mask R-CNN, processes images at 3× the speed without a drop-in accuracy. The addition of the segmentation sub-network introduces a small overhead (2M parameters) that lowers the processing speed of the RetinaNet by less than 1 fps. The difference in speeds is more pronounced on CPU since the Mask R-CNN has several small operations that exploit the speed-up provided by a GPU. It should be mentioned that the original publications had reported faster speeds for both Mask R-CNN and RetinaNet, but here the results obtained from our own experiments were used for the sake of consistency. The code to run these models were obtained from the TensorFlow Object detection GitHub repository [5].

10.2.7 Comparison of CSD with Theoretical Distribution

Size distribution in particle growth processes tends to follow a right-skewed lognormal distribution [19, 20]. One explanation for this phenomenon is that at every instant in time, the particles can either grow in size by collision or stay the same. Since the latter is more likely, the bulk of the particles have small sizes with a tiny fraction attaining the larger, improbable values. The results of the model-generated

TABLE 10.1
Processing Speeds of Deep Learning Models with a ResNet-50 Backbone

Model	Parameters	FPS on GPU (Time per Frame)	FPS on CPU (Time per Frame)
Mask R-CNN	44M	3.0 (0.33)	0.077 (12.98)
RetinaNet	34M	8.9 (0.11)	0.33 (3.03)
Our model	36M	8.1 (0.12)	0.27 (3.7)

histogram were validated by comparing it to a log-normal distribution fitted to the manually determined mean and standard deviation. First, the log-normal distribution parameters are calculated from the mean and standard deviation using the formula described as follows:

$$\mu = ln\left(\frac{m^2}{\sqrt{s^2 + m^2}}\right), \tag{10.4}$$

$$\sigma = \sqrt{ln\left(\frac{s^2}{m^2} + 1\right)}. \tag{10.5}$$

In Figure 10.17, the log-normal distributions fitted to the manually determined parameters are shown for three different sampling times during the reaction. The histograms are generated by identifying the crystals in the images at each sampling time, measuring their size and recording the values. It is seen that the predictions closely match the expected log-normal distribution. In Figure 10.18, the distributions from the various sampling times are overlaid to show both the increase in mean size and spread of the crystal sizes.

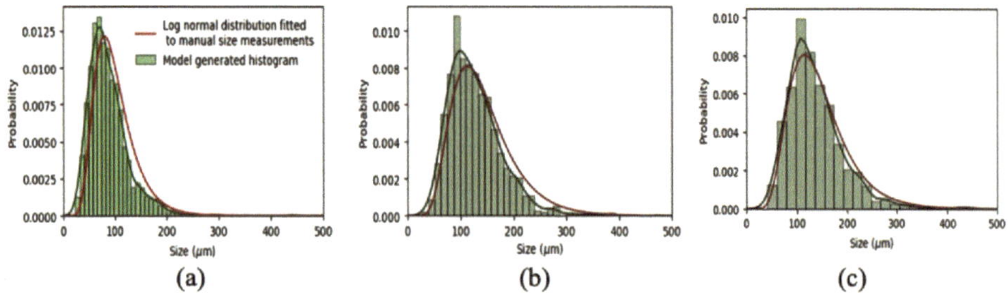

FIGURE 10.17 Comparing the results of the model-generated size distribution with a log-normal distribution fitted to manually measured crystal sizes at different sampling times of a single experimental batch. (a) 30 min, (b) 180 min, and (c) 420 min.

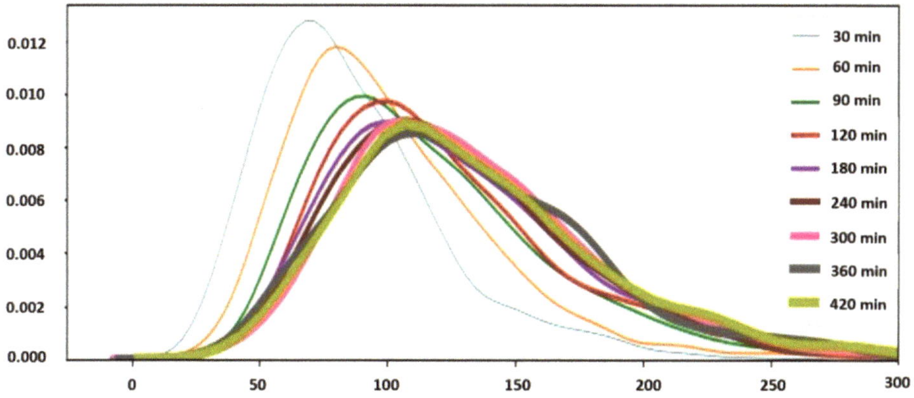

FIGURE 10.18 Time evolution of distribution during reaction.

10.3 CASE STUDY 3: DIGITAL TWIN FOR A DEMETHANIZER COLUMN

10.3.1 Process Description

Natural gas stands as a global reservoir of both cleaner and affordable energy, with its production rapidly on the rise, particularly in North America. Significant technological strides in drilling and multi-stage fracturing have facilitated access to vast reserves of natural gas trapped within shale formations. Given the heightened market value of heavier hydrocarbons (C2+), the extraction of natural gas liquids (NGLs) frequently occurs from the original feed gas. Within this context, the utilization of turboexpanders emerges as one of the predominant strategies in the gas industry for effectively separating these valuable hydrocarbon liquids [21].

Among the various structures for extracting NGLs, the gas-subcooled process (GSP) and the cold residue recycle (CRR) process are two of the most extensively used in gas processing plants [22]. Each of these recovery methods exhibits distinct advantages and drawbacks, with their suitability hinging on a multitude of factors that range from operational requisites and process constraints to the technological orientation of the operator. In this particular case study, our focus centers on the CRR unit.

In the CRR unit (see Figure 10.19), the initial step involves cooling the raw gas through the utilization of both the heat exchanger (E-100) and the chiller (E-101). Once cooled, this leads to the separation of the resultant liquid component using the flash tank (TK-100), strategically positioned ahead of the demethanizer column (T-100). A portion of this liquid is then combined with the vapor exiting the flash tank (TK-100). Subsequently, this combined mixture undergoes cooling via the heat exchanger (E-102), followed by expansion through the Joule-Thompson valve (JTV-101). The resultant output is then introduced into the column's second stage. Simultaneously, another segment of the liquid is depressurized using the JTV-100 before being fed into the 26th stage. Meanwhile, the remaining vapor proceeds to the

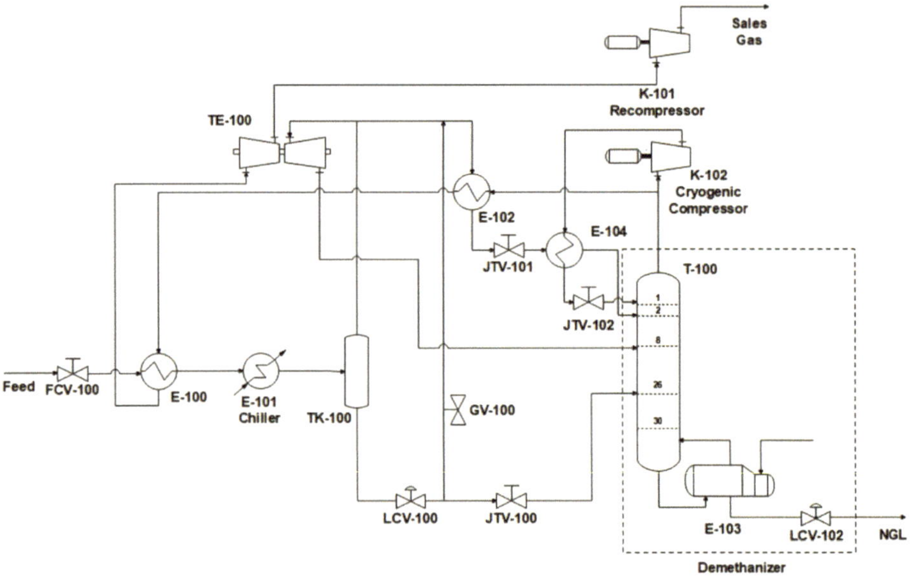

FIGURE 10.19 CRR scheme for NGL recovery.

turboexpander (TE-100) and is introduced into the column's eighth stage. To achieve the necessary separation, a fraction of the demethanizer's upper output is compressed by the cryogenic compressor (K-102). This compression yields ethane-free reflux, which then interacts with the heavier compounds within the vapor emerging from the column's upper section, effectively carrying them into the descending liquid phase [23]. Before being reintroduced, this reflux is subjected to cooling through the heat exchanger (E-104) and subsequent expansion via the JTV-102, adjusting it to the desired column top pressure. The remaining product from the column requires further processing for commercial applications, following its role as a refrigeration medium for the heat exchangers E-100 and E-102. Finally, the NGL bottom product originating from the demethanizer column undergoes an additional round of separation within the fractionation train.

Central to the separation process is the demethanizer column (T-100), which stands as the core unit. It operates as a 30-stage distillation column, incorporating a reboiler (E-103). The primary objective of this unit is to effectuate the separation of methane from the raw gas. To achieve this, the unit functions under cryogenic conditions, facilitating a liquid-vapor equilibrium between methane and the denser hydrocarbon fraction. The nominal temperature at the top tray of this unit hovers around −116°C.

A dynamic process model was developed in HYSYS® for the selected CRR scheme and used for controllability analysis [24–28]. The process model was constructed based on realistic operating conditions. A complete description and operating conditions can be found in the original publications. In this section, a digital surrogate model (digital twin) of the column was developed using the concepts introduced in Chapter 9.

10.3.2 Demethanizer Digital Twin Architecture (DDT)

To acquire a digital surrogate model for the demethanizer column, the task was divided into three primary objectives [29, 30]. The first objective focused on estimating the compositions of both the top and bottom segments of the column as well as its internal and external flows. This was accomplished by creating two distinct Long Short-Term Memory (LSTM) models specifically designed for this purpose. Subsequently, a neural model architecture was developed and juxtaposed to accurately simulate the internal separation processes taking place within the column.

To determine the compositions of the demethanizer's top and bottom products, LSTM neural networks were utilized to construct a model named NMPC (Neural Model for Product Composition Estimation). For estimating the top and bottom product streams, along with the reflux and boil-up streams, another model called NMCF (Neural Model for Column Flow Estimation) was formulated. Additionally, a model referred to as the Column Neural Model (CNM) was developed to simulate the internals of the column. This CNM model utilized a dedicated LSTM cell to represent the operations at each stage of the distillation column and the reboiler. By employing this approach, the dynamic aspects of the separation process occurring within each stage of the column were effectively captured. Moreover, bidirectional connections were incorporated to consider interactions between adjacent stages, thus simulating the internal flows within the distillation column.

The individual models were amalgamated into a comprehensive architecture, forming the foundation of the final data-driven model for the demethanizer column known as the Demethanizer Digital Twin (DDT). The DDT integrated the structure of the CNM model with the NMPC and NMCF models. Specifically, the NMCF-derived estimations of reflux and boil-up flows were utilized as inputs for the model, and the composition estimations obtained from the NMPC model replaced the composition measurements required by the NM2 architecture. This elimination of dependence on composition analyzers in the CNM architecture allowed for a complete model of the column to be developed, enabling the estimation of demethanizer column operations using only readily available measurements. Subsequent sections provide comprehensive insights into the model architectures employed to effectively realize each of the discrete components.

10.3.3 Product Composition LSTM Neural Model

The NMPC architecture, depicted in Figure 10.20, integrates two distinct LSTM cells, each one responsible for computing the compositions of one of the product streams. The input data shared by these cells encompasses the pressure and temperature values of all column feed streams, in addition to those of the top and bottom product streams. This input also includes variables such as the reboiler duty, cryogenic compressor power, and the pressure and temperature settings of the flash tank (TK-100). The selection of input data for each network was guided by a knowledge-based approach, considering the variables that exhibit the most robust correlation with the desired outputs. Furthermore, we took into consideration variables that are easily measurable within real plant scenarios.

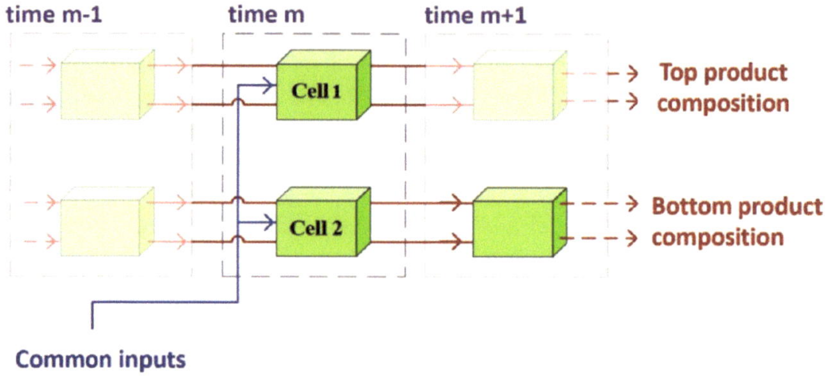

FIGURE 10.20 Schematic representation of the NMPC recurrent structure [30].

10.3.4 Column Flows LSTM Model

The architecture of the NMCF, schematized in Figure 10.21, involves the use of three interacting LSTM cells. The first cell is responsible for calculating the material balance around the cryogenic compressor and, as shown in Figure 10.21, has the goal of estimating the reflux stream. The second cell is responsible for calculating the total column material balance and gives as outputs the flows of the top and bottom column products. The third cell is responsible for estimating the material balance around the reboiler and is responsible for the estimation of boil-up flow. The input data considered in the model was chosen with a knowledge-based approach and differed from cell to cell. The cells' outputs are also provided as input in the other cells to simulate the behavior of flows in the column. The first cell receives as input the reflux ratio in addition to the estimation of boil-up and distillate. The second cell receives as input the measurement of the input flowrates and reflux and boil-up streams' estimations. The third cell receives as input column bottom temperature and bottom product temperature and bottom product and reflux streams' estimations.

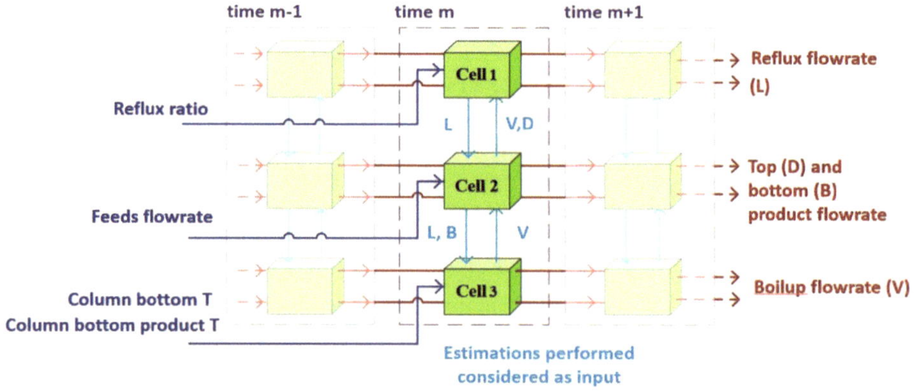

FIGURE 10.21 Schematic representation of the NMCF recurrent structure [30].

10.3.5 Column LSTM Model

Figures 10.22 and 10.23 provide schematic representations of the CNM layout and a specific inner LSTM cell. Illustrated in Figure 10.23, the examined cell (denoted by the subscript "s") utilizes a fusion of the current hidden states and internal cell states ($h_{t-1,s}$ and $c_{t-1,s}$), along with the hidden states and internal cell states computed by neighboring LSTM cells (indicated by subscripts $s-1$ and $s+1$) at the current time step. Furthermore, for the network terminal cells simulating the uppermost tray of the column and the reboiler, the computed hidden states from the absent adjoining cell ($h_{t,0}$ and $h_{t,32}$) are substituted with real measurements of corresponding variables: the reflux for the initial cell and the incoming liquid stream for the final cell within the network.

10.3.6 Results

10.3.6.1 Datasets and Model Training

The DDT training and testing datasets were constructed based on a two-month operational period of the CRR separation unit, utilizing simulations within Aspen HYSYS®. These datasets were generated by introducing varying amplitude and duration ramp changes to the plant feed flowrate, mimicking the fluctuations in daily natural gas demand. These fluctuations exhibit an ascending trend during daylight hours and a descending trend during nighttime hours, with peak variations reaching 10% of the nominal plant feed value. Among all the recorded datasets, 80% were allocated for network training, while the remaining 20% were designated for the validation phase.

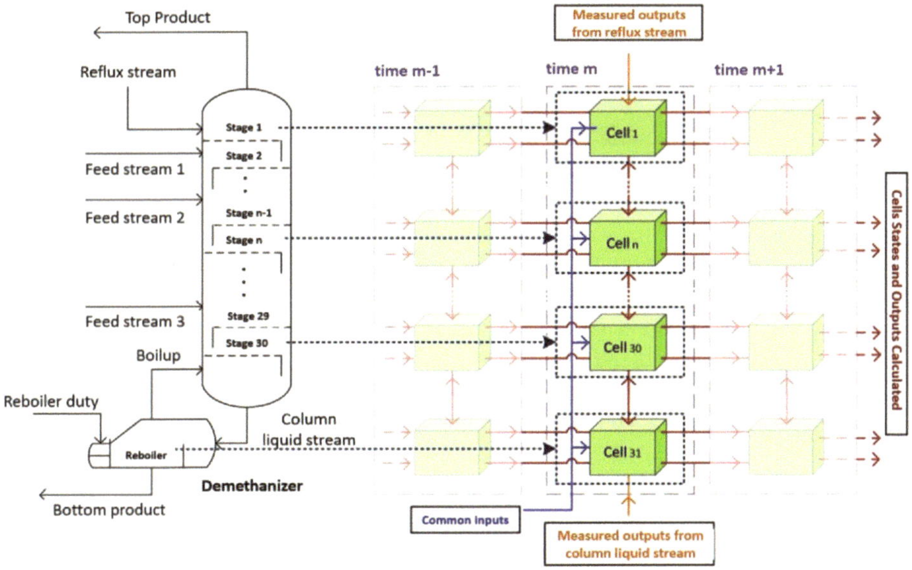

FIGURE 10.22 Schematic representation of the model recurrent structure [30].

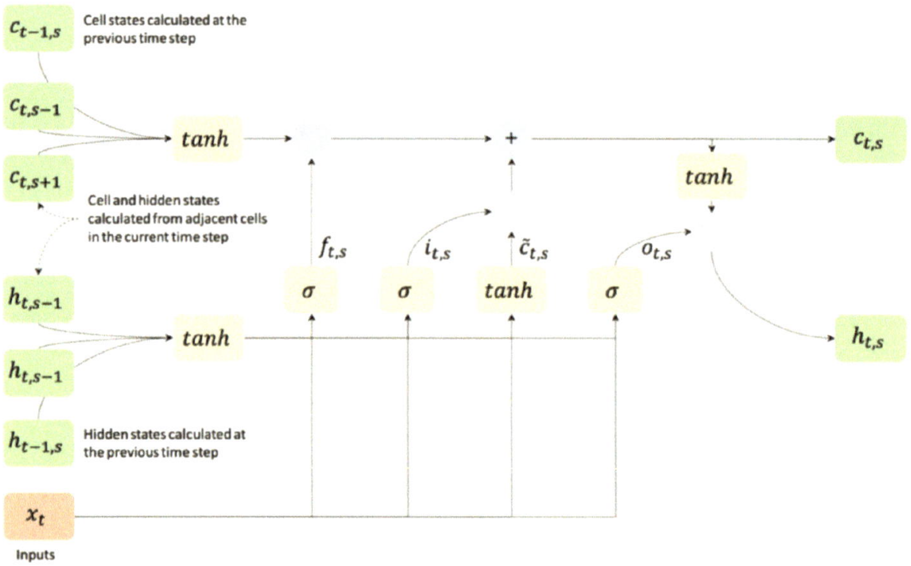

FIGURE 10.23 Schematic representation of an inner LSTM cell.

To assess the performance of the proposed models, a test process involved simulating 2 days of plant operation with ramp variations of 5% in the nominal plant feed flowrate. All datasets were acquired using a sampling interval of 20 seconds. Additionally, to enhance realism, measurement noise was introduced, incorporating standard deviations (further details available in [29, 30]).

The loss function considered in the training optimization problem is given by the Mean Squared Error (MSE) defined as reported in Eq. 10.6.

$$MSE = \frac{1}{N_b} \sum_{j=1}^{O} \sum_{n=1}^{N} \sum_{m=1}^{M} (\hat{y}_{j,n,m} - y_{j,n,m})^2, \tag{10.6}$$

where N_b corresponds to the number of observations in the current iteration; O is the number of considered outputs in a given cell, while the general output is denoted by j; N represents the length of the selected time sequence; M is the number of cells employed by the different models; $\hat{y}$ and y denote a general hidden state calculated by the considered model and the corresponding actual value, respectively.

The loss function is computed during each iteration, employing the backpropagation through Time algorithm to calculate the corresponding gradient concerning weights and biases. This gradient is subsequently applied in the parameter update process facilitated by the Adaptive Moment Estimation optimizer (ADAM) as introduced by Kingma and Ba [31]. Determination of the network hyperparameters involved a multi-objective optimization approach employing the Non-dominated Sorting Genetic Algorithm (NSGA-II) algorithm found within the pymoo package, developed by Blank and Deb [32]. The optimized hyperparameters encompass batch size, sequence length, hidden size, and learning rate. The specific hyperparameters utilized are detailed in [29].

10.3.6.2 DDT Estimation Performance

Results for the DDT model's performance on the test data are presented and discussed in this section. Figure 10.24 provides a visual representation of DDT's effectiveness in reconstructing steady-state profiles for key components, such as ethane and temperature, along the columns. This illustration highlights the accuracy of the DDT model in reconstructing steady states profiles within the column. Importantly, this performance extends to other variables, including pressure and temperature along the column, as detailed in reference [30].

Figure 10.25 illustrates the top and bottom methane composition as well as temperature time evolution under a 5% decrease in the plant feed flowrate. These results demonstrate the model's capability to closely approximate methane and temperature responses with an estimation error of less than 0.5°C. Moreover, both methane and temperature profiles exhibit low levels of noise in the estimated data, attributed to the excellent measurement noise reduction abilities of the NMPC concentration estimations used in the DDT model.

To have a complete overview of the results obtained by the DDT model, developed for the prediction of demethanizer column operations, the comparison of the 3D transient profiles obtained for column key components composition and the actual transient profiles is depicted in Figure 10.26. As shown in the figure, the estimated transient profiles obtained by the DDT show that the model can predict not only the steady state condition but also accurately reconstruct the time evolution of the profiles with no detectable deviations from the actual column profiles.

The model is also able to predict the time evolution of the other components in the column. Thus, from a qualitative point of view, the model developed correctly approximates the target variables. It succeeds in extrapolating the information contained in the inputs provided, and those provided by the outputs of the adjacent

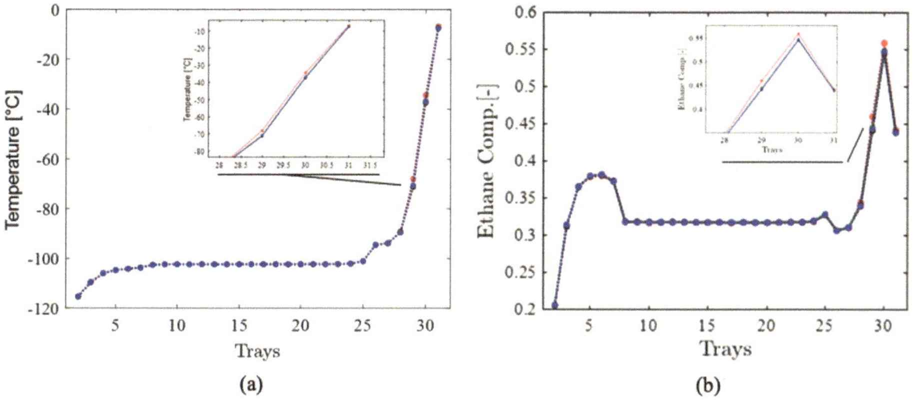

FIGURE 10.24 Column steady state profiles: (a) temperature and (b) ethane at the new steady state after 5% decrease in the plant feed nominal value.

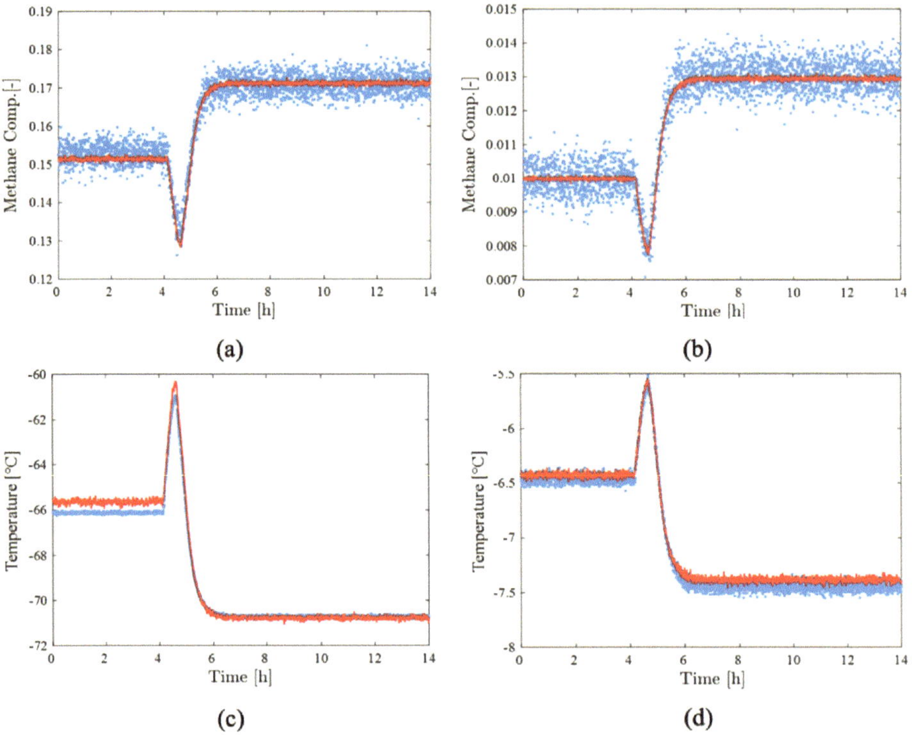

FIGURE 10.25 Comparison of (a) top and (b) bottom transient profiles of methane composition and (c) top and (d) bottom tray temperature under a decreasing variation of 5% in the plant feed nominal value. Blue – measured profiles, red – predicted.

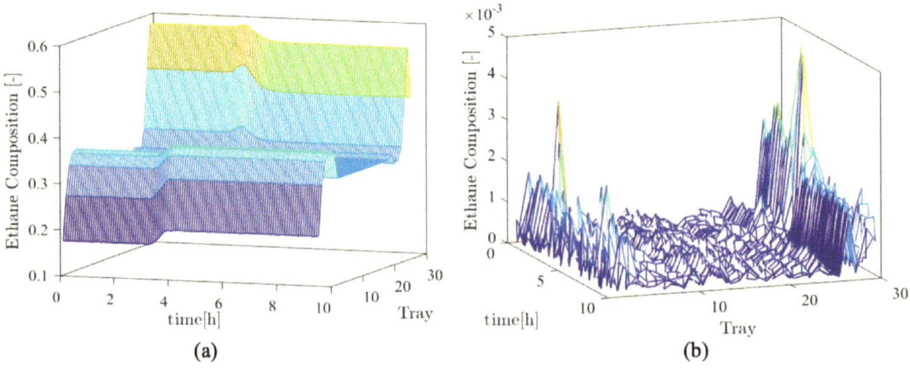

FIGURE 10.26 Transient column profiles obtained under a decreasing variation of 5% in the plant feed nominal value for: (a) estimated ethane composition and (b) error transient profiles [30].

cells of the model, to produce a mathematical model that succeeds in estimating the separation that occurs in each column tray. With regard to the ethane components, the estimation error of the column transient profiles of ethane composition is shown in Figure 10.26b. Here, it is shown that for this component, the DDT can accurately predict the actual composition transient profiles for all the column trays. The estimation errors increase in correspondence to the column region where most of the variations for the considered component occur. Indeed, the estimation error for ethane increases around the column feed trays, and for both the components, it is higher in the column terminal regions, where most of the separation takes place. In general, for both the key components, the deviations of the predicted transient column profiles from the actual profiles are always lower than the maximum assumed measurement error, leading to the achievement of estimates more accurate than the measures achievable with measurement sensors considering the assumed standard deviations.

10.4 CONCLUSIONS

In this chapter, we introduced a monitoring framework tailored for pyrolysis reactor operations for dealing with high temperatures and limited sensor capabilities. Leveraging infrared thermography and advanced image processing techniques, the methodology enables real-time detection of operational anomalies. The segmentation architecture, combining a pre-trained ResNet-50 with U-Net decoding, accurately identifies tube regions within infrared images. Integrated temperature and shape information allow for automated anomaly detection, supported by multivariate statistical methods and historical data.

We also explored crystal growth and characterization, where subtle variations in conditions profoundly influence product quality, from pharmaceuticals to food processing. We showed how CNNs can be used to automate the process of identifying and measuring crystal properties, circumventing challenges associated with manual methods and traditional image processing algorithms. By training the network on labeled images and employing transfer learning from existing datasets, the framework achieved efficient model training and inference, demonstrating the potential of machine learning approaches to revolutionize crystal characterization and pave the way for advancements in materials science and industrial applications.

Finally, we investigated a CRR unit within an NGL processing plant. The development of the DDT using RNNs showcases the benefits of a comprehensive model that integrates various neural network architectures to accurately simulate column operations.

REFERENCES

1. Zhu W., Ma Y., Benton M. G., Romagnoli J. A., Zhan Y. (2019), Deep learning for pyrolysis reactor monitoring: From thermal imaging toward smart monitoring system. AIChE Journal, 65(2), pp. 582–591, doi: 10.1002/aic.16452
2. He K., Zhang X., Ren S., Sun J. (2016), Deep residual learning for image recognition. 2016 IEEE Conference on Computer Vision and Pattern Recognition (CVPR), doi: 10.1109/cvpr.2016.90

3. Shelhamer E., Long J., Darrell T. (2017), Fully convolutional networks for semantic segmentation. IEEE Transactions on Pattern Analysis and Machine Intelligence, 39(4), pp. 640–651, doi: 10.1109/TPAMI.2016.2572683
4. Ronneberger O., Fischer P., Brox T. (2015), U-net: Convolutional networks for biomedical image segmentation. In Navab, N., Hornegger, J., Wells, W., Frangi, A. (Eds), Medical Image Computing and Computer-Assisted Intervention – MICCAI 2015. MICCAI 2015. Lecture Notes in Computer Science, vol 9351. Springer, doi: 10.1007/978-3-319-24574-4_28
5. Abadi M., Barham P., Chen J., Chen Z., Davis A., Dean J., Devin M., Ghemawat S., Irving G., Isard M., Kudlur M., Levenberg J., Monga R., Moore S., Murray D. G., Steiner B., Tucker P., Vasudevan V., Warden P., … Zhang X. (2016), TensorFlow: A system for large-scale machine learning. ArXiv, doi: 10.48550/arxiv.1605.08695
6. Chen L.-C., Papandreou G., Kokkinos I., Murphy K., Yuille A. L. (2018), Deeplab: Semantic image segmentation with deep convolutional nets, atrous convolution, and fully connected CRFs. IEEE Transactions on Pattern Analysis and Machine Intelligence, 40(4), pp. 834–848, doi: 10.1109/TPAMI.2017.2699184
7. Zhu W., Sun W., Romagnoli J. (2018), Adaptive k-nearest-neighbor method for process monitoring. Industrial & Engineering Chemistry Research, 57(7), pp. 2574–2586, doi: 10.1021/acs.iecr.7b03771
8. Khadka P., Ro J., Kim H., Kim I., Kim J. T., Kim H., Cho J. M., Yun G., Lee J. (2014), Pharmaceutical particle technologies: An approach to improve drug solubility, dissolution and bioavailability. Asian Journal of Pharmaceutical Sciences, 9(6), pp. 304–316, doi: 10.1016/j.ajps.2014.05.005
9. Hartel R. W. (2019), Crystallization in foods. In A. S. Myerson, D. Erdemir, and A. Y. Lee (Eds.), Handbook of Industrial Crystallization. Cambridge University Press, pp. 460–478.
10. Larsen P. A., Rawlings J. B., Ferrier N. J. (2007), Model-based object recognition to measure crystal size and shape distributions from in situ video images. Chemical Engineering Science, 62(5), pp. 1430–1441, doi: 10.1016/j.ces.2006.11.018
11. Calderon De Anda J., Wang X. Z., Roberts K. J. (2005), Multi-scale segmentation image analysis for the in-process monitoring of particle shape with batch crystallisers. Chemical Engineering Science, 60(4), pp. 1053–1065, doi: 10.1016/j.ces.2004.09.068
12. Chen J., Wang X. Z. (2005), A wavelet method for analysis of droplet and particle images for monitoring heterogeneous processes. Chemical Engineering Communications, 192(4), pp. 499–515, doi: 10.1080/00986440590477908
13. Larsen P. A., Rawlings J. B., Ferrier N. J. (2006), An algorithm for analyzing noisy, in situ images of high-aspect-ratio crystals to monitor particle size distribution. Chemical Engineering Science, 61 (16), pp. 5236–5248, doi: 10.1016/j.ces.2006.03.035
14. Presles B. (2010), Novel image analysis method for in situ monitoring the particle size distribution of batch crystallization processes. Journal of Electronic Imaging, 19(3), p. 031207, doi: 10.1117/1.3462800
15. Dharmayat S., Calderon De Anda J., Hammond R. B., Lai X., Roberts K. J., Wang X. Z. (2006), Polymorphic transformation of l-glutamic acid monitored using combined on-line video microscopy and X-ray diffraction. Journal of Crystal Growth, 294(1), pp. 35–40, doi:10.1016/j.jcrysgro.2006.05.072
16. Wang X. Z., Roberts K. J., Ma C. (2008), Crystal growth measurement using 2D and 3D imaging and the perspectives for shape control. Chemical Engineering Science, 63(5), pp. 1173–1184, doi: 10.1016/j.ces.2007.07.018
17. Manee V., Zhu W., Romagnoli J. A. (2019), A deep learning image-based sensor for real-time crystal size distribution characterization. Industrial & Engineering Chemistry Research, 58(51), pp. 23175–23186, doi: 10.1021/acs.iecr.9b02450

18. Zhang B., Willis R., Romagnoli J. A., Fois C., Tronci S., Baratti R. (2014), Image-based multiresolution-ANN approach for online particle size characterization. Industrial & Engineering Chemistry Research, 53(17), pp. 7008–7018, doi: 10.1021/ie4019098
19. Manee V. (2022), A data-based framework for monitoring and controlling particulate systems. Doctoral Dissertations, Louisiana State University, p. 5846.
20. Söderlund J., Kiss L. B., Niklasson G. A., Granqvist C. G. (1998), Lognormal size distributions in particle growth processes without coagulation. Physical Review Letters, 80(11), pp. 2386–2388, doi: 10.1103/PhysRevLett.80.2386
21. Li Y., Xu F., Gong C. (2017), System optimization of turbo-expander process for natural gas liquid recovery. Chemical Engineering Research and Design, 124, pp. 159–169, doi: 10.1016/j.cherd.2017.06.001
22. Kherbeck L., Chebbi R. (2015), Optimizing ethane recovery in turboexpander processes. Journal of Industrial and Engineering Chemistry, 21, pp. 292–297, doi: 10.1016/j.jiec.2014.02.035
23. Getu M., Mahadzir S., Long N. V. D., Lee M. (2013), Techno-economic analysis of potential natural gas liquid (NGL) recovery processes under variations of feed compositions. Chemical Engineering Research and Design, 91(7), pp. 1272–1283, doi: 10.1016/j.cherd.2013.01.015
24. Chebeir J., Salas S. D., Romagnoli J. A. (2019), Operability assessment on alternative natural gas liquids recovery schemes. Journal of Natural Gas Science and Engineering, 71, p. 102974, doi: 10.1016/j.jngse.2019.102974
25. Zhu W., Chebeir J., Romagnoli J. A. (2020), Operation optimization of a cryogenic NGL recovery unit using deep learning based surrogate modeling. Computers & Chemical Engineering, 137, p. 106815, doi: 10.1016/j.compchemeng.2020.106815
26. Mandis M., Chebeir J., Baratti R., Romagnoli J. A., Tronci S. (2021), Machine learning for monitoring and control of NGL recovery plants. Chemical Engineering Transactions, 86, pp. 997–1002.
27. Mandis M., Chebeir J. A., Tronci S., Baratti R., Romagnoli J. A. (2021), Control of a natural gas liquid recovery plant in a GSP unit under feed and composition disturbances. IFAC-Papers OnLine, 54(3), pp. 182–187, doi: 10.1016/j.ifacol.2021.08.239
28. Mandis M., Baratti R., Chebeir J., Tronci S., Romagnoli J. A. (2022), Performance assessment of control strategies with application to NGL separation units. Journal of Natural Gas Science and Engineering, 106, p. 104763, doi: 10.1016/j.jngse.2022.104763
29. Mandis M. (2023), Monitoring and control for NGL recovery plant, PhD Thesis, University of Cagliari
30. Mandis M., Baratti R., Chebeir J., Tronci S., Romagnoli J. A. (2024), Exploring nontraditional LSTM architectures for modeling demethanizer column operations. Computers & Chemical Engineering, 183, April 2024, p. 108591
31. Kingma D. P., Ba J. (2014), Adam: A method for stochastic optimization. arXiv preprint arXiv:1412.6980, doi: 10.48550/arXiv.1412.6980
32. Blank J., Deb K. (2020), Pymoo: Multi-objective optimization in python. IEEE Access, 8, pp. 89497–89509, doi: 10.1109/ACCESS.2020.2990567

Part III

Reinforcement Learning and Generative AI

Equipping machines with the capabilities to perform tasks using machine learning often requires the combination of multiple techniques and approaches. In most cases, the need goes beyond making a prediction about the state of a system and involves taking actions or making decisions based on such predictions. Reinforcement learning and generative artificial intelligence (AI) are two archetypes of these solutions, where multiple modalities of machine learning (ML) come into play and new, unique methods emerge to further expand the reach of AI. The final part of this book is devoted to introduce the main notions and ideas around reinforcement learning and generative AI and showcase how these methods can be applied to Chemical Engineering.

DOI: 10.1201/9781003455905-13

11 Reinforcement Learning

Reinforcement learning (RL) is a branch of artificial intelligence that has gained significant attention in recent years for its potential applications in various domains, including chemical engineering. RL provides a framework for training intelligent agents to make sequential decisions in dynamic environments, with the aim of maximizing cumulative rewards. In the context of chemical engineering, RL can be utilized to optimize complex processes, control systems, and design efficient strategies for resource allocation. The next two chapters explore the application of RL in chemical engineering and highlight its potential benefits and challenges.

11.1 INTRODUCTION

To this point, we have discussed methods and strategies to train a computer agent so it can *describe* the state of the system. Given some information about the current configuration of a system (inputs, independent variables), we ask the model to complete it and return new, previously unknown information (outputs, dependent variables). This approach then enables us to make decisions and determine the best course of action. But what if we were to train a computer so that it can *react* to the current state of the system? Such is the realm of RL, which will be discussed in Chapters 11 and 12.

Along with supervised and unsupervised learning, RL is another subset of machine learning methods with important applications in Chemical Engineering. In RL, an agent is allowed to explore an environment to find the best possible behavior to achieve a specific task. Key to RL is the agent's ability to perceive and interpret the environment, and it is in this aspect that supervised learning advancements, such as computer vision via convolutional neural networks (CNNs) and sequence analysis via recurrent neural networks (RNNs), have propelled RL developments. A great example of these methods is AlphaGo [1], a combination of RL and deep learning designed to learn how to play the board game Go, that defeated human world champions in the game. The intuition behind RL, where the agent first perceives the environment, reacts to it, and learns from the feedback, resembles the notions of process control. Because of this, RL poses great promise in applications in the chemical process industry.

Example 11.1: Chemical Reaction Optimization

To illustrate the notion of RL, imagine you are a newly hired chemical engineer in a chemical company. You are tasked with improving the yield of a chemical reaction by changing two reaction conditions: temperature and pH. As you start to analyze this task, your supervisor pops into your office again, hands you a sheet of reactions that have been tried in the past (shown in Table 11.1 and Figure 11.1),

DOI: 10.1201/9781003455905-14

TABLE 11.1
Historical Reaction Data

Experiment No	Temperature	pH	Yield (%)
1	50	4	30
2	75	7	80
3	60	3	40
4	80	11	70

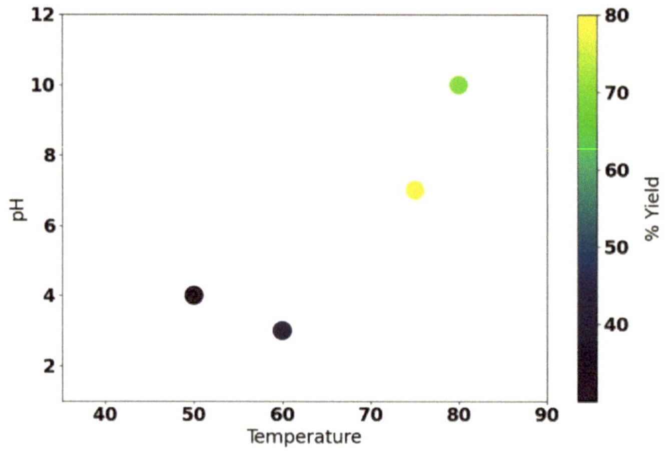

FIGURE 11.1 Data for example 11.1.

and mentions that there is limited time to solve this problem. You ask her how much time and she says she wants an answer in 3 days. You mentally calculate the time it will take per experiment and conclude that you can try at most nine experiments.

Based on the historical data in Table 11.1, which experiment will you try first and why? Will you choose all nine experiments to run at once or choose experiments sequentially based on the results of previous reactions?

One possibility is to choose conditions that are similar to the ones in experiment number 2. Since this set of conditions (75°C and pH of 7) has resulted in high yield, maybe higher yields could be discovered by exploring similar set of conditions. Another equally valid solution could be to explore areas in the parameter space that have been unexplored in Table 11.1. For example, the region in the plot above with low temperature and high pH or the region with high temperature and low pH has not been explored.

If you predetermine the nine experiments to run and then run them one by one irrespective of their results, your decision will only be supported by the data in Table 11.1. Such approach could use the methods described in the previous chapters to obtain a predictive model to inform which experiments to choose. However, given the size of the dataset in Table 11.1, it is unlikely that this approach would solve the problem right away. If, on the other hand, you choose each experiment based on the result of the previous experiment (in addition to the historical data in Table 11.1), your decisions will be informed by the data you are collecting and likely lead to better choices. This paradigm of active engagement with the environment is the key idea of RL.

RL employs an actor that iteratively interacts with the environment and modifies its actions to maximize rewards received from the environment [2]. The main advantage of the RL algorithm is that it learns to optimize strategies by exploration of the environment, independent of the linearity or multi-variability of the system [3]. The learned strategy is obtained from data about the system, and it does not require parameter-tuning or real-time optimization, which allows RL to be easily adaptable to different tasks once the framework is established.

For instance, an RL-based controller can be easily generalized from single input, single output (SISO) systems to multiple inputs, multiple outputs (MIMO) systems, which was examined by Spielberg et al. [3] in simple linear systems of chemical processes in both SISO and MIMO cases. In this chapter, we introduce the theoretical aspects of RL and its related algorithms, while Chapter 12 will present two applications of these methods to chemical processes.

11.2 REINFORCEMENT LEARNING FUNDAMENTALS

11.2.1 Data Requirements for Reinforcement Learning

It is worth starting our discussion about RL by identifying the information needed to successfully deploy an RL application. This will also help framing the kind of problems RL is useful for in the context of Chemical Engineering. The main aspect of RL is the agent's access to information about the environment. Since safety and cost limitations prevent us from giving the computer access to a real plant, we must rely on simulations or bench-scale experiments to train an RL model. If we use a simulation as the environment, then we have the same data needs and opportunities as in the problems described in Chapters 6–8. This means, we would need either a knowledge-based model or labeled data to build a data-driven model. If, on the other hand, we allow the agent to interact with a physical system, deep learning methods as those presented in Chapters 9 and 10 would be necessary for the agent to perceive the environment. In other words, we must provide the learning agent either with numerical descriptions of the environment or with the means to access those descriptions by itself.

11.2.2 Components of the RL Framework

As stated before, the center of the RL framework is an *environment* which holds the dynamics of the task being solved. Relevant information about the constantly changing status of the environment is held within the *state* (s_t). An agent uses this information to interact with the environment by taking an *action* (a_t). In response to the agent's action, the environment transitions into a new state (s_{t+1}) and sends out a *reward* (r_{t+1}) signaling the quality of the action (see Figure 11.2).

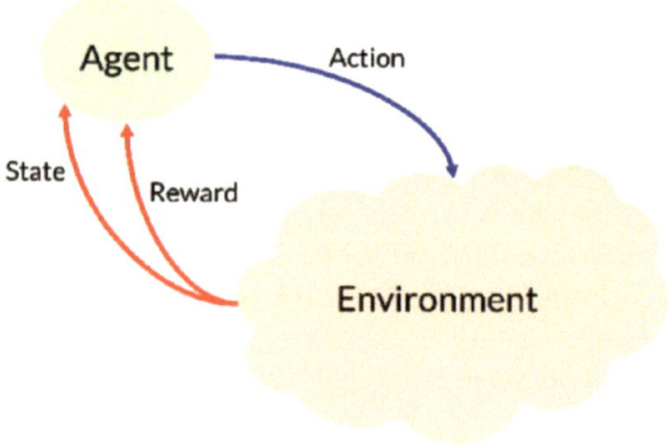

FIGURE 11.2 Relationship between the learning agent and the environment in the reinforcement learning framework.

The playbook that the agent uses to make its decisions is termed the *policy* of the agent. The agent keeps a record of its interactions and the rewards it sees at each state. Using this information, it builds out a map of preference or value for every state in the environment. This map is called the *value function* in the RL framework. If the map has a preference or value associated with both states and actions, it is called the *action-value* function or the *Q-function*. The agent then uses either one of these maps to guide its policy by making better decisions until it ultimately discovers the optimal policy to solve the task.

Example 11.2: Chemical Reaction Optimization with RL

Let us think about problem introduced in Example 11.1 using the RL framework. If we replace the human decision maker with an RL agent, what are the actions that this RL agent can take to improve the yield of the reaction? Among yield, temperature, and pH, which is the quantity that the RL agent cannot directly manipulate? Why do you think this quantity cannot be manipulated directly? What can be considered as the environment in the reaction optimization problem? In the RL framework, the agent takes actions, and the environment sends out reward signals indicating the quality of the actions. For this problem, when should an action be rewarded highly and when should an action be rewarded poorly?

Like the human decision maker, the RL agent can change the temperature and pH of the chemical reaction to influence the yield. The RL agent cannot manipulate the yield of the reaction directly because the effect of temperature and pH on the yield of the reaction is unknown and possibly complex. The environment holds the dynamics of the task being solved. In this problem, the dynamics refers to the effect of temperature and pH on the yield, which is unknown, but can be learnt through constant interaction between the agent and the environment. If an action taken by the RL agent results in discovering a set of conditions that produce higher yield than previously observed, then the action should be rewarded highly. On the other hand, if the action results in finding a yield that is lower than or similar to what has previously been observed, then it should be rewarded negatively.

The task to be solved is framed as a Markov Decision Process [4] with a state space (a list of all possible states in the problem), an action space (a list of all possible actions), reward distribution, and a stationary transition dynamics distribution with conditional density $p(s_{t+1}, r_{t+1}|s_t, a_t)$ following the Markov property $p(s_{t+1}, r_{t+1}|s_t, a_t) = p(s_{t+1}|s_1, a_1, \ldots, s_t, a_t)$.

The Markov property ensures that the current state and reward is dependent only on the previous state and action but, given that information, independent of all earlier states and actions. This property is not meant to restrict the types of environments that can be solved with RL but instead meant to define what qualifies as an appropriate state. If the state does not include relevant information about the past interactions between the agent and the environment that could influence future decisions, it is said to not have the Markov property.

11.2.3 Solving an RL Problem

The agent samples actions in the environment by following a policy $\pi(a \mid s)$, where π holds the probability of choosing a particular action a while being in a given state s. By following the policy, the agent generates a sequence of states, actions, and rewards. While the reward that the agent receives is an indication of the quality of a particular action, it says very little about the overall effectiveness of the agent's policy. It is easy to imagine a policy that favors actions which maximize immediate reward but simultaneously takes the agent to a state that has low potential for future rewards. This policy would likely not be a good way to solve the problem.

RL algorithms are therefore set up to maximize not just the immediate rewards but the *sum of future rewards*. This quantity, called the return R_t (Eq. 11.1), is usually used in a manner that discounts the value of future rewards and gives more weight to immediate rewards. Unlike rewards, the returns are not an inherent property of the state and action, but it is a function of the policy that the agent is using. Agents beginning from the same starting state, but following different policies, can generate very different trajectories of states, actions, and rewards and therefore alter the returns.

$$R_t = \sum_k \gamma^k r_{t+k}. \tag{11.1}$$

Here, $\gamma \in [0,1]$ is a discount factor that assigns more value to the most immediate rewards.

Value functions are used to store the expected returns associated with each state (or combination of state and action). The expected return is used instead of the sample return because the rewards and returns are not always deterministic but can be stochastic. Since the returns are a function of the policy, each value function is also associated with a particular policy. There are two types of value functions, one that is only defined over states (V^π) and one that is defined over both states and actions (Q^π).

$$V^\pi = E(R_t \mid s_t = s, \pi), \tag{11.2}$$

$$Q^\pi = E(R_t \mid s_t = s, A_t = a_t, \pi). \tag{11.3}$$

The agent's goal, in this setting, is to learn a policy π^* that maximizes the expected return. The basic RL algorithm using Q-learning is presented in Algorithm 11.1. The procedure starts with an arbitrary initialization of the Q-function. The first state is observed, and an action either random or one that maximizes the current Q-function is taken. The motivation for introducing randomness in the selection of actions is to promote *exploration*, that is, the consideration of new actions, and minimize *exploitation*, that is the use of only the currently known best actions. The ε parameter regulates the exploration/exploitation balance. Then, the state after the action taken is observed and a reward is recorded. The Q-function is updated using α, a learning rate, and γ, the discount factor. The process continues until all steps in the episode are exhausted. The learning agent approaches the actual Q-function over many episodes.

Algorithm 11.1. *Q*-learning principle.

1. Initialize $Q(s, a)$ arbitrarily.
2. For each episode:

$$s \leftarrow s_t$$

For each step:

$$a \leftarrow \begin{cases} \underset{a}{\operatorname{argmax}}\, Q(s, a_t), & \textit{with probability } 1-\varepsilon \\ \textit{random } a, & \textit{with probability } \varepsilon \end{cases}.$$

Take action a and observe.

$$r' \leftarrow r_{t+1},\ s' \leftarrow s_{t+1}.$$

$$Q(s, a) \leftarrow Q(s, a) + \alpha\left[r' + \gamma \max Q(s', a_{t+1}) - Q(s, a)\right].$$

$$s \leftarrow s'.$$

11.2.4 Deep Reinforcement Learning (DRL)

Value functions are integral to the performance of RL algorithms. Historically, the value functions were either plain tables that stored data or linear function approximators. While they were limited in their capacity, they were attractive because they guaranteed convergence in some scenarios. Nonlinear function approximators have also been explored in the past, but it was not pursued because of divergence issues.

With the revival of interest in deep learning, neural networks have been used as function approximators. Neural networks can learn a much larger class of functions than their linear counterparts. They learn by minimizing a loss function between their predictions and the target labels using the stochastic gradient algorithm. The problems of divergence associated with the earlier generation of nonlinear function approximators were solved by [5] to develop a model that learned to play video games. The model eventually surpassed the performance of a human expert on some of the games. Since then, neural networks have been used to model not just the value function but also the policy and the environment. In the recent past, DRL algorithms have surpassed human-level performance in complex games like AlphaGo [1]. Although these DRL algorithms still require a lot of learning iterations to arrive at the optimal policy, they can now solve a much wider class of problems than was previously possible.

11.2.5 RL Solution Methods

There are three major types of approaches to solve the RL problem. The first type is the Q-learning method in which the agent tries to learn the optimal Q function [6, 7]. With access to the optimal Q function, the agent can sample all possible actions in a state and pick the optimal action. The algorithms that use Q-learning typically tend to be off-policy, meaning that they can learn from trajectories generated by older policies. This increases the efficiency of learning. The second type of approach is the policy gradient algorithms [8] where the goal of the agent is to maximize a performance objective by iteratively improving its policy. The algorithms that use policy gradient tend to be on-policy since the performance objective requires trajectories generated by the current policy. This in turn makes these algorithms less efficient since they require a lot of experience to find the optimal policy. The final type is the actor-critic approach [9, 10] which combines aspects of Q-learning and policy gradient. Here, the agent learns both a Q-function and a policy that maximizes a performance objective.

11.3 REINFORCEMENT LEARNING ALGORITHMS

There exists a wide variety of RL algorithms, each of them with advantages in certain applications. To help navigate the many options available, perhaps the most important distinction is whether a model of the environment is used to estimate both states and rewards or if that information is extracted directly from the environment. This divides RL algorithms between *model free* and *model based*. Among model-based RL algorithms, the model of the environment can be either provided directly to the learning agent or learned by it through iterations. For the cases of model-free algorithms, there are two main types: policy learning and Q-learning. Because of their relevance to process control and Chemical Engineering applications, policy learning algorithms are presented here. The reader is encouraged to explore other algorithms with potential applications for the chemical process industry [11].

11.3.1 Actor-Critic Algorithms

RL was initially studied only with discrete action space, but practical problems sometimes require control actions in a continuous action space [12]. The actor-critic algorithm is a widely known architecture based on the policy gradient theorem which allows applications in continuous space [13]. The actor-critic network has two components: an actor network that is updated by the policy gradient theorem and a critic network that estimates the action-value function $Q(s, a)$ (Figure 11.3).

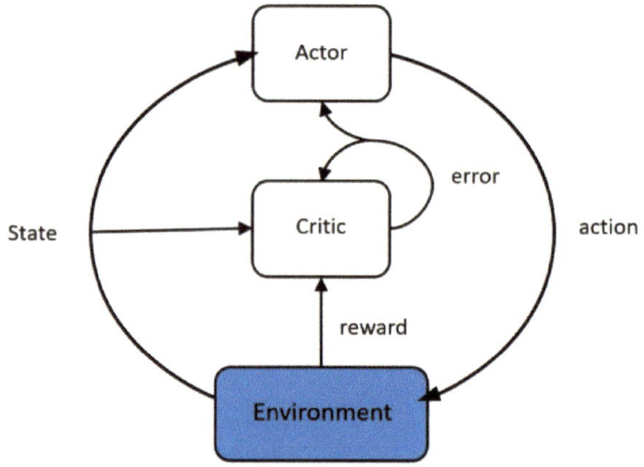

FIGURE 11.3 Schematic diagram of the actor-critic algorithm: actor network performs an action and is updated by critic network.

11.3.2 Deep Deterministic Policy Gradient (DDPG) Method Algorithm

To improve the trainability of the existing policy gradient method [14], Lillicrap et al. [12] developed Deep Deterministic Policy Gradient (DDPG) method including the feature of learning in mini-batches [15] rather than learning online. Mini-batch learning uses a replay buffer to store tuples of historical samples in the form of $[s_t, a_t, r_t, s_{t+1}]$, which stabilizes training and makes efficient use of the hardware optimization. During training, mini-batches are randomly drawn from the replay buffer to update the weights of the networks [12]. Additionally, separate target networks are created as copies of the original actor network and critic network. The weights of the target networks are constrained to change slowly, which significantly enhances stability of the network.

Moreover, for deep neural networks, the distribution of every layer's input changes with training [16]. This phenomenon, defined as internal covariate shift, leads to extreme difficulty in training models with large nonlinearities. Networks with batch normalization have shown extraordinary performance in speeding up training and improving the performance of the original setup by reducing internal covariate shift. DDPG networks are usually extremely difficult to train due to the high degree of freedom from exploration. Therefore, for training a DDPG network, batch normalization layers are placed before the activation function at each layer of the neural network to enhance training efficiency [12]. With the combination of actor-critic algorithm, replay buffer, target networks, and batch normalization, DDPG is able to perform continuous control, while the training efficiency and stability are improved from original actor-critic network [12]. These features make DDPG an ideal candidate for application in industrial settings, where the control actions are often continuous.

11.3.3 A3C Algorithm

The algorithm implemented in one of our applications was Asynchronous Advantage Actor-Critic (A3C) algorithm in DRL [17]. A3C is a variant of the standard RL, where it takes advantage of actor-critic algorithm and asynchronous learning by parallel learning agents. Actor-critic algorithm utilizes separated policy and value function approximated by neural networks with weights θ_p and θ_v, respectively, where the policy network $\pi_{\theta_p}(s_t)$ generates a probability distribution of favorable actions given current state s_t, and the value network $V_{\theta_v}(s_t)$ evaluates the expected return from state s_t following Bellman equation similar to the Q-function in standard RL setup. During training, both policy and value network are updated continuously, where the weight of the value network θ_v is updated with loss function to minimize the difference between the value function and the observed reward [18]:

$$L_v(\theta_v) = E\left(R_t - V_{\theta_v}(s_t)\right)^2, \tag{11.4}$$

and the weight of the policy network θ_p is updated based on policy gradient theorem [17] in the direction of $\nabla_{\theta_p} \log \pi_{\theta_p}(a_t \mid s_t)$ aiming to favor actions that maximize expected reward [18]. The asynchronous aspect of A3C corresponds to the use of CPU multi-threading, allowing parallel learners to explore policies simultaneously, which significantly enhances the efficiency and stability of training. Notably, the parallel learners in A3C communicate through a global network, where A3C uses the experience of state transitions $\left[s_t, a_t, r_t, s_t+1, a_t+1, r_t+1, \ldots, s_t+n, a_{t+n}, r_{t+n}\right]$ gathered by all the parallel learners to update the global θ_p and θ_v every n time steps. When the global network is updated, the weights of the parallel learners are also synchronized with the global θ_p and θ_v, which allows each parallel learner to incorporate experience generated by all learners.

In the case study of fed-batch control, the A3C algorithm was constructed with four parallel learners where each learner interacted with a separate environment simultaneously. At each time step, a policy network generated a probability distribution of control actions of nitrate nutrient additions during interaction with the simulated reaction system, where each policy network consisted of a neural network with one hidden layer with 200 neurons. The control action with highest probability was selected to feedback to the reaction system, where the value network estimated the expected reward based on the response of the reaction system. The value network was composed of a neural network with one hidden layer of 100 hidden neurons that calculated the expected reward calculated according to the yield of the desired bio-product.

11.3.4 PPO Algorithm

Proximal Policy Optimization (PPO) [19] is a robust RL algorithm with policy gradient family that has shown good generalization ability and data efficiency. In policy gradient methods, the gradient of policy is computed with an estimator with stochastic gradient ascent algorithm. The loss function of policy gradient algorithm is shown as follows:

$$L_{PG} = \hat{E}_t\left[\log \pi_\theta(a_t \mid s_t)\hat{A}_t\right], \tag{11.5}$$

where π_θ is a stochastic policy, and $\hat{A}_t$ is the estimator of the advantage function at time t, which is the difference between the value function $V(s) = E_\pi\left[R_t \mid s_t = s\right]$ and Q function.

PPO is an improved algorithm over the Trust Region Policy Optimization (TRPO) [20] under the policy gradient family, which inherits benefits including good data efficiency and reliable performance from TRPO but shows simpler implementation and sample complexity. In trust region methods, it maximizes the objective function with penalty coefficient β and the Kullback-Leibler (KL) divergence of the new policy update.

$$\max_{\theta} \hat{E}_t\left[\frac{\pi_\theta(a_t \mid s_t)}{\pi_{\theta_{old}}(a_t \mid s_t)}\hat{A}_t - \beta KL\left[\pi_{\theta_{old}}(\cdot \mid s_t), \pi_{\theta(\cdot \mid s_t)}\right]\right]. \tag{11.6}$$

Therefore, the TRPO maximizes the "surrogate objective". Let $y_t(\theta)$ be the probability ratio between the policy π_θ and the old policy $\pi_{\theta_{old}}$.

$$y_t(\theta) = \frac{\pi_\theta(a_t \mid s_t)}{\pi_{\theta_{old}}(a_t \mid s_t)}. \tag{11.7}$$

The TRPO objective function refers to conservative policy iteration (CPI) and can be written as

$$L_{CPI}(\theta) = \hat{E}_t\left[\frac{\pi_\theta(a_t \mid s_t)}{\pi_{\theta_{old}}(a_t \mid s_t)}\hat{A}_t\right] = \hat{E}_t\left(y_t(\theta)\hat{A}_t\right). \tag{11.8}$$

The key difference between PPO to TRPO is that PPO uses a clipped surrogate objective (CLIP) to avoid huge policy updates without a constraint. The PPO proposes a clipped objective to avoid excessive changes that moves $y_t(\theta)$ away from 1:

$$L_{CLIP}\ (\theta) = \hat{E}_t\left[\min\left(r_t(\theta)\hat{A}_t\left(y_t(\theta), (1-\epsilon, 1+\epsilon)\right)\right)\right]. \tag{11.9}$$

In the clipped surrogate objective, the clip $\left(y_t(\theta), (1-\epsilon, 1+\epsilon)\hat{A}_t\right)$ term constrains the ratio y_t from moving outside the boundary $[1-\epsilon, 1+\epsilon]$, where the default clipping ratio is $\epsilon = 0.2$. With the clipped surrogate objective, PPO shows improved computational efficiency and better sample complexity and thus demonstrates better generalized ability and simpler for implementation than many other RL algorithms.

11.3.5 Twin Delayed Deep Deterministic Policy Gradient (TD3) Algorithm

TD3 is the successor to the DDPG [21]. Up until recently, DDPG was one of the most used algorithms for continuous control problems such as robotics and autonomous driving. Although DDPG is capable of providing excellent results, it has its drawbacks. Like many RL algorithms, training DDPG can be unstable and heavily reliant on finding the correct hyper parameters for the current task. This is caused by the algorithm continuously over estimating the Q values of the critic (value) network. These estimation errors build up over time and can lead to the agent falling into a local optimum or experience catastrophic forgetting. TD3 addresses this issue by focusing on reducing the overestimation bias seen in previous algorithms. This is done with the addition of three key features:

1. Using a pair of critic networks
2. Delayed updates of the actor
3. Action noise regularization

11.3.5.1 Twin Critic Networks

The first feature added to TD3 is the use of two critic networks. This was inspired by the technique seen in deep reinforcement learning (DRL) with Double Q-learning which involved estimating the current Q value using a separate target value function, thus reducing the bias. However, the technique doesn't work perfectly for actor critic methods. This is because the policy and target networks are updated so slowly that they look very similar, which brings bias back into the picture. Instead, an older implementation seen in Double Q-learning [22] is used. TD3 uses clipped Double Q-learning where it takes the smallest value of the two critic networks (the lesser of two evils if you will). This method favors underestimation of Q values. This underestimation bias isn't a problem as the low values will not be propagated through the algorithm, unlike overestimate values. This provides a more stable approximation, thus improving the stability of the entire algorithm.

11.3.5.2 Delayed Updates

Target networks are a great tool for introducing stability to an agent training; however, in the case of actor critic methods, there are some issues to this technique. This is caused by the interaction between the policy (actor) and critic (value) networks. The training of the agent diverges when a poor policy is overestimated. Our agent's policy will then continue to get worse as it is updating on states with a lot of error.

In order to fix this, we simply need to carry out updates of the policy network less frequently than the value network. This allows the value network to become more stable and reduce errors before it is used to update the policy network. In practice, the policy network is updated after a fixed period of time steps, while the value network continues to update after each time step. These less frequent policy updates will have value estimate with lower variance and therefore should result in a better policy.

11.3.5.3 Noise Regularization

The final portion of TD3 looks at smoothing the target policy. Deterministic policy methods tend to produce target values with high variance when updating the critic. This is caused by overfitting to spikes in the value estimate. In order to reduce this variance, TD3 uses a regularization technique known as target policy smoothing. Ideally there would be no variance between target values, with similar actions receiving similar values. TD3 reduces this variance by adding a small amount of random noise to the target and averaging over mini-batches. The range of noise is clipped in order to keep the target value close to the original action. By adding this additional noise to the value estimate, policies tend to be more stable as the target value is returning a higher value for actions that are more robust to noise and interference.

In the second case study, the TD3 algorithm was utilized to implement RL to control the size distribution of particles during anti-solvent crystallization by adjusting the anti-solvent flowrate and the temperature. This approach significantly accelerates the control loop by resolving the optimization problem offline, thereby allowing ample time for image processing and executing control actions.

11.4 DRL FOR CHEMICAL PROCESS CONTROL AND OPTIMIZATION

Despite the huge success of DRL algorithms in chess and video games, whether DRL can achieve intelligent control behaviors like the game simulation remains a question. There exist many intrinsic differences between game environments and chemical reactions that make naive application of DRL for chemical reaction control troublesome.

The first challenge is that the real-time exploration of DRL within a plant system is not achievable due to the cost of time, material, and safety issues. For the game environments, the observation and reward are always clearly given by the game simulator, while the rules of the game are usually relatively straightforward, where the game either provides a final score or a win/lose situation [21]. Such traits make game environments ideal for DRL exploration and control optimization.

Chemical systems, however, do not have a clear definition of the control objective, while the design of the objective function can be directly related to the controller performance. Additionally, since the optimization can only be performed via simulation, an accurate and fast-responding reaction model is critical for the development of the controller. Another problem with chemical systems optimization is the time delay problem. In most game environments, once the control action is performed, the effect onto the environment is immediate. The time delay in many chemical systems makes training of the controller more difficult than video game environments. In chemical systems, the introduction of an action can often result in a time delay that leads to minimal or even imperceptible immediate effects within the environment. This delay presents a considerable challenge when it comes to optimizing RL policies in DRL.

Numerous challenges exist in the realm of chemical process control. One prominent concern is the computational efficiency of the underlying model, which is intricately linked to the training time. To illustrate, in cases involving reaction systems with a multitude of measured variables and intricate reaction dynamics, the simulation process may prove to be prohibitively slow for practical use as an interactive environment for DRL optimization. Moreover, the abundance of both measured and unmeasured variables, coupled with potential inaccuracies in the simulation, collectively magnifies the complexities inherent in the application of DRL techniques for chemical process control.

In the following chapter, we will focus on two chemical engineering case studies: the experimental application for optimal operation of a fed-batch growth reaction and the control of an experimental crystallization reactor.

11.5 CONCLUSIONS

This chapter provided a comprehensive overview of RL and its applications in Chemical Engineering. By positioning RL within the broader context of machine learning and detailing its practical implications, the chapter underscores its importance in decision-making processes within complex systems like chemical reactions. Furthermore, the integration of deep learning techniques into RL, particularly DRL, has expanded its capabilities to address more intricate challenges, paving the way for enhanced optimization in chemical processes and beyond.

The chapter has also elucidated the key components of the RL framework, emphasizing the significance of value functions in guiding the learning agent toward maximizing expected returns. It has explored various RL algorithms, ranging from model-free approaches like Q-learning to advanced methods such as actor-critic architectures and PPO. Notably, the discussion highlights the potential of algorithms like A3C and TD3 in addressing real-time control tasks and continuous action spaces, offering promising avenues for application in industrial settings.

Despite the notable achievements of RL algorithms, their application to chemical process control presents unique challenges, including safety concerns, time delays, and computational efficiency. However, through case studies like the optimal operation of fed-batch growth reactions and the control of experimental crystallization reactors, the potential of RL in addressing these challenges becomes evident. By exploring practical examples, we can further uncover the capabilities and limitations of RL algorithms in real-world chemical engineering applications, driving innovation and optimization in the field.

REFERENCES

1. Silver D., Schrittwieser J., Simonyan K., Antonoglou I., Huang A., Guez A., Hubert T., Baker L., Lai M., Bolton A., Chen Y., Lillicrap T., Hui F., Sifre L., van den Driessche G., Graepel T., Hassabi D. (2017), Mastering the game of go without human knowledge. Nature, 550(7676), pp. 354–359, doi: 10.1038/nature24270
2. Mitchell T. M. (1997), Machine Learning. McGraw-Hill, Inc.
3. Spielberg S., Tulsyan A., Lawrence N. P., Loewen P. D., Gopaluni R. B. (2019), Toward self-driving processes: A deep reinforcement learning approach to control. AIChE Journal, 65(10), p. e16689, doi: 10.1002/aic.16689
4. Bellman R. (1957), A Markovian decision process. Journal of Mathematics and Mechanics, 6(5), pp. 679–684. http://www.jstor.org/stable/24900506
5. Mnih V., Kavukcuoglu K., Silver D., Graves A., Antonoglou I., Wierstra D., Riedmiller M. (2013), Playing Atari with deep reinforcement learning, Advances in Neural Information Processing Systems. http://arxiv.org/abs/1312.5602
6. Watkins C. (1989), Learning from delayed rewards. PhD Thesis, University of Cambridge.
7. Watkins C. J. C. H., Dayan P. (1992), Q-learning. Machine Learning, 8(3–4), pp. 279–292, doi: 10.1007/BF00992698
8. Schulman J., Wolski F., Dhariwal P., Radford A., Klimov O. (2017), Proximal policy optimization algorithms, arXiv e-prints, arXiv:1707.06347 [Online]. https://ui.adsabs.harvard.edu/abs/2017arXiv170706347S
9. Fujimoto S., van Hoof H., Meger D. (2018), Addressing function approximation error in actor-critic methods, arXiv:1802.09477. https://ui.adsabs.harvard.edu/abs/2018arXiv180209477F

10. Lillicrap T. P., Hunt J. J., Pritzel A., Heess N., Erez T., Tassa Y., Silver D., Wierstra D. (2015), Continuous control with deep reinforcement learning, arXiv:1509.02971. https://ui.adsabs.harvard.edu/abs/2015arXiv150902971L
11. Achiam J. (2018), Spinning up in deep reinforcement learning. OpenAI, October 13. https://spinningup.openai.com/en/latest/spinningup/spinningup.html
12. Lillicrap T., Hunt J., Prit A. (2015), Continuous control with deep reinforcement learning. CoRR, 1509, p. 02971.
13. Sutton R. S., McAllester D., Singh S., Mansour Y. (1999), Policy gradient methods for reinforcement learning with function approximation. Proceedings of the 12th International Conference on Neural Information Processing Systems, MIT Press.
14. Silver D., Lever G., Heess N., Degris T., Wierstra D., Riedmiller M. (2014), Deterministic policy gradient algorithms. Proceedings of the 31st International Conference on International Conference on Machine Learning, 32, JMLR. Docx: https://dl.acm.org/doi/10.5555/3044805.3044850.
15. Mnih V., Kavukcuoglu K., Silve D., Rusu A., Veness J., Bellemare M. G., Graves A., Riedmiller M. A., Fidjeland A., Ostrovski G., Petersen S., Beattie C., Sadik A., Antonoglou I., King H., Kumaran D., Wierstra D., Legg S., Hassabis D. (2015), Human-level control through deep reinforcement learning. Nature, 518(7540), pp. 529–533, doi: 10.1038/nature14236
16. Ioffe S., Szegedy C. (2015), Batch normalization: accelerating deep network training by reducing internal covariate shift. Proceedings of the 32nd International Conference on International Conference on Machine, JMLR.org. doi: 10.5555/3045118.3045167
17. Mnih V., Puigdomenech Badia A., Mirza M., Graves A., Lillicrap T., Harley T., Silver D., Kavukcuoglu K. (2015), Asynchronous methods for deep reinforcement learning. Proceedings of the 33rd International Conference on International Conference on Machine Learning, 48. https://dl.acm.org/doi/10.5555/3045390.3045594
18. Li H. Collins C. R., Ribelli T. G., Matyjaszewski K., Gordon G. J., Kowalewski T., Yaron D. G. (2018), Tuning the molecular weight distribution from atom transfer radical polymerization using deep reinforcement learning. Molecular Systems Design & Engineering, 3(3), pp. 496–508, doi: 10.1039/C7ME00131B
19. Schulman J., Wolski F., Dhariwal P., Radford A., Klimov O. (2017), Proximal policy optimization algorithms. ArXiv, 1707.06347.
20. Schulman J., Levine S., Moritz P., Jordan M., Abbeel P. (2015), Trust region policy optimization. Proceedings of the 32nd International Conference on International Conference on Machine Learning, 37. doi: 10.5555/3045118.3045319
21. Shin J., Badgwell T. A., Liu K. H., Lee J. H. (2019), Reinforcement learning – Overview of recent progress and implications for process control. Computers & Chemical Engineering, 127, pp. 282–294, doi: 10.1016/j.compchemeng.2019.05.029
22. Hasselt, H. van. (2010), Double Q-learning. Proceedings of the 23rd International Conference on Neural Information Processing Systems, 2, pp. 2613–2621, doi: 10.5555/2997046.2997187

12 Reinforcement Learning Case Studies

12.1 CASE STUDY 1: OPTIMIZATION OF FED-BATCH GROWTH REACTION

12.1.1 The Process

To maximize the yield of valuable bio products, cyanobacteria are cultured in a reaction system. These blue-green algae have a high nutritional value and are widely cultivated worldwide as a dietary supplement due to their antioxidant properties [1]. C-phycocyanin (C-PC) is a blue phycobiliprotein found in the light-harvesting complexes of cyanobacteria. It has anti-inflammatory, neuroprotective, and antioxidant properties and is therefore considered a high-value bio-product [2]. Moreover, its blue color makes it an excellent candidate to replace synthetic pigments commonly used in the food and cosmetics industries [1]. As C-PC is extracted and purified from cyanobacterial cultures, increasing its content per gram of biomass (i.e., yield) has become essential for its industrialization with high efficiency and productivity [3].

This case study focuses on the examination of Plectonema sp. UTEX 1541, a filamentous cyanobacterium commonly referred to as Leptolyngbya sp. PCC 6402 [4], hereafter simply referred as Plectonema. This particular cyanobacterium exhibits high yields of C-PC production. However, unlike the extensively researched Spirulina, Plectonema lacks comprehensive kinetic studies, posing challenges in optimizing product yield through conventional methods. To address this limitation, an artificial neural network (ANN) is employed to simulate the growth rate and C-PC accumulation in Plectonema. This data-driven approach eliminates the need for mechanistic models or assumptions. Subsequently, deep reinforcement learning (DRL) is utilized to select an efficient reaction strategy, thus bypassing the laborious process of determining the reaction mechanism, optimizing first principles models, or exhaustively searching through all potential control profiles.

12.1.2 Experimental Setup

The University of Texas at Austin culture collection provided the cyanobacterium Plectonema sp. UTEX 1541. Plectonema was cultivated in sterile 250 mL Erlenmeyer flasks covered with foam plugs, using modified BG11 medium containing varying concentrations of sodium nitrate ($NaNO_3$) as a key nutrient (300 mg/L, 600 mg/L, 900 mg/L, and 1200 mg/L). The flasks were kept in an illuminated New Brunswick scientific Innova 4340 incubator shaker at 28°C with 130 RPM agitation. A constant illumination of 100 $\mu mol/m^{-2}s^{-1}$ was maintained using four 20 W white fluorescent lamps. The effective culture volume

DOI: 10.1201/9781003455905-15

was 110 mL, with 56% v/v of empty space to facilitate uptake of atmospheric carbon dioxide, the sole carbon source for C-PC synthesis.

Plectonema growth was monitored for at least 17 days and in one experiment lasting 55 days. To begin, the optical density (measured as absorbance at 730 nm) of all cyanobacterial cultures was adjusted to 0.1, and throughout the growth period, measurements were taken for biomass density, C-PC concentration, and nitrate ion levels. Liquid culture absorbance at 730 nm was used to estimate biomass generation at different points during Plectonema growth, using a Beckman Coulter DU730 Life science UV-visible spectrophotometer. We constructed an empirical relationship between dry Plectonema biomass and liquid culture absorbance at 730 nm based on 54 data pairs (available in the Supplementary Materials). To determine C-PC concentration, cyanobacterial cultures were digested with lysozyme in TE buffer (2.7 mg mL^{-1}), following a modified protocol based on standard methods [5]. Nitrate ion concentration was determined using a modified nitrate reduction test adapted from nitrate tests from Millipore Sigma and Dalynn Biologicals.

12.1.3 Raw Data Pretreatment

As discussed earlier in Chapter 2, missing data is a common issue both in laboratory and industrial settings [6]. In this case, a third-order polynomial fitting to the original experimental data was used to approximate the missing points. This approach not only solved the missing data issue but also served as a data smoothing pretreatment to reduce experimental fluctuations. It is important to note that each polynomial function is only an approximation of the variable trajectory under a specific experimental condition and does not represent the physical mechanism or dynamic model of the reaction. Thus, it cannot be used as a dynamic interactive environment for controller development.

To train the ANN, pretreated data with polynomial imputation [7] and smoothing was used. The training dataset consisted of a total of 86 data pairs, and a testing dataset was used that included 16 data pairs. Each data pair consisted of four data points: current time, biomass concentration, nitrate concentration, and C-PC concentration.

12.1.4 The Environment

12.1.4.1 NN Implementation

Using TensorFlow [8] in Python 3.6, a feedforward neural network with 64 neurons in hidden layer 1, 64 neurons in hidden layer 2, and 4 neurons in the output layer (Figure 12.1) was constructed. The input to the neural network included the current time (t), current biomass concentration ($[x]$), current nitrate concentration ($[N]$), and C-PC concentration ($[C-PC]$). The neural network output was the gradient of time (Δt), biomass concentration ($\Delta[x]$), nitrate concentration ($\Delta[N]$), and C-PC concentration ($\Delta[C-PC]$) at the next time step. The recorded output can then be used as input for the next time step, creating a continuous profile of Plectonema growth, nitrate consumption, and C-PC production. The Mean Squared Error (MSE) was used to compute the training loss, and the weight of the neural network was optimized using the Adam Optimizer [9].

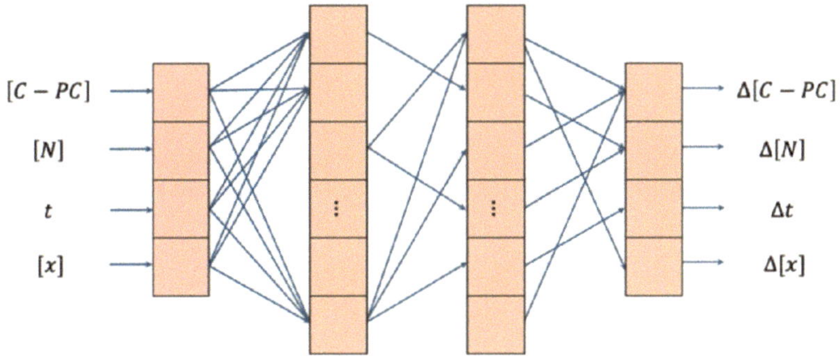

FIGURE 12.1 Diagram of the feedforward neural network. The input is the measurements of current time, biomass concentration, nitrate concentration, and C-PC concentration; the output is the gradients of the measurements.

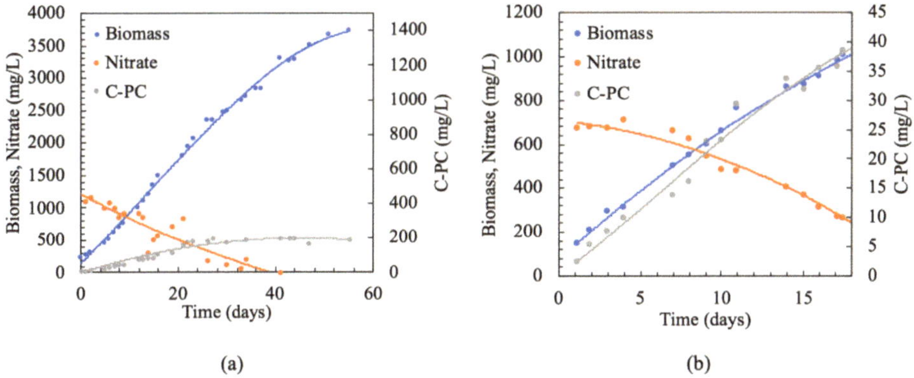

FIGURE 12.2 Experimental results L: (a) initial nitrate concentration of 1200 mg/L and (b) initial nitrate concentration of 900 mg/L. Dots represent experimental data and solid lines represent polynomial fit.

In total, four sets of experimental data were gathered under different conditions, and out of these, three sets (300 mg/L, 900 mg/L, and 1200 mg/L) were utilized for training the network, while the remaining set (600 mg/L) was used for testing. Figure 12.2a and Figure 12.2b illustrates the training and testing data, respectively.

12.1.4.2 Minimizing Overfitting Effect

It is often difficult and expensive to collect sufficient experimental data to train an ANN, leading to the possibility of overfitting, where the network becomes too closely fitted to the limited training data points [10]. To mitigate this issue, various approaches have been explored. One such approach is to perform data augmentation with noise injection [10]. This approach was also employed in del Rio-Chanona's work, which involved training an ANN using data generated from a mechanistic growth model [11]. By duplicating the original training data with ±5% random noise injection, the training data size was expanded to 172 data points.

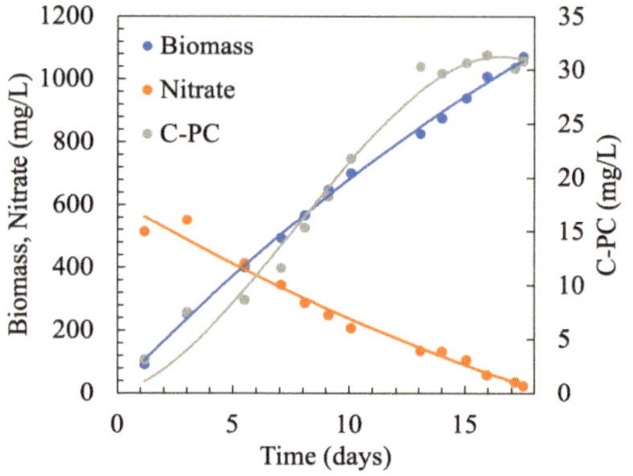

FIGURE 12.3 Validation result from the ANN growth model. The experimental condition for validation is different from the training data.

Another strategy to prevent overfitting is regularization. In this study, the dropout algorithm [12] was implemented. Dropout randomly deletes hidden nodes in the neural network with a probability of $1 - p$ during each training iteration. The remaining nodes are then trained with backpropagation as usual [13]. By introducing a probabilistic approach to dropping nodes, the dropout algorithm adds noise to the training process and promotes the development of higher generalization ability in the network [12]. In our implementation of the ANN growth model, we added a dropout layer with $p = 0.85$ to the second hidden layer. To determine if the ANN was overfitting, we compared the total MSE loss of the test dataset with the implementation of noise injection and dropout algorithm. A low training error and high validation error would indicate overfitting.

Results show that using the dropout algorithm alone outperforms other candidates, while noise injection combined with augmented data negatively impacts training stability. Although both the dropout algorithm and noise injection aim to reduce overfitting by introducing noise into the training process, the former shows better generalization ability by probabilistically neglecting nodes during training, while the latter harms training data accuracy and results in higher validation error. Therefore, the dropout algorithm was chosen for training the ANN.

Figure 12.3 shows the validation results of the ANN simulation predicting the growth profile with 600 mg/L nitrate concentration. The ANN generated highly accurate predictions when compared to the experimental data, indicating its ability to predict growth rates with previously unseen conditions. The ANN was saved and used as the training environment for the DRL algorithm.

12.1.5 DRL of Fed-Batch Growth Reaction

12.1.5.1 DRL Implementation

The A3C algorithm was utilized with four parallel learners. Each learner simultaneously interacted with a separate environment. At each time step, a policy network generated a probability distribution of control actions for nitrate nutrient additions

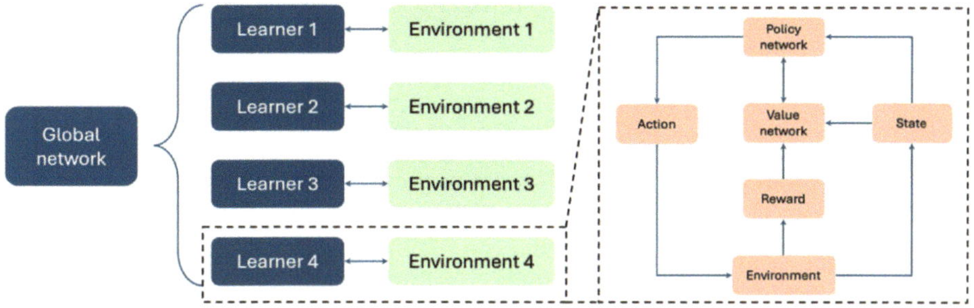

FIGURE 12.4 Flowchart showing the A3C reinforcement learning algorithm: four parallel learners that explore policies within the environment simultaneously.

during interaction with the simulated reaction system. Each policy network had a neural network consisting of one hidden layer with 200 neurons and a SoftMax activation function in the output layer. The control action with the highest probability was selected and fed back to the reaction system. Simultaneously, the value network estimated the expected reward based on the C-PC yield obtained at the end of the fed-batch reaction in the simulation. The value network consisted of a neural network with one hidden layer of 200 neurons and a linear output layer. To prevent overfitting, a dropout layer with $p = 0.85$ was implemented in both the policy network and value network before the output layer. The parallel learners are synchronized periodically with the global network updated every n training iteration ($n = 2$). Figure 12.4 illustrates the schematic diagram of the Asynchronous Advantage Actor-Critic (A3C) reinforcement learning (RL) algorithm.

For training the controller, the state input for the network was a tuple including time t, biomass $[x]$, nitrate concentration $[N]$, $C-PC$ concentration $[C-PC]$, and total nitrate allowed $[N_{tot}]$ at each time step. The output from the policy network was the nutrient to be added to the reactor. In training the networks, the allowed nitrate nutrient N_{tot} was set to a fixed number according to the operator's demand. To test the validity of the controller, we used 1000 mg/L, 1500 mg/L, 2000 mg/L, and 2500 mg/L as the total nitrate amount to be added to the reactor. The action output at each time step was chosen from six different nutrient dosages from 0 mg/L to 250 mg/L to the reactor each day. After nitrate amount met the constraint level, further nutrient addition was prohibited until the end of each run and would be reset to the initial value for the next run.

12.1.5.2 Reward

To train the controller, the control agent was rewarded by a reward function, which was directly associated with the yield of C-PC production at the end of the simulation shown in Eq. 12.1:

$$r_t = \begin{cases} c.Y_{C-CP} \; if \; t = t_{max} \\ -0.1 \quad otherwise \end{cases}, \tag{12.1}$$

where c is a constant and $Y_{C\text{-}PC}$ is the yield of $C-PC$ and Y_{C-CP} is defined as follows:

$$Y_{C-CP} = \frac{[C-PC]}{[Biomass]}. \tag{12.2}$$

The positive reward was provided at the terminal state of the experiment, $t = t_{max}$, as an incentive to maximize the yield of C-PC production at the end of the experiment, whereas a small negative reward was given at every time step before the reaction reached terminal state. The discount factor to calculate cumulative reward is $\gamma = 0.99$. In order to improve exploration of the policy network by discouraging premature convergence to suboptimal policy, the β, which controls the entropy regularization term, is set to be 0.005.

12.1.5.3 Training of A3C Controller

Training of the DRL controller was completed in an Intel Core i7-3770K CPU, with 30,000 iterations. Training took approximately 300 seconds to complete. The fast training speed can be credited to the parallel computation with the asynchronous aspect in the A3C algorithm, where it employs multiple parallel actor networks which learns to maximize cumulative rewards simultaneously [14].

12.1.5.4 RL Control Profiles

Thanks to the data-driven nature and high computational speed of the ANN and A3C algorithm, a fed-batch control strategy can be achieved even without fully understanding the reaction mechanism for the microorganism metabolism. It is suggested by the controller that without economic analysis or feasibility constraints, a maximized level of nitrate concentration in the reactor should be maintained to obtain the highest level of C-PC production. However, economic profitability comes with maximizing the yield of the desirable product while minimizing the operational cost of the reactants. Maintaining maximal level of nutrient concentration is not a feasible solution to facilitate the industrialization of the production process due to the cost of nitrate.

For the industrialization of C-PC production, the key factor to be considered is the amount of nitrate nutrient that comes per unit cost of nitrate. DRL has a high degree of flexibility; therefore, it can be adjusted to meet the operational constraints and varying control objectives conveniently. Additionally, training the A3C controller is computationally efficient, allowing the control strategy to be optimized with varying market requirement on a routine basis in real operation. With the ANN simulation based on the laboratory scale fed-batch growth reactor, we operated the controller based on the economic constraints including growth time-span and nutrient limit to validate A3C's ability in simulation. For real-world applications, the constraints can be tuned to meet more specific economic considerations, such as the market price of the nitrate and the desired product, but for the laboratory-scale reactor, we only considered total nitrate limit and growth time-span for simplicity. With the aforementioned constraints, the control profiles were obtained from the A3C control agent under nitrate limit from 1000 mg/L to 2500 mg/L and growth period of 20 days and

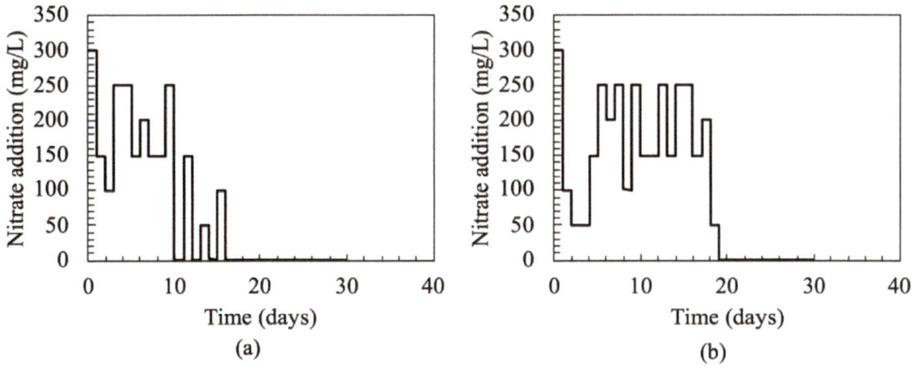

FIGURE 12.5 Control profiles generated by the A3C controller for the 30-day fed-batch process with different total nitrate amount (a) 2000 mg/L and (b) 2500 mg/L.

30 days. Two of the control profiles suggested by the A3C controller are shown in Figure 12.5.

From the control profiles for varying conditions, the A3C controller generally suggests to add nitrate nutrient during the early phase of the cyanobacterial growth. The controller output indicates that maintaining a high nitrate concentration during the early phase of algae growth is essential for C-PC production. Also, with longer growth period and higher total nitrate amount, the nitrate concentration peak is delayed which is observed from the control profile of 2500 mg/L nitrate limit. This observation was also suggested in the optimization of C-PC in *Spirulina sp.* by del Rio-Chanona et al. [11] that C-PC was mainly accumulated during early phase of cell growth. They suggested that the light attenuation effect by the increasing biomass can affect the C-PC accumulation in later phase of the growth period [11].

Therefore, high C-PC content can be achieved by maintaining a high nitrate concentration in the earlier phase of cyanobacterial growth. The biomass concentration is not considerably affected by the varying amount of nitrate, where the C-PC production significantly enhanced with increased amount of nitrate addition.

12.1.5.5 Experimental Validation

To validate the performance of the reaction simulation and A3C controller, triplicates of parallel experiments were performed following an action profile generated from the controller for the 30-day growth with 2000 mg/L nitrate limit. The results from the A3C control profile were compared to another experimental group, where the nitrate was replenished to the cell culture on day 7 and day 14 of the cell growth, and each addition replenished the nitrate concentration by 1000 mg/L. Nitrate additions to the cell culture are shown in Figure 12.6, the total nitrate amount for the validation experiment is 2000 mg/L, and the C-PC is harvested after 30 days of culture.

The experimental results for the two different nitrate input profiles for the 30-day fed-batch growth are shown in Figure 12.7 with standard deviation shown with error

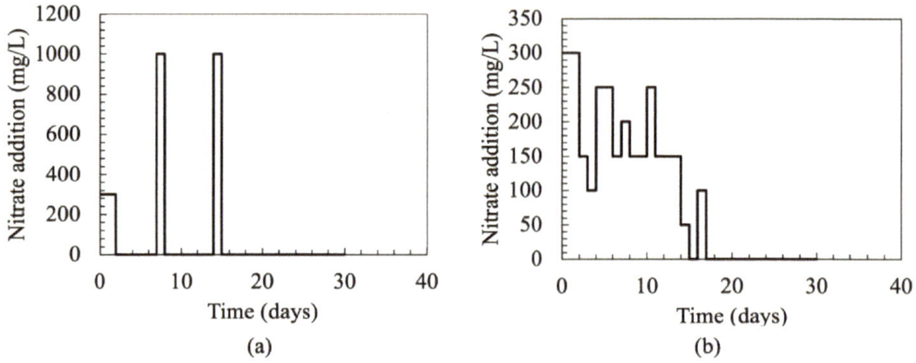

FIGURE 12.6 Control profiles for the validation experiment. Total amount of nitrate addition is limited to 2000 mg/L; the initial concentration of nitrate in the culture is 300 mg/L. (a) 2-step replenishment and (b) control profile generated by the A3C agent.

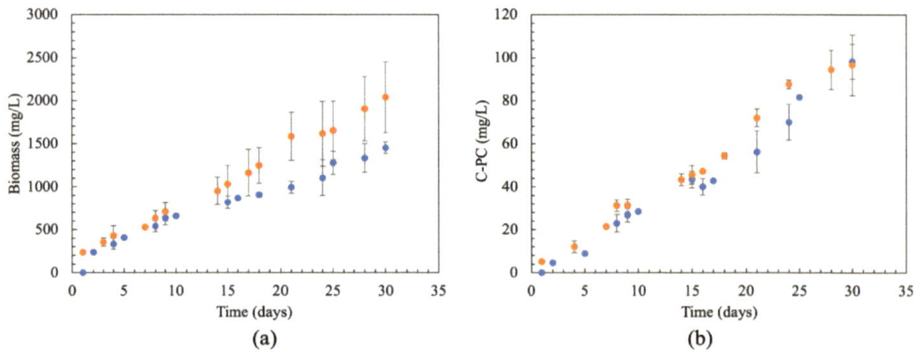

FIGURE 12.7 Experimental results obtained in the two conditions plotted over time, with standard deviation shown with error bars calculated from the triplicates of experimental data (blue: A3C controller, orange: two-step addition of nitrate nutrient). (a) Biomass accumulation and (b) C-PC concentration.

bars. From the experimental results, the concentration of C-PC in the A3C controller experiments (116.0 mg/L) was 20.1% higher than the control group (96.6 mg/L) at the end of the 30-day fed-batch experiment. The results demonstrate that the C-PC production can be significantly boosted with an improved control profile with same total nitrate input.

The results show that C-PC concentration and yield increased significantly by following the control profile suggested by the A3C controller for the 30-day fed-batch reaction with 2000 mg/L total nitrate addition. It can be observed that the A3C controller group had higher nitrate utilization to produce the desired product. In the A3C controlled experiment, the added nitrate nutrient was used to produce more C-PC; in contrast, in the two-step addition experiment, the cyanobacteria accumulated higher biomass level and lower amount of C-PC with same amount of nitrate nutrient. The results suggested that the C-PC production could be enhanced by altering the feed rate of the nutrient into the culture, even when the total amount of nutrient was constrained to be equal. Although the current reactor was built based on a laboratory scale, the results demonstrate great economic potential.

Following the control profile suggested by the A3C controller significantly improves the yield of the desired product compared to a non-optimized control group with same amount of nitrate. The experimental results showcase the potential of applying machine learning algorithms in both simulation and control of a fed-batch reaction system, which combines ANN reaction simulation and A3C control strategies that improved the C-PC production in *Plectonema*. The A3C controller also demonstrates robust performance in proposing control strategies, given various experimental constraints, including total nutrient amount and culture duration. The proposed approach shows great potential as a solution to improve the productivity and profitability, given economic constraints with a reaction system with unknown mechanism. Therefore, this approach can be further employed for a wide range of applications for process control with reaction systems without *a priori* knowledge about the reaction kinetics.

12.2 CASE STUDY 2: DRL CONTROL OF AN EXPERIMENTAL CRYSTALLIZATION REACTOR

12.2.1 The Process

As discussed previously in Chapter 10, in crystallization operations, the size and shape of crystals can determine the surface properties of the final product and impact downstream operations such as filtering, drying, and blending. Since large-scale manufacture of the final products is contingent on the consistency of the crystal properties, the control of crystal growth is an important problem that has been widely studied in literature.

Crystals are grown from supersaturated solution by the random collision of solute molecules in the host solvent under favorable conditions. The growth begins with the formation of stable nuclei and proceeds by the repeated addition of solute molecules to their surface. If the solution is overly supersaturated, the rate of nucleation increases rapidly resulting in proliferation of the solution with crystal nuclei and impeding their subsequent growth. On the other hand, if the solution is under-saturated, the conditions do not favor either nuclei formation or crystal growth and can even lead to dissolution of the existing crystals. Controlled growth occurs somewhere between these two extremities. While the saturation conditions that promote such growth cannot be measured directly, process variables like solution temperature, anti-solvent weight fraction in the solution, and crystal properties like size and shape can be monitored to make an estimate. They can also be manipulated to alter the degree of supersaturation and control the growth of target crystals.

The size distribution of crystals becomes important in applications like medical inhalers which deliver drug particles to the lungs [15]. If the particles to be delivered are too large, they collide with the bifurcating passages in the respiratory tract and are blocked from reaching the blood vessels in the deep lung region. On the other hand, if the particles are too small, they are likely to be blown away during exhalation. In such situations, a tight control over the size distribution of particles is necessary for efficient drug delivery. In addition to this example, size distribution has important consequences for downstream pharmaceutical operations such as filtration, washing, drying, milling, mixing, and tableting.

Control frameworks for particle size distribution leverage knowledge of the underlying dynamics to effectively manage particle sizes. This knowledge is often represented through population balance [16–18] or stochastic models [19–22], which describe how the crystal population evolves based on external variables such as temperature, anti-solvent weight fraction, and others. Model-based control frameworks utilize this information to guide the crystal population toward the desired size distribution. They achieve this by solving optimization problems in real-time to compute the optimal sequence of control actions. In contrast, as discussed in Chapter 11, RL operates as a learning-from-experience framework.

In the following sections, we present an approach that utilizes RL to control the size distribution of particles during anti-solvent crystallization (see Figure 12.8). This approach significantly accelerates the control loop by solving the optimization problem offline, thereby allowing ample time for image processing and executing control actions. For this application, we employ the Twin Delayed Deep Deterministic Policy Gradients (TD3) algorithm, which excels in handling problems with continuous state and action spaces.

The approach discussed in Chapter 10 is utilized to monitor the size distribution of crystals during crystallization. It addresses several challenges that cannot be effectively handled by classic pattern recognition algorithms. These challenges encompass identifying crystal boundaries in images acquired from medium- to high-concentration solutions [18], distinguishing boundaries of adjacent or partially overlapping crystals, and managing varying lighting conditions, among others.

By employing these tools, we first solve the open-loop optimization problem to attain the desired crystal size distribution from a random initial state. Next, we utilize the trained RL policy to implement closed-loop control, calculating the optimal control action at each sampling time.

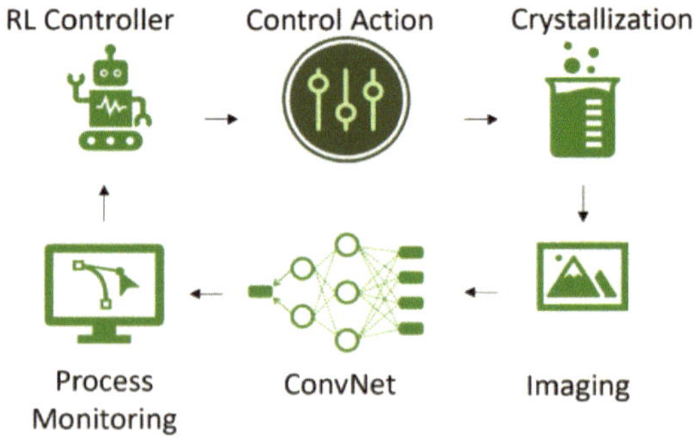

FIGURE 12.8 Machine learning-based control framework for batch crystallization.

12.2.2 The Environment

In non-isothermal anti-solvent crystallization, the addition of an anti-solvent and the temperature of the solution are simultaneously manipulated to control the supersaturation of the solution. The effect of these variables on the growth of crystals can be described by a stochastic modeling approach [22]. According to this approach, the variation of the crystal size distribution ψ can be described by the Fokker Planck Equation (Eq. 12.3) in Stratonovich form.

$$\frac{\partial \psi(L,t)}{\partial t} = \frac{\partial}{\partial L}\left\{-\left[r(q,T)L\left(1-\frac{L}{K(q,T)}\right)-D(q,T)L\right]\psi(L,t)+D(q,T)L^2\frac{\partial}{\partial L}\psi(L,t)\right\}, \tag{12.3}$$

with a log-normal initial condition and boundary conditions given by Eqs. 12.4–12.6.

$$\psi(L,0)=\frac{exp\left\{-\frac{\left[ln(L)-\mu_0\right]}{2\sigma_0^2}\right\}}{\sigma_0 L\sqrt{2\pi}}, \tag{12.4}$$

$$-\left[r(q,T)L\left(1-\frac{L}{K(q,T)}\right)-D(q,T)L\right]\psi(L,t)+D(q,T)L^2\frac{\partial}{\partial L}\psi(L,t), \tag{12.5}$$
$$at\ L=0\ and\ \forall t$$

$$\frac{\partial \psi(L,t)}{\partial L}=0\ at\ L\to\infty\ and\ \forall t. \tag{12.6}$$

The parameters in the equation r, K, and D can be expressed as functions of the anti-solvent flowrate (q) and the temperature (T) given by the following equations.

$$r(q,T)=\gamma_{0r}+\gamma_{1r}q+\gamma_{2r}T, \tag{12.7}$$

$$K(q,T)=\gamma_{0K}+\gamma_{1K}\,q^2+\gamma_{2K}T^2, \tag{12.8}$$

$$D(q,T)=\gamma_{0D}+\gamma_{1D}q+\gamma_{2D}T^2. \tag{12.9}$$

$\gamma_{0r},\gamma_{1r},\gamma_{2r},\gamma_{0K},\gamma_{1K},\gamma_{2K},\gamma_{0D},\gamma_{1D}$, and γ_{2D} can be estimated by fitting the equation to experimental data. The values for these parameters are obtained from the work done by Ghadipasha et al. [23].

12.2.3 Implementation Details

In order to use the TD3 algorithm to solve the Fokker Planck Equation, the state definition and the reward function need to be specified.

12.2.3.1 State

The state of the system encapsulates information about the crystal size distribution. The simplest case would contain the mean size of the crystals and the standard deviation in the state definition. This has the advantage of being easy to interpret and also satisfies the Markov property. But in order to guide the agent toward the desired crystal size distribution, the target mean size and target standard deviation were also included in the state definition. These targets could alternatively be included in the reward formulation, but it was included here in order to learn a general optimal control law.

The range of possible mean sizes is between 120 microns and 165 microns while the range of the standard deviation is between 48 and 64 microns. The two cannot be chosen independently. At training time, the target mean size and target standard deviation were randomly sampled from a uniform distribution of possible values. The state values were normalized between −1 and 1 before they were passed to the actor and critic networks for training.

$$State = \left[Mean,\ Target\ \ Mean,\ Variance,\ Target\ \ Variance\right]. \qquad (12.10)$$

12.2.3.2 Action

The actions that can influence the state are the temperature and the anti-solvent flowrates. Both these actions were bounded with the temperature having an upper and lower bound of 30°C and 10°C, respectively, while the anti-solvent flowrate was bounded between 3 mL/min and 0.7 mL/min. Similar to the state definition, the action values were normalized between −1 and 1 before they were passed to the actor neural network for training.

12.2.3.3 Reward

A simple partial rewarding scheme was used. The agent was rewarded (+1 for mean size and +0.1 for standard deviation) when the state values were sufficiently close to the target values. This was arbitrarily set to be 0.5 microns for the mean size and 0.3 units for the standard deviation. For all other scenarios, the agent was not rewarded.

The environment takes in the value of the current state and the actions executed by the agent in that state to generate the subsequent state. This was done by solving Eq. 12.3 using a finite volume method. The FiPy library in python was used to perform this numerical approximation. The state values (mean and standard deviation) were used to generate the corresponding log-normal distribution, and this was passed as the initial condition to Eq. 12.3. The FiPy code then solves the equation for one period of sampling time and returns the next state.

12.2.4 Results

12.2.4.1 Open-Loop Optimization

The Fokker Planck Equation (Eq. 12.3) used to describe the crystallization kinetics was solved using the FiPy package. This was used as the environment for the TD3 algorithm, and it was solved in three stages. In the first stage, the RL agent was rewarded only when the mean size and standard deviation of the crystal population reached a fixed set-point from a fixed initial state. Once this was accomplished, the task was expanded to teach the agent to reach any arbitrary set-point from the range of possible set-points (i.e. between 120 μm and 160 μm). Here too the initial state was fixed. In the final stage of training, both the set-point and the initial state were randomly assigned, and the agent was rewarded based on its ability to reach the set-point.

The optimal policies learnt by the agent is shown in Figure 12.9. It is seen that with access to a simple reward function, the agent discovers the competing effects of the two actions. For example, high temperature and low anti-solvent flowrate lead to higher asymptotic mean sizes, while low temperature and high anti-solvent flowrates produce lower asymptotic mean sizes. The policy prefers changes in the anti-solvent flowrate for increasing mean size at smaller values, and as the mean sizes increases,

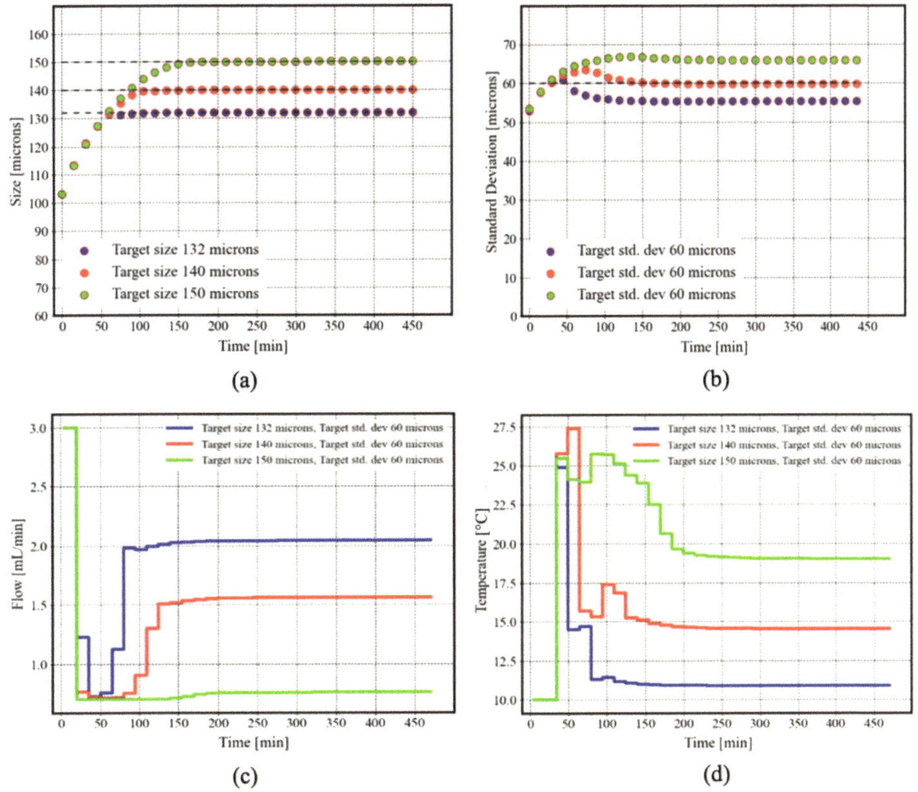

FIGURE 12.9 Open-loop optimal trajectories generated by following DRL policy. (a) Size, (b) standard deviation, (c) flow, and (d) temperature.

it uses a combination of both anti-solvent flowrate and temperature to do this. Once the anti-solvent flowrate is saturated at the lowest value, it relies completely on temperature for further increase in mean sizes. The algorithm used to learn this policy is shown in **Algorithm 12.1**.

Algorithm 12.1. Policy for open-loop optimization.

1. Pick an initial start state (mean value and standard deviation) and target state from a uniform distribution.
2. Run the policy (i.e.. execute the action in the environment).
3. Get the reward and the next state.
4. Repeat steps 2 and 3 until the target state is reached.
5. Update the agent and the critic using the experience gained thus far.

12.2.4.2 Optimal Feedback Control

Finally, the optimized trajectories were experimentally tested under feedback environment on the experimental system described in Chapter 8. Three sets of experiments were run with different mean size set-points (132 μm, 140 μm, and 150 μm) and a single standard deviation set-point (60 μm). To train the TD3 model to respond to disturbances, **Algorithm 12.1** was modified. Random disturbances sampled from a Gaussian distribution were introduced at each new sampling time. The reward function was left unchanged, and the agent was forced to respond to these disturbances and drive the system back to the desired set-point. The updated algorithm is shown in **Algorithm 12.2**.

Feedback control using the RL controller was employed to drive the crystal population toward the set-point and minimize variation. At every sampling time, 25 images were captured by the microscopic camera and sent to the CNN for analysis. The resulting measurements were fed as inputs to the RL agent, and through a single forward pass of the optimal policy, the appropriate control action was determined. Since the RL policy was determined offline, no additional time was necessary to retrain the algorithm.

The results of feedback control are shown in Figures 12.10 and 12.11 both for the controlled and manipulated variables, respectively. For all three experiments, the controller manages to reach the mean size target-set-point through a smooth trajectory. The manipulated variable profiles illustrate the effect of the feedback action to achieve the desired set-point. This can be seen in the adjustments to the off-line optimal trajectory brought about by the feedback action of the RL controller. The standard deviation, on the other hand, is harder to control (specially for this system in which the solubility is weakly dependent on temperature) but stays within 75% of the set-point. Again, the results show important improvements as compared to the open-loop case. Finally, a histogram of the crystal size distribution obtained at the end of the experiment for all three scenarios is shown in Figure 12.12.

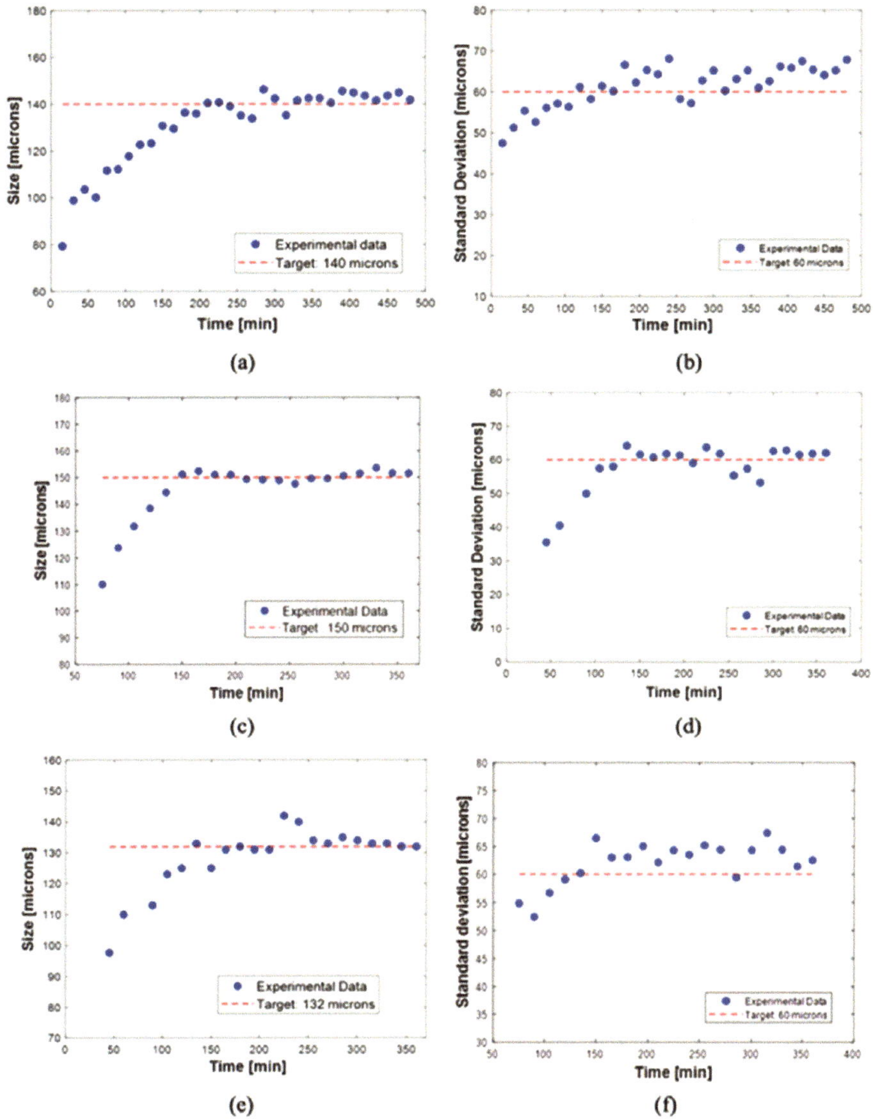

FIGURE 12.10 Results of using the DRL policy to guide the reaction toward the set-point. (a, c, e) These figures show the experimentally observed average crystal size trajectories for three different setpoints (140, 150, 132 microns). (b, d, f) These figures show the observed crystal size standard deviation for a setpoint of 60 microns.

Algorithm 12.2. Policy for closed-loop general control law.

1. Pick an initial start state and target state (mean value and standard deviation) from a uniform distribution.
2. Run the policy (i.e., execute the action selected by the agent in the environment).

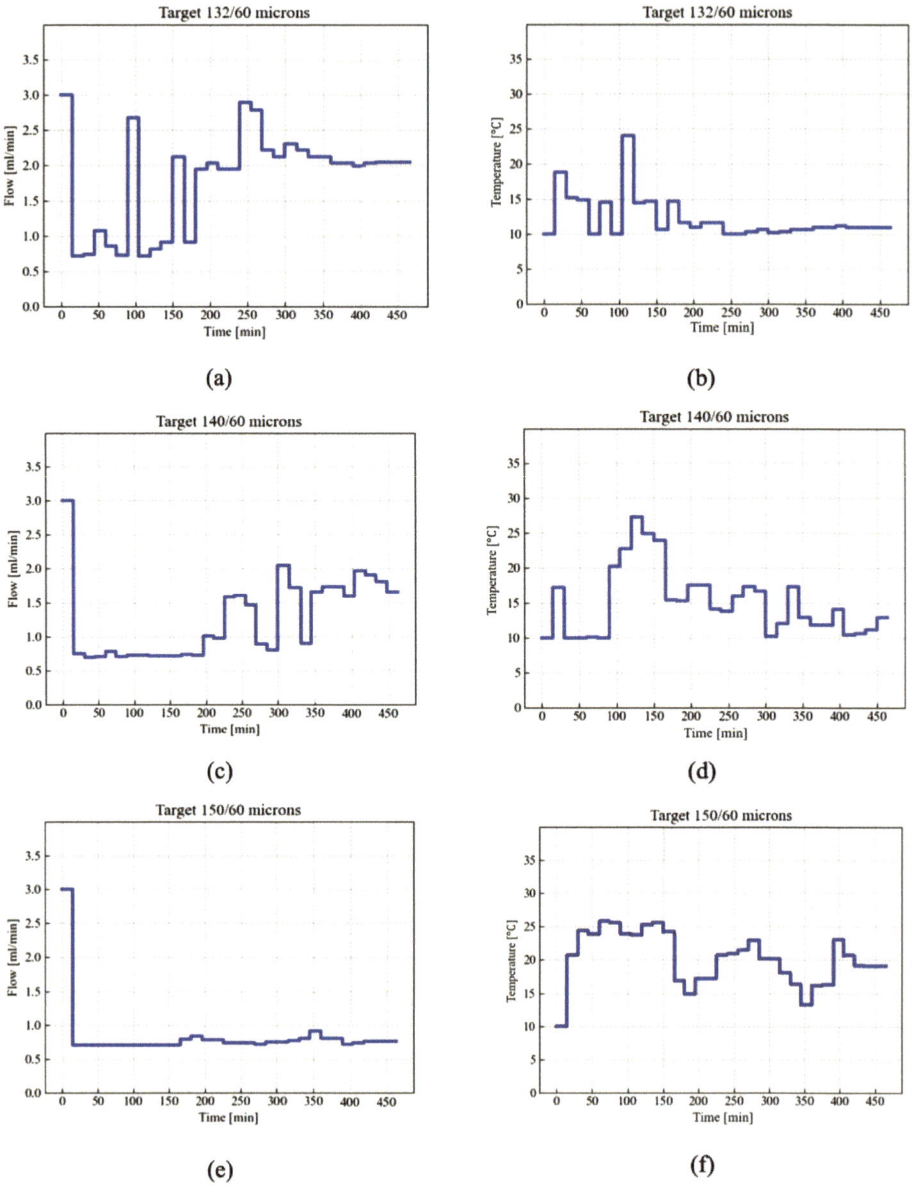

FIGURE 12.11 Manipulated variables using the DRL policy to guide the reaction toward the set-point. (a, c, e) This show the anti-solvent flow rate profile for three different crystal size/standard deviation setpoints (132/60 microns, 140/60 microns, 150/60 microns) while (b, d, f) show the corresponding temperature profiles.

3. Get the reward and the next state.
4. Introduce a disturbance with probability ϵ.
5. Repeat steps 2–4 until the target state is reached.
6. Update the agent and the critic using the experience gained thus far.

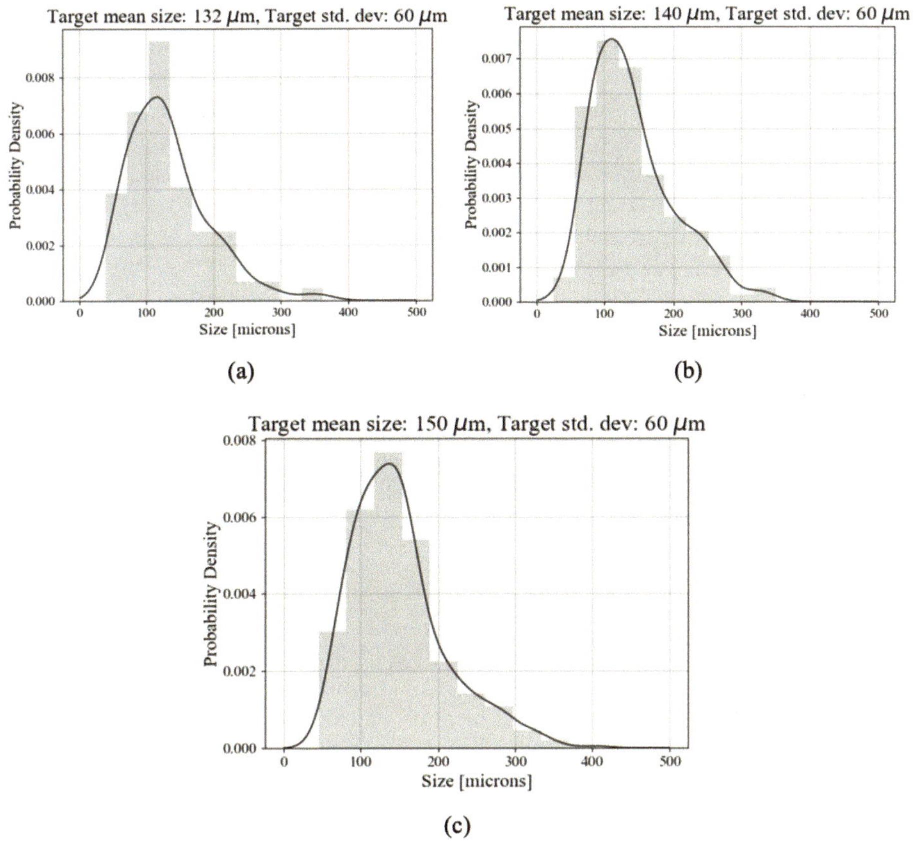

FIGURE 12.12 Histogram of crystal size distribution at the end of batch from closed-loop experiments. (a–c) These figure show the experimentally observed crystal size histograms for three different crystal size/standard deviation setpoints (132/60 microns, 140/60 microns, 150/60 microns).

12.3 CONCLUSIONS

In this chapter, we have explored two approaches to enhance the performance of high-value processes: C-PC yield in bioreactors and crystal size distribution in anti-solvent crystallization. By integrating ANN and DRL, we aimed to optimize C-PC yield in Plectonema sp. UTEX 1541 and control crystal size distribution effectively. The significance of C-PC as a bio-product with diverse applications underscores the need for novel optimization strategies, especially in lesser-studied cyanobacterial species like Plectonema. Through experimental setups and data preprocessing techniques, we demonstrated the application of ANN and DRL in simulating growth dynamics and selecting efficient reaction strategies, resulting in significant improvements in C-PC concentration and yield.

In the realm of anti-solvent crystallization, we explored the fundamental principles governing crystal growth and the critical role of size and shape in downstream operations. Leveraging insights from population balance and stochastic models, coupled with RL techniques like TD3, we presented a novel framework for optimizing crystal size distribution. The meticulous description of implementation details, including state definition, action formulation, and reward scheme, offers practical insights into deploying RL algorithms for process control. Through numerical approximation and experimental validation, we showcased the efficacy of the TD3 algorithm in steering crystal populations toward desired set-points, even in the presence of disturbances, thereby laying the foundation for enhanced control strategies in complex chemical processes.

Overall, these case studies show how by combining theoretical understanding with practical implementations of machine learning algorithms, we can not only advance the optimization of bio-product yield and crystal size distribution but also open avenues for further research in process control and optimization. These approaches hold promise for enhancing productivity and profitability in various industrial applications, particularly in pharmaceuticals and beyond, paving the way for future advancements in optimizing complex reaction systems with unknown kinetics.

REFERENCES

1. del Rio-Chanona E. A., Zhang D., Xie Y., Manirafasha E., Jing K. (2015), Dynamic simulation and optimization for *Arthrospira platensis* growth and c-phycocyanin production. Industrial & Engineering Chemistry Research, 54(43), pp. 10606–10614.
2. Romay C., Gonzalez R., Ledon N., Remirez D., Rimbau V. (2003), C-phycocyanin: A biliprotein with antioxidant, anti-inflammatory and neuroprotective effects. Current Protein and Peptide Science, 4(3), pp. 207–216.
3. Eriksen N. T. (2008), Production of phycocyanin—a pigment with applications in biology, biotechnology, foods and medicine. Applied Microbiology and Biotechnology, 80(1), pp. 1–14.
4. Rippka R., Deruelles J., Waterbury J. B., Herdman M., Stanier R. Y. (1979), Generic assignments, strain histories and properties of pure cultures of cyanobacteria. Microbiology, 111(1), pp. 1–61.
5. Boussiba S., Richmond A. E. (1980), C-phycocyanin as a storage protein in the blue-green alga *Spirulina platensis*. Archives of Microbiology, 125(1–2), pp. 143–147.
6. Jo B., Ginexi E. M., Ialongo N. S. (2010), Handling missing data in randomized experiments with noncompliance. Prevention Science, 11(4), pp. 384–396.
7. Junninen H., Niska H., Tuppurainen K., Ruuskanen J., Kolehmainen M. (2004), Methods for imputation of missing values in air quality data sets. Atmospheric Environment, 38(18), pp. 2895–2907.
8. Abadi M., Agarwal A., Barham P., Brevdo E., Chen Z., Citro C., Corrado G. S., Davis A., Dean J., Devin M., Ghemawat S., Goodfellow I., Harp A., Irving G., Isard M., Jia Y., Jozefowicz R., Kaiser L., Kudlur M., Levenberg J., Mane D., Monga R., Moore S., Murray D., Olah C., Schuster ́M., Shlens J., Steiner B., Sutskever I., Talwar K., Tucker P., Vanhoucke V., Vasudevan V., Viegas F., Vinyals O., Warden P., Wattenberg M., Wicke M., Yu Y., Zheng X. (2015), TensorFlow: Large-scale machine learning on heterogeneous systems. Software available from tensorflow.org. doi: 10.48550/arXiv.1603.04467
9. Kingma D. P., Ba J. (2014), Adam: A method for stochastic optimization. arXiv:1412.6980v9. doi: 10.48550/arXiv.1412.6980

10. Zur R. M., Jiang Y., Pesce L. L., Drukker K. (2009), Noise injection for training artificial neural networks: A comparison with weight decay and early stopping. Medical Physics, 36(10), pp. 4810–4818.
11. del Rio-Chanona E. A., Manirafasha E., Zhang D., Yue Q., Jing K. (2016), Dynamic modeling and optimization of cyanobacterial c-phycocyanin production process by artificial neural network. Algal Research, 13, pp. 7–15.
12. Srivastava N., Hinton G., Krizhevsky A., Sutskever I., Salakhutdinov R. (2014), Dropout: A simple way to prevent neural networks from overfitting. The Journal of Machine Learning Research, 15(1), pp. 1929–1958.
13. Baldi P., Sadowski P. J. (2013), Understanding dropout. Advances in Neural Information Processing Systems, pp. 2814–2822. doi: 10.5555/2999792.2999926
14. Mnih V., Badia A. P., Mirza M., Graves A., Lillicrap T., Harley T., Silver D., Kavukcuoglu K. (2016), Asynchronous methods for deep reinforcement learning. International Conference on Machine Learning, pp. 1928–1937. doi: 10.5555/3045390.3045594
15. Tsuda A., Henry F. S., Butler J. P. (2013), Particle transport and deposition: Basic physics of particle kinetics. Comprehensive Physiology, 3, pp. 1437–1471.
16. Nagy Z. K., Fujiwara M., Braatz R. D. (2007), Recent advances in the modelling and control of cooling and antisolvent crystallization of pharmaceuticals. IFAC Proceedings Volumes (IFAC-PapersOnline); IFAC Secretariat, 40, pp. 29–38, doi: 10.3182/20070606-3-MX-2915.00054
17. Nowee M., Abbas A., Romagnoli J. A. (2008), Model-based optimal strategies for controlling particle size in antisolvent crystallization operations. Crystal Growth & Design, 8, pp. 2698–2706, doi: 10.1021/cg700720t
18. Trifkovic M., Sheikhzadeh M., Rohani S. (2008), Kinetics estimation and single and multi-objective optimization of a seeded, anti-solvent isothermal batch crystallizer. Industrial & Engineering Chemistry Research, 47(5), pp. 1586–1595. doi: 10.1021/ie071125g
19. Grosso M., Galan O., Baratt R., Romagnoli J. A. (2009), A stochastic formulation for the description of the crystal size distribution in antisolvent crystallization processes. AIChE Journal, 56(8), pp. 2077–2087, doi: 10.1002/aic.12130
20. Galán O., Grosso M., Baratti R., Romagnoli J. A. (2010), Stochastic approach for the calculation of anti-solvent addition policies in crystallization operations: An application to a bench-scale semi-batch crystallizer. Chemical Engineering Science, 65(5), pp. 1797–1810, doi: 10.1016/j.ces.2009.11.023
21. Cogoni G., Tronci S., Mistretta G., Baratti R., Romagnoli J. A. (2013), Stochastic approach for the prediction of PSD in nonisothermal antisolvent crystallization processes. AIChE Journal, 59(8), pp. 2843–2851, doi: 10.1002/aic.14089
22. Baratti R., Tronci S., Romagnoli J. A. (2017), A generalized stochastic modelling approach for crystal size distribution in antisolvent crystallization operations. AIChE Journal, doi: 10.1002/aic.15372
23. Ghadipasha N., Romagnoli J. A., Tronci S., Baratti R. (2018), A model-based approach for controlling particle size distribution in combined cooling-antisolvent crystallization processes. Chemical Engineering Science, doi: 10.1016/j.ces.2018.06.032

13 Generative AI

Machine learning (ML) algorithms, together with ideas from predictive modeling and optimization, can help us solve complex problems in the chemical process industry. In most cases, we apply ML methods to classical problems where classical solution methods are not efficient enough to provide a reasonable answer to the problem at hand. In this chapter, we explore the question of whether ML can help us tackle new kinds of problems, thus unlocking completely new avenues of research and discovery. We will do this by introducing key ideas – and an example – about what is known as generative ML.

13.1 INTRODUCTION

As humans, we not only want to understand the world around us but also want to *imagine* what it could be. Generative machine learning (ML) methods incorporate methods from different subfields of ML to mimic creative processes. To do this, generative models are trained to first learn about a topic and then simultaneously try out new ideas and critique them based on that knowledge. When a human agent asks for something *new*, line an image of a picnic scene, a generative model can give an answer with a chance of being good based on its training, often using optimization methods for the identification. The combination of methods and applications related to generative ML methods is what has come to be called Generative AI.

13.2 DEFINITION AND IMPORTANCE OF GENERATIVE METHODS

Generative ML methods can produce new instances of data based on prior training. The ML algorithms introduced in the previous chapters are all designed to learn the underlying distribution of a given dataset or, as we explained in Chapter 6, find the hypothesis that better explains the outcomes of a phenomenon by observing the data available. Generative models take this knowledge and add a new component – another algorithm – that uses it to produce new, unseen data points while maximizing the likelihood that it would belong to the learn distribution or that it would agree with the hypothesis found.

The capability of producing unseen data instances opens a wide range of applications, such as:

- Data augmentation by generating additional synthetic samples. This helps to alleviate the data scarcity problem and improves the generalization and robustness of ML models. By expanding the training set, generative methods enable better model performance and accuracy.
- Imputation and missing data handling by generating plausible replacements based on the learned data distribution. This is particularly useful in scenarios where data is incomplete or corrupted, enabling better analysis and decision-making.

DOI: 10.1201/9781003455905-16

- Creative output generation by generating novel and compelling outputs, such as realistic images, music compositions, and textual content. They enable machines to exhibit creative intelligence and empower artists, designers, and content creators with new tools and possibilities.
- Domain adaptation and style transfer: Generative methods can learn the characteristics of one domain and transfer them to another. This enables tasks such as style transfer in images, where the style of a particular artwork can be applied to different images while preserving their content.
- Reinforcement learning (RL) by simulating realistic environments and generating diverse scenarios; generative methods facilitate training and exploration in complex RL setups.

Overall, generative methods play a significant role in expanding the capabilities of ML systems. They provide the means to understand, create, and manipulate data in creative and practical ways, with implications across various domains, including healthcare, finance, entertainment, and scientific research. As generative modeling techniques continue to advance, their impact on AI-driven systems and human creativity is expected to grow exponentially.

13.3 SELF-SUPERVISED LEARNING

As stated before, the first step in a generative ML implementation is learning the data distribution. Generally speaking, this can be achieved using the methods described in the previous chapters. However, in most applications involving generative ML, large amounts of labeled data are needed to obtain a meaningful model. The primary problem with this requirement is that labels for these datasets are often inaccessible. This can be due to the cost involved in obtaining labels for the dataset or because we lack clarity on what the labels should be. Hence, a different branch of ML that we have not explored yet, called self-supervised learning (SSL), is needed.

SSL is useful for problems in which we have a lot of unlabeled data but do not have the labels needed to provide supervision to train a supervised model. In such a scenario, SSL algorithms rely on the structure of the data to define a "pretext" learning objective. If the data is text, then the learning objective could be as simple as masking a word in a sentence and having the model predict the masked word based on its surrounding context. This objective forces the model to learn the relationships among the words in the sentence without having any explicit labels. This technique also works with images where the learning objective could involve masking a portion of the image and having the model predict the masked pixels. Once the model has been trained on these "pretext" objectives, the representations learned by the model can be used as a starting point for finetuning models on more relevant tasks [1]. Below, we provide a brief introduction of some of the most widely used methods for SSL. The reader is encouraged to learn more about SSL from the references provided at the end of this chapter.

Autoencoders (AEs): AEs are a type of SSL method that consists of an encoder and a decoder. The encoder takes an input data sample and maps it to a lower-dimensional latent space representation. The decoder then takes the latent representation and reconstructs the original input. By training the AE to minimize the reconstruction error, it learns a compressed representation of the data and can generate new samples by sampling from the latent space [2, 3].

Variational Autoencoders (VAEs): VAEs build upon the concept of AEs. In addition to reconstructing the input, VAEs also learn a probability distribution in the latent space. This distribution allows the model to generate new samples by sampling from the latent space. VAEs use techniques from variational inference to learn the parameters of the latent distribution [4, 5].

Generative Adversarial Networks (GANs): GANs are another popular class of SSL that involve two neural networks: a generator and a discriminator. The generator network takes random noise as input and generates new samples, while the discriminator network tries to distinguish between real and generated samples. The two networks are trained in a competitive manner, with the generator trying to fool the discriminator and the discriminator trying to accurately classify the samples. GANs have been successful in generating high-quality images, audio, and even video [6].

Normalizing Flows: Normalizing flows are a family of SSL method that aim to learn a bijective mapping between a simple probability distribution – the latent space – and a complex target distribution – the original input space. By applying a series of invertible transformations to the simple distribution, normalizing flows can transform it into the target distribution. This allows for efficient sampling from the target distribution and computation of its density [7, 8].

Autoregressive models: Autoregressive models model the joint distribution of the data by factorizing it into a product of conditional probabilities in which the probability of each variable is depending on the previous variables. These models learn to generate samples by sequentially generating each component of the data, conditioned on the previously generated components [9].

13.4 VARIATIONAL AUTOENCODERS IN GENERATIVE ML

The architecture of a generative ML implementation is illustrated here using VAEs. As stated before, the basic structure of an AE consists of two main components: encoder and decoder.

The encoder takes the input data and maps it to a lower-dimensional representation, often referred to as the latent space or bottleneck. The encoder network typically consists of an artificial neural network, with each layer transforming the input data into a more compact representation. The final layer of the encoder outputs the *latent* representation. The latent space is a lower-dimensional representation of the input data. It serves as a compressed representation that captures the most important features or characteristics of the input. The dimensionality of the latent space is usually much smaller than the dimensionality of the input data.

The decoder takes the latent representation and reconstructs the original input data. Like the encoder, the decoder network consists of multiple layers, with each layer progressively expanding the dimensionality of the representation. The final layer of the decoder produces the output, which should match the input data as closely as possible.

13.4.1 Model Training

The training process of an AE involves minimizing a reconstruction loss, which quantifies the difference between the original input and the reconstructed output. The most commonly used reconstruction loss is the Mean Squared Error (MSE), but other loss functions such as binary cross-entropy can also be used depending on the nature of the data.

During training, the AE learns to encode the input data into the latent space and then decode it back to reconstruct the original input. By minimizing the reconstruction loss, the AE learns to capture the essential features of the input data in the latent space. The idea is that a compressed representation in the latent space can capture the most salient aspects of the data, discarding redundant or noisy information.

13.4.2 Inference and Generation

Once the AE is trained, it can be used for generating new data samples. This is done by interpolating samples in the latent space and passing them through the decoder to produce the generated outputs. By sampling different points in the latent space, different outputs can be generated, allowing the AE to generate diverse samples.

13.4.3 Probability Distribution of the Latent Space

An improvement on the AE approach shown before is to set up the model to learn a probability distribution over the latent space. While the encoder in a deterministic AE would map the input to a point in the latent space, here the approach is to map the input to the parameters of a known distribution. For example, if a normal distribution is the distribution of choice, then the encoder would map the input to the mean and variance of the normal distribution. By constraining the latent space to follow an assumed distribution, we effectively regularize the model. This is achieved by implementing the following modifications to the AE:

1. Reparameterization trick: To ensure that the model remains differentiable, the VAE uses the reparameterization trick during training. Instead of directly sampling from the estimated distribution, the trick involves sampling from a standard Gaussian distribution and then transforming the samples using the mean and variance parameters predicted by the encoder.
2. Loss function: VAEs are trained using a loss function that consists of two terms: the reconstruction loss and a regularization term. The reconstruction loss measures the similarity between the original input and the reconstructed output, encouraging the model to faithfully reproduce the input data. The regularization term encourages the learned latent distribution to resemble a prior distribution, typically a standard Gaussian. This term is

computed using the Kullback-Leibler (KL) divergence between the estimated latent distribution and the prior distribution.

3. Sampling: Once the VAE is trained, it can generate new samples by drawing points from the latent space distribution and passing them through the decoder.
4. Control over the latent space representation: In a VAE, constraints can be enforced over the latent space representation. By introducing the constraint that the dimensions of the latent space are independent, the VAE has been shown to learn semantically meaningful representations.

13.5 GENERATIVE AI MOLECULAR DESIGN

A complete implementation of generative ML is presented in this section to showcase not only the capabilities of these methods for chemistry and chemical engineering applications but also how different subfields of ML can come together to produce an effective solution [10]. The main components of the procedure are shown in Figure 13.1, and the details of the implementation are provided below.

13.5.1 Feature Engineering

The concept of feature engineering was briefly introduced in Chapter 6. In short, feature engineering can serve two purposes: on the one hand, it translates data that would otherwise be too complex for the computer to consume – such as text or graphs, and on the other hand, it allows the embedding of domain knowledge into the data by using mechanistic models to produce the engineering features. For applications involving molecular representations, translating such representations into machine readable data is a key step.

A widespread methodology, the Simplified Molecular-Input Line-Entry System (SMILES) [11] uses ASCII strings, for representing 2D or 3D molecular representations on a large scale [12]. However, despite the ubiquity and proven utility of SMILES in computational chemistry, the method lacks the syntactic and semantic

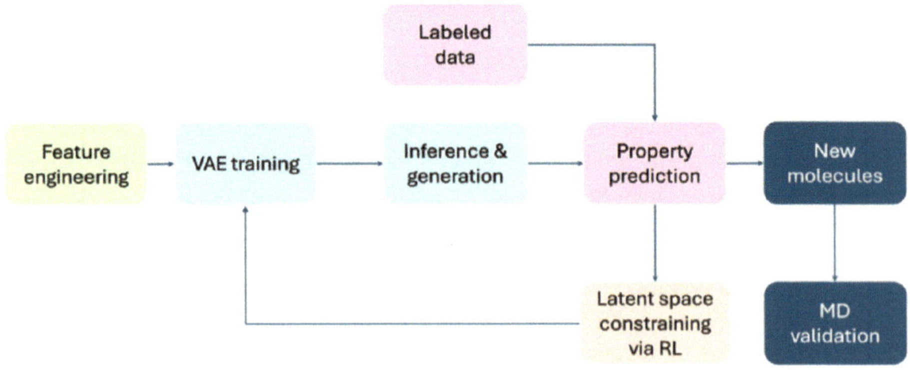

FIGURE 13.1 Overall generative machine learning implementation for molecule discovery.

robustness necessary for the representation of complex molecules and their properties [13]. SELF-referencing embedded string (SELFIES), an improvement upon SMILES, introduces an approach for representing molecular graphs as character strings, enabling the unique identification of any given molecule [13]. Because of its complete robustness, every combination of symbols in the SELFIES alphabet maps to a chemically valid graph, which makes the methodology much more powerful.

13.5.2 Learning the Latent Space and Generating New Molecules

A VAE model was used to learn the latent space of a database of molecules. The encoder network takes a molecular structure, which is represented by atom types, connectivity, and coordinates – encoded using SELFIES, as its input and then maps it to the distribution within the latent space. Within this process, the encoder predicts the parameters, namely the mean and variance, of the approximate posterior multivariate Gaussian distribution. The latent space itself serves as a low-dimensional continuous representation of molecular structures, with each point in this space corresponding to a potential molecular structure. The decoder network takes a point from the latent space and maps it back to a molecular structure. In this step, the decoder predicts the parameters pertaining to atom types, connectivity, and coordinates, which collectively define the reconstructed molecule.

To generate new samples, we sample the latent space and use the decoder to produce the corresponding molecules (see Figure 13.2). Although the training process of the VAE together with the regularization provided by the variational aspect promotes a latent space containing only chemically valid molecules, such validity must still be verified [14]. Moreover, when the goal of the implementation is that of generating molecules with specific target properties, additional steps need to be taken to maximize the likelihood of generating not only valid molecules but also molecules with the desired characteristics. The following section outlines a method to address this challenge.

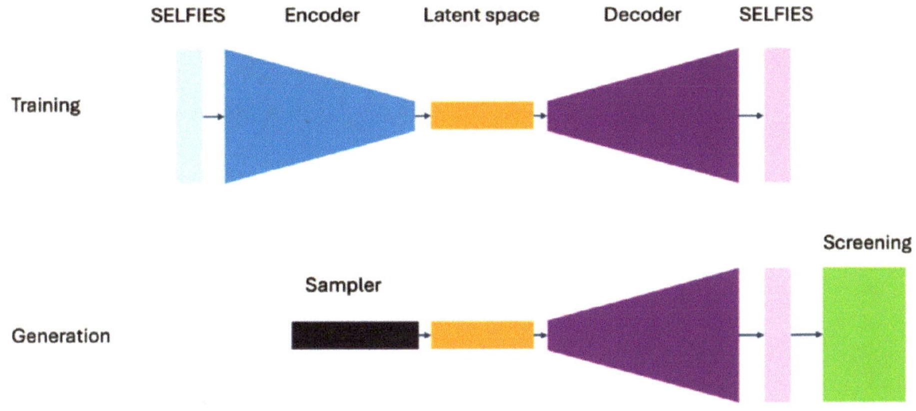

FIGURE 13.2 Schematic representation of the molecules' generation using a variational autoencoder.

13.5.3 Generating Molecules with Specific Properties

The continuous representations of molecules obtained through the VAE allows for the exploration of the chemical space in search of molecules with desired properties [15, 16, 17]. However, although this search space might be reduced through the training of the VAE and the further constraint of learning a probability distribution, it is still an extensive search space with many candidate molecules that do not meet the specify criteria. To address this, a combination of representation and RL is introduced here to further reduce the search space, thus increasing the likelihood of finding molecules with the desired properties.

A graphical convolutional neural network (GCNN) model [18] is used to estimate the value of the target property for a molecule drawn from the latent space and decoded. This predictive model is trained offline using a database of SELFIES with their corresponding property values. With this, the difference between the desired value for the target property and the predicted values estimated from the GCNN model can be used as fitness criterion for the search.

Once we have a way of evaluating the fitness of a given molecule, we can use RL to train an agent to draw from the latent space only molecules that are likely to have the desired properties. In this context, an action is a point in the latent space, the environment is the GCNN model, and the reward is based on the distance from a desired target property value.

13.6 EXAMPLE: NON-IONIC SURFACTANT MOLECULES WITH LMC

The feasibility of this approach is demonstrated in a case study focused on non-ionic surfactant molecules [10]. Specifically, it targets the generation of new molecules with low Critical Micelle Concentration (CMC), a crucial surfactant characteristic.

In colloidal and surface chemistry, the CMC is defined as the concentration of surfactants above which micelles form and all additional surfactants added to the system will form micelles [19]. The CMC is an important characteristic of a surfactant. Before reaching the CMC, the surface tension changes strongly with the concentration of the surfactant. In this study, 288 surfactant molecules were considered, and the aim is to generate new molecules with low CMC which will improve efficiency, solubility, and stability.

13.6.1 Data

The VAE model was pre-trained on a large polymer database (over 20,000 molecules) obtained from Kaggle, an open-sourced platform for data science competitions and datasets. Training of the VAE model was done on a dataset sourced from Gil [20]. The non-ionic surfactant dataset with CMC property used for fine-tuning the pre-trained VAE model was obtained from literature [21–25]. The predictive model based on graph convolution network also leveraged data from literature. Here we trained the GNN with the same experimental CMC data.

13.6.2 VAE Finetuning Using Transfer Learning and Reinforcement Learning

To leverage the knowledge gained from the pretrained VAE model, a transfer learning approach was adopted. The weights of compatible layers, that is, those layers whose dimensions align, are transferred from the pretrained VAE model to the VAE model intended for fine-tuning. This process is crucial for initializing the model effectively, providing a starting point that already captures valuable features from the larger dataset. Following the initialization through transfer learning, the VAE model is trained on the experimental dataset of non-ionic surfactants, which consists of 285 molecules. As explained in Chapter 8, some form of regularization is needed for transfer learning. In this case, a smaller learning rate than that of the pretraining step was implemented. The loss function encompasses a reconstruction loss, which measures the difference between the generated SELFIES and the original input SELFIES, and the Kullback-Leibler divergence, which enforces a regularization on the latent space.

Following the pretraining and fine-tuning phase, the RL component is introduced. The VAE generates molecules, and the property prediction model evaluates their properties. If a generated molecule surpasses the property threshold, a positive reward of 100 is assigned; otherwise, a penalty of –100 is imposed. The policy network is then updated using policy gradients, optimizing the VAE's latent representation to maximize rewards. The RL algorithm, operating over 2000 episodes, effectively optimizes the VAE's latent space representation using policy gradients. The seamless integration of RL significantly enhances the VAE's ability to generate molecules that align with the desired chemical properties.

13.6.3 Hyperparameter Tuning

Hyperparameters were tuned one at a time through inspection. The parameters adjusted included batch size, number of epochs, and learning rate. Experimentation involved varying batch sizes to find the optimal balance between computational efficiency and model convergence. Smaller batch sizes facilitated faster convergence, while larger ones ensured more stable gradients. After experimenting with various sizes, a batch size of 64 was selected as it demonstrated a favorable compromise between convergence speed and model stability. The model's performance was tracked using VAE loss, and the epoch value at which performance improvement plateaued was chosen.

After extensive experimentation, 1000 epochs were chosen as it provided a favorable balance between training duration and reaching a satisfactory outcome. Various learning rates were explored aiming to find one that ensured stable convergence without overshooting or sluggish progress. Subsequently, through thorough testing, a learning rate of 0.001 emerged as optimal, facilitating efficient model training and delivering promising results in terms of performance. he chosen hyperparameter combination (64, 1000, 0.001) played a crucial role in attaining the desired equilibrium between model convergence, computational efficiency, and overall efficacy in molecule generation

13.6.4 Results

Figure 13.3 depicts the latent space of molecules generated using the VAE alone (without fine-tuning). The scatter plot displays a set of points in a 2D space using a principal component analysis (PCA) projection, each representing a molecule. These molecules were generated based on the VAE's learned distribution in latent space. Fine-tuning the pre-trained VAE using transfer learning led to a more extensive and dispersed distribution of points in the latent space, compared to directly training the VAE with smaller dataset. This is because by encompassing a wide range of molecular features in a pre-trained model, the VAE model is better equipped to generate meaningful representations for a broad array of molecular structures. This also leads to lower VAE loss and faster convergence during fine-tuning, as the model has already learned some useful features. A wider spread in the latent space indicates that the model has learned to represent a broader range of chemical features and structures and would enhance the robustness and generalization capabilities of the model. This would facilitate the generation of novel molecules with varied properties and characteristics, contributing to a more diverse chemical space. This diverse latent space allows for more effective exploration of potential chemical designs and structures, such that by integrating with RL, we were able to navigate the latent space to identify promising molecular candidates that may exhibit desired properties for specific application.

In Figure 13.4, we observe a visualization depicting two principal components of the 64-dimensional latent space. The cluster represented by the red points corresponds to molecules generated exclusively by the VAE model. These molecules are produced based on the learned latent space representation, with the objective of reconstructing input molecules. On the other hand, the cluster represented by the black points, denoted as VAE+RL, comprises molecules generated by the VAE model with RL guidance. RL assists in refining the generated molecules according to specific criteria or objectives, thereby yielding more desirable or optimized molecules. The discernible separation of the VAE+RL cluster from the larger VAE

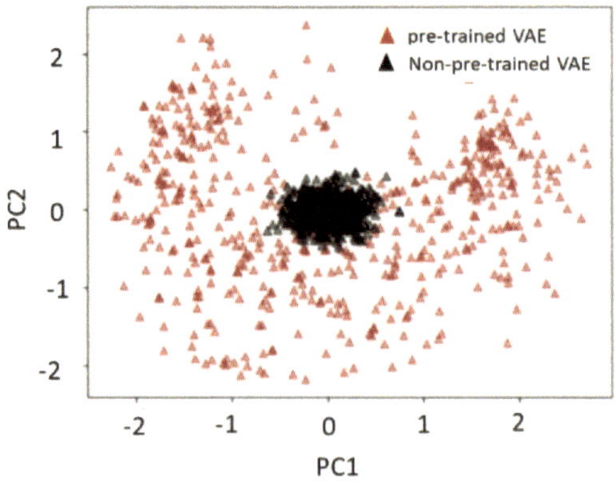

FIGURE 13.3 Latent space visualization of pre-trained VAE and the non-pre-trained. Pretraining shows a widespread distribution of generated molecules in an unconstrained space.

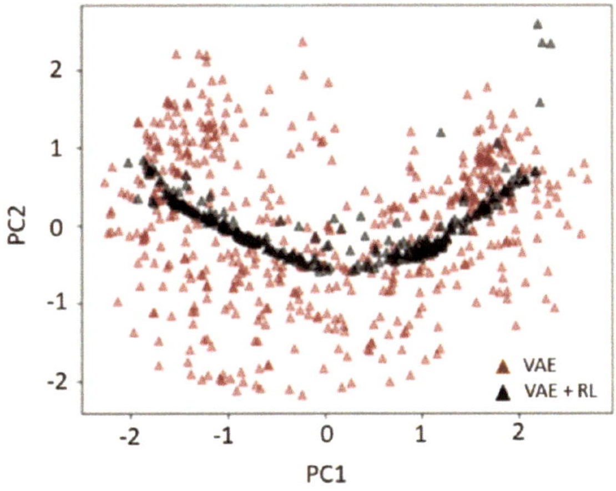

FIGURE 13.4 Latent space visualization of the RL-generated molecules in the unconstrained space. RL-generated molecules are in a region within diverse space. The VAE + RL-generated molecules exhibited a distinctive cluster implying that the RL strategy successfully influenced the VAE.

cluster implies that the RL-guided approach significantly influences the generated molecules. This refinement results in molecules that meet specified property thresholds, leading to a more compact and distinct region within the latent space.

13.6.5 Validation

Figure 13.5 provides a visualization of a collection of surfactants and their SMILES string. RDKit [26] was used to convert each SMILES string into its corresponding molecular form, providing an accessible insight into the results of the generative process. The RDKit library was also used to check the validity of the generated molecules. Using RDKit's molecular structure parser that checks atoms' valency and consistency of bonds in aromatic rings, all new generated molecules were shown to be valid. Each of these surfactants has a CMC value less than 0.1 mM aligning perfectly with the set threshold.

Furthermore, a molecular dynamics (MD)-informed validation was conducted to ensure the structural integrity of the generated molecules, confirming their adherence to fundamental chemical principles. The focus was on preventing over- or under-coordination of atoms, which might occur due to unusual valences. This validation included an assessment of cyclic structures and structures with cyclic branches, aimed at verifying the expected number of rings and assessing the chemical plausibility of their sizes. Additionally, to gauge the practical relevance of these molecules in surfactant design, Tanimoto similarity coefficients were computed (Figure 13.5) [10]. This involved comparing the generated molecules with original surfactant molecules from the training dataset to quantitatively assess their similarity. This approach effectively screened the generated molecules based on their resemblance to existing surfactant molecules within the dataset. The validation

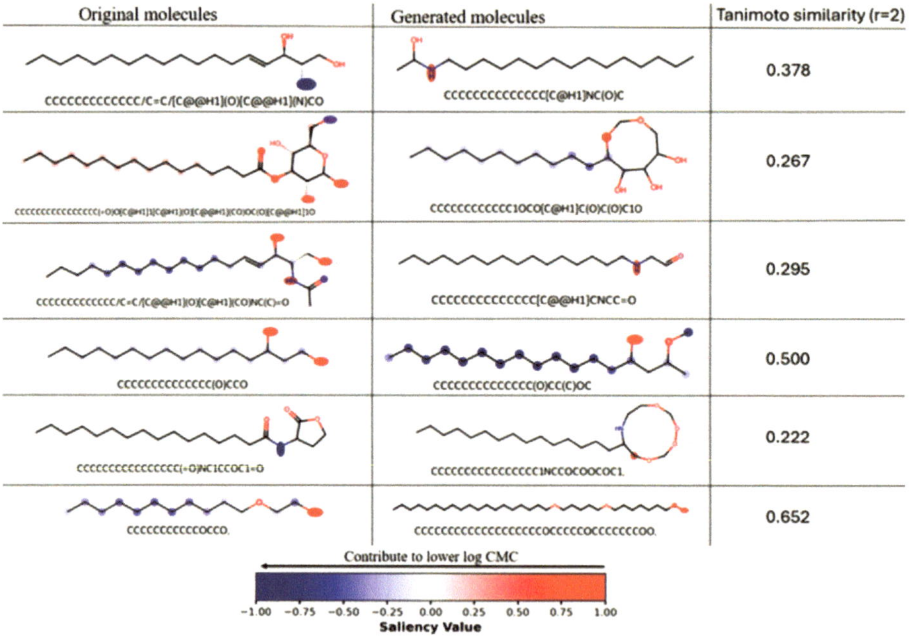

FIGURE 13.5 Using the Tanimoto similarity to compare the molecular structures of the generated molecules to the molecules in the training dataset.

results suggest that the generated molecules are not just theoretically viable but also potentially capable of being synthesized. This stems from the observation that these molecules exhibit a remarkable proximity to the most similar structures within the surfactant training dataset. In essence, they appear to represent only a few chemical mutations away from their real-world counterparts. To ascertain the reliability of the generated surfactant molecules, MD simulations were conducted to calculate the free energy of solvation [10]. The ΔGsolv values for the surfactants analyzed showed negativity across the studied temperatures, indicating the spontaneous micellization process in aqueous solutions. Therefore, the resulting micelle solutions exhibit thermodynamic stability.

13.6.6 Model Explainability

A fundamental requirement for applying generative methods to solve Chemical Engineering problems is the capacity to provide insightful explanations regarding model behavior. One approach to achieving this in ML involves analyzing the features utilized by the model to make predictions. In the context of understanding the features of generated surfactant molecules affecting CMC values, saliency maps were employed. Saliency maps, commonly used in fields like representation learning for image classification, reveal key features in images [27, 28]. To generate a saliency map, gradients of input atom features were computed using backpropagation, then multiplied element-wise with the input [27]. Multiplying input features by their gradients unveils their relative importance, particularly for categorical features such as the presence or absence of specific atoms like carbon. This study adopted the saliency analysis method as outlined by Qin et al [21].

Figure 13.5 also illustrates the saliency maps for several chosen surfactants, comparing both original and newly generated molecules. It's important to note that the original surfactants selected for this comparison meet the specified CMC threshold value. This comparison allows us to assess how the models' decisions impact the significance of molecular features. In these maps, atoms (represented as nodes in the graph) are color-coded according to their normalized gradients: red indicates a more positive contribution to CMC, while blue signifies a more negative contribution. These saliency maps confirm that polar atoms such as O and N contribute to higher CMC values, while nonpolar atoms like C contribute to lower CMC values, consistent with our qualitative expectations. Additionally, the maps reveal that as surfactants branch out more, lighter blue hues appear, indicating smaller CMC values compared to unbranched tail nodes in samples. These observed patterns are in line with our intuition that surfactants with long, unbranched tail groups tend to have lower CMC values, while those with increasing polarity tend to have higher values.

13.7 CONCLUSIONS

This chapter has provided an exploration of generative ML methods and their applications in molecular design and chemical engineering. By synthesizing new data points based on learned distributions or hypotheses, generative models enhance ML systems' capabilities, enabling data augmentation, missing data handling, creative output generation, domain adaptation, style transfer, and RL. Methods such as AEs, VAEs, GANs, and autoregressive models were discussed, showcasing their ability to learn complex data distributions and generate diverse outputs.

Furthermore, the chapter demonstrated the practical implementation of VAEs in generative ML-driven molecular design. Through transfer learning, fine-tuning, and RL, VAEs were utilized to generate novel molecules with specific properties, such as low CMC in non-ionic surfactants. The integration of RL facilitated the exploration of the latent space to identify molecules meeting desired criteria, highlighting the potential of generative ML methods in accelerating molecular discovery and design. Overall, this chapter underscores the transformative potential of generative ML methods in advancing scientific research, innovation, and problem-solving across diverse domains.

REFERENCES

1. Balestriero R., Ibrahim M., Sobal V., Morcos A., Shekhar S., Goldstein T., Bordes F., Bardes A., Mialon G., Tian Y., Schwarzschild A., Wilson A. G., Geiping J., Garrido Q., Fernandez P., Bar A., Pirsiavash H., LeCun Y., Goldblum M. (2023), A cookbook of self-supervised learning. arXiv preprint arXiv:2304.12210v2, doi: 10.48550/ARXIV.2304.12210
2. Rumelhart D. E., Hinton G. H., Williams R. J. (1986), Learning internal representations by error propagation [Online]. https://api.semanticscholar.org/CorpusID:62245742

3. Bank D., Koenigstein N., Giryes R. (2021), Autoencoders, Deep Learn. Sci. [Online]. https://api.semanticscholar.org/CorpusID:212717965
4. Kingma D. P., Welling M. (2019), An introduction to variational autoencoders. now [Online]. http://ieeexplore.ieee.org/document/9051780
5. Kingma D. P., Welling M. (2022), Auto-Encoding Variational Bayes. arXiv:1312.6114v11, doi: 10.48550/arXiv.1312.6114
6. I. Goodfellow I., Pouget-Abadie J., Mirza M., Xu B., Warde-Farley D., Ozair S., Courville A., Bengio Y. (2020), Generative adversarial networks. Communications of the ACM, 63(11), pp. 139–144, doi: 10.1145/3422622
7. Tabak E. G., Vanden-Eijnden E. (2010), Density estimation by dual ascent of the log-likelihood. Communications in Mathematical Sciences, 8(1), pp. 217–233.
8. Kobyzev I., Prince S. J. D., Brubaker M. A. (2021), Normalizing flows: An introduction and review of current methods. IEEE Transactions on Pattern Analysis and Machine Intelligence, 43(11), pp. 3964–3979, doi: 10.1109/tpami.2020.2992934.
9. Van Den Oord A., Kalchbrenner N., Kavukcuoglu K. (2016), Pixel recurrent neural networks, Proceedings of the 33rd International Conference on International Conference on Machine Learning, Volume 48. JMLR.org, pp. 1747–1756.
10. Nnadili M., Okafor A. N., Olayiwola T., Akinpelub D., Kumar R., Romagnoli J. A. (2024), Surfactant-specific AI-driven molecular design: Integrating generative models, predictive modeling, and reinforcement learning for tailored surfactant synthesis, Industrial & Engineering Chemistry Research, doi: 10.1021/acs.iecr.4c00401
11. Weininger D. (1988), SMILES, a chemical language and information system. 1. Introduction to methodology and encoding rules. Journal of Chemical Information and Computer Sciences, 28(1), pp. 31–36.
12. Campos D., Ji H. (2021), Img2smi: Translating molecular structure images to simplified molecular-input line-entry system. arXiv preprint arXiv:2109.04202, doi: 10.48550/arXiv.2109.04202
13. Krenn M., Ai, Barthel S., Carson N., Frei A., Frey N.C., Friederich P., Gaudin T., Gayle A.A., Jablonka K.M., Lameiro R.F., Lemm D., Lo A., Moosavi S.M., Nápoles-Duarte J.M., Nigam A., Pollice R., K., Schatzschneider U., Schwaller P., Skreta M., Smit B., Strieth-Kalthoff F., Sun C., Tom G., Falk von Rudorff G., Wang A., Andrew White, Young A., Yu R., Aspuru-Guzik Krenn A.(2022), SELFIES and the future of molecular string representations. Patterns, 3(10), p. 100588.
14. An J., Cho S. (2015), Variational autoencoder based anomaly detection using reconstruction probability. Special Lecture on IE, 2(1), pp. 1–18.
15. Cheng Y., Gong Y., Liu Y., Song B., Zou Q. (2021), Molecular design in drug discovery: A comprehensive review of deep generative models. Briefings in Bioinformatics, 22(6), p. bbab344.
16. Jørgensen P. B., Schmidt M. N., Winther O. (2018), Deep generative models for molecular science. Molecular Informatics, 37(1–2), p. 1700133.
17. Vanhaelen Q., Lin Y. C., Zhavoronkov A. (2020), The advent of generative chemistry. ACS Medicinal Chemistry Letters, 11(8), pp. 1496–1505.
18. Kipf T. N., Welling M. (2016), Semi-supervised classification with graph convolutional networks. arXiv preprint arXiv:1609.02907. https://doi.org/10.48550/arXiv.1609.02907
19. Cifuentes A., Bernal J. L., Diez-Masa J. C. (1997), Determination of critical micelle concentration values using capillary electrophoresis instrumentation. Analytical Chemistry, 69(20), pp. 4271–4274.
20. Gil V. S. (2022), Polymers - Molecules for machine learning. version 1. https://www.kaggle.com/datasets/victorsabanzagil/polymers/data
21. Qin S., Jin T., Van Lehn R. C., Zavala V. M. (2021), Predicting critical micelle concentrations for surfactants using graph convolutional neural networks. The Journal of Physical Chemistry B, 125(37), pp. 10610–10620.

22. Saunders R. A., Platts J. A. (2004), Correlation and prediction of critical micelle concentration using polar surface area and LFER methods. Journal of Physical Organic Chemistry, 17(5), pp. 431–438.
23. Lee S., Lee J., Yu H. (2016), Synthesis of environment friendly nonionic surfactants from sugar base and characterization of interfacial properties for detergent application. Journal of Industrial and Engineering Chemistry, 38, pp. 157–166.
24. Chaveriat L., Gosselin I., Machut C., Martin P. (2013), Synthesis, surface tension properties and antibacterial activities of amphiphilic D-galactopyranose derivatives. European Journal of Medicinal Chemistry, Vol. 62, April 2013, pp. 177–186.
25. Yan C., Li G. (2023), The rise of machine learning in polymer discovery. Advanced Intelligent Systems, 5(4), p. 2200243.
26. Landrum G. (2016), Rdkit: Open-source cheminformatics software. Landrum2016RDKit2016_09_4, https://github.com/rdkit/rdkit/releases/tag/Release_2016_09_4
27. Adebayo J., Gilmer J., Muelly M., Goodfellow I., Hardt M., Kim B. (2018), Sanity checks for saliency maps. Advances in Neural Information Processing Systems31. NeurIPS 2018, Montreal, Canada
28. Simonyan K., Vedaldi A., Zisserman A. (2013), Deep inside convolutional networks: Visualizing image classification models and saliency maps. arXiv preprint arXiv:1312, doi: 10.48550/arXiv.1312.6034

Appendix A: FASTMAN-JMP Tool Architecture

The FASTMAN Python environment, officially known as FAST MANual Data Classification (FASTMAN), was developed by the Process Systems Engineering group at Louisiana State University (PSE@LSU) [1]. Its primary purpose is to facilitate the rapid and straightforward application of pattern recognition techniques derived from data mining and unsupervised machine learning for the analysis of historical datasets. FASTMAN equips users with a collection of Python-based algorithms, alongside data visualization tools and variable contribution calculations. These tools are designed to aid users in validating clustering outcomes for a specific dataset and the chosen hyperparameters for the applied algorithms.

In addition to FASTMAN, a new tool called FASTMAN-JMP has been introduced [2]. FASTMAN-JMP combines the statistical data analysis software JMP [3] with the FASTMAN Python environment. This combination is packaged as an add-in utilizing JMP Scripting Language (JSL), as demonstrated in Section A.1. Once the user's Python environment is correctly configured, this framework offers a user-friendly and practical application interface (Figure A.1), integrating powerful analysis methods.

Notably, FASTMAN-JMP eliminates the need for users to possess an in-depth understanding of the intricate implementation of each offered method, including tasks such as data cleaning, normalization, sampling, dimensionality reduction, and data clustering. This tool serves as the offline, model-building component within the process monitoring workflow. In this workflow, clustering is performed on a historical database to train a model, which can then be deployed for real-time process monitoring. The subsequent section delves into the workflow provided by FASTMAN-JMP in detail.

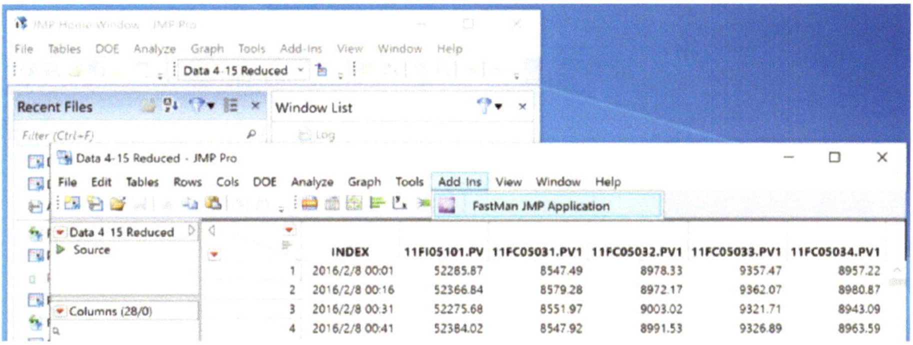

FIGURE A.1 FASTMAN-JMP add-in easily accessible from any JMP window or data table.

A.1 FASTMAN-JMP DATA PROCESSING WORKFLOW

FASTMAN-JMP offers users the convenience of performing various preprocessing steps, such as data cleaning, sampling, and normalization, following the selection of a datafile. The available preprocessing options are detailed in Figure A.2 of the tool. Users are empowered to clean non-numeric data, address missing data, and eliminate outliers. Furthermore, to streamline data analysis and reduce its size, users can opt to sample the data and conduct subsequent analysis on this subset. This sampling can be accomplished through random selection of data points within a specified interval or by computing the average of all data points within a given interval to represent it. Additionally, since this tool is implemented as an add-in for JMP, users can harness the extensive built-in capabilities of JMP for data preprocessing prior to utilizing the Python algorithms provided within the tool.

At the time of preparing this book, FASTMAN-JMP has implemented several dimensionality reduction methods, including principal component analysis (PCA), independent component analysis (ICA), Spectral Embedding, t-distributed stochastic neighbor embedding (t-SNE), uniform manifold approximation and projection (UMAP), Triplet Manifold Approximation (TriMAP), and most recently, Pairwise Controlled Manifold Approximation (PaCMAP). Users do not need to have a deep understanding of each method before using them, as FASTMAN-JMP will employ default parameters if none are specified (refer to Figure A.3).

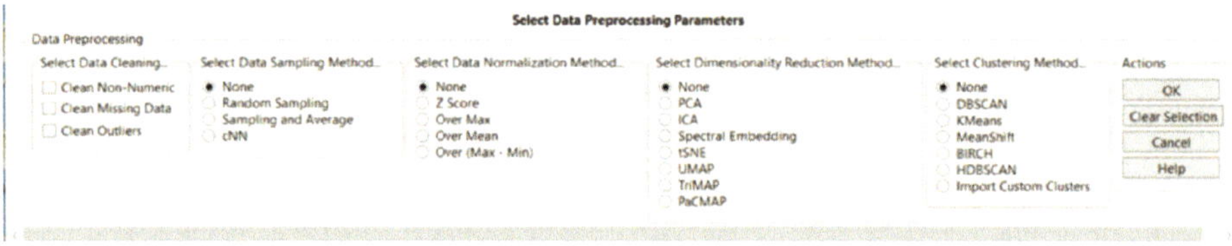

FIGURE A.2 FASTMAN-JMP preprocessing step with available selections.

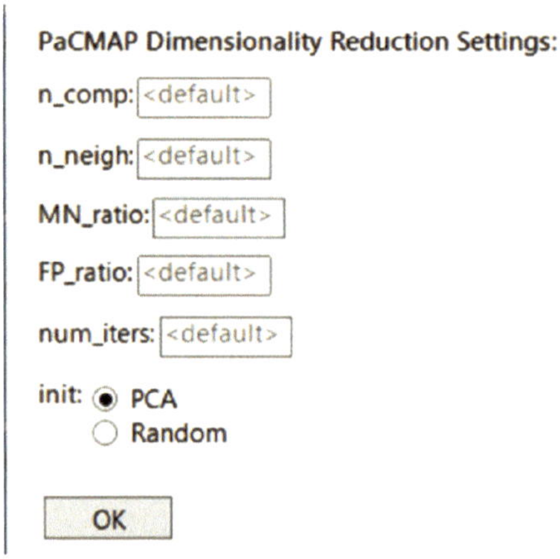

FIGURE A.3 FASTMAN-JMP hyperparameter selection pop-up window.

This feature allows users to become more familiar with the methods and later fine-tune parameters as needed.

Furthermore, FASTMAN-JMP offers a range of clustering methods, including K-means, MeanShift, Balanced Iterative Reducing and Clustering using Hierarchies (BIRCH), density-based spatial clustering of applications with noise (DBSCAN), and hierarchical density-based spatial clustering of applications with noise (HDBSCAN). Additionally, users have the option to upload custom cluster labels to overlay on the preprocessed and reduced data projection. Just like with the dimensionality reduction methods, FASTMAN-JMP does not demand that users fully comprehend each clustering method upfront, and default parameters are used in the absence of user specifications.

A.2 FASTMAN-JMP ANALYSIS FEATURES

Once a user has made a specific selection in any of the previous steps, if the algorithms need extra information for their calculations, such as a desired value for a hyperparameter in one of the chosen steps, a new window will appear. In this window, the user will be prompted to provide the necessary information, as illustrated in Figure A.4. It's important to note that this tool does not demand that the user comprehends every method beforehand; it will automatically employ default parameters if none are specified.

This window not only displays the selected parameters used in the analysis for reference but also offers additional features. These features include the ability to toggle the graph axes between date/time and row indexing, switch the graph coloring to project time-series colors onto the clustered data (which is particularly helpful for understanding the time periods comprising different clusters), and the option to rename, merge, or hide clusters from the graphs. FASTMAN-JMP empowers users to select new parameters and rerun the analysis in a new window, enabling the simultaneous opening of multiple screens for easy reference and comparison.

When clicking the "Export" button, the user can choose the file path location to save the data labeled with the final clusters, images of any auxiliary graphs generated during the analysis, and the trained model file. This model file can then be uploaded into an online process monitoring platform. Multiple visualizations are readily available for analysis, as shown in Figure A.5. These include a 3D scatterplot on the "3D Visualization" tab and the "Alt. Visualizations" tab, which provides insights into alternative dimensionality reduction and clustering methods' outcomes using the same data and preprocessing selections without requiring you to repeat each step from the beginning.

All graphs generated using this tool are dynamic and interconnected with the source data table and other relevant graphs. When you select a cluster on one graph, it automatically highlights the corresponding data points in the data table and time-series graph. This feature proves invaluable for visualization purposes and for scenarios where the user needs to exclude specific clusters or data points from further analysis, a task easily accomplished by clicking on the selected rows in the data table. Because the platform is designed as an add-in for JMP, users benefit from a high degree of flexibility when customizing their experience.

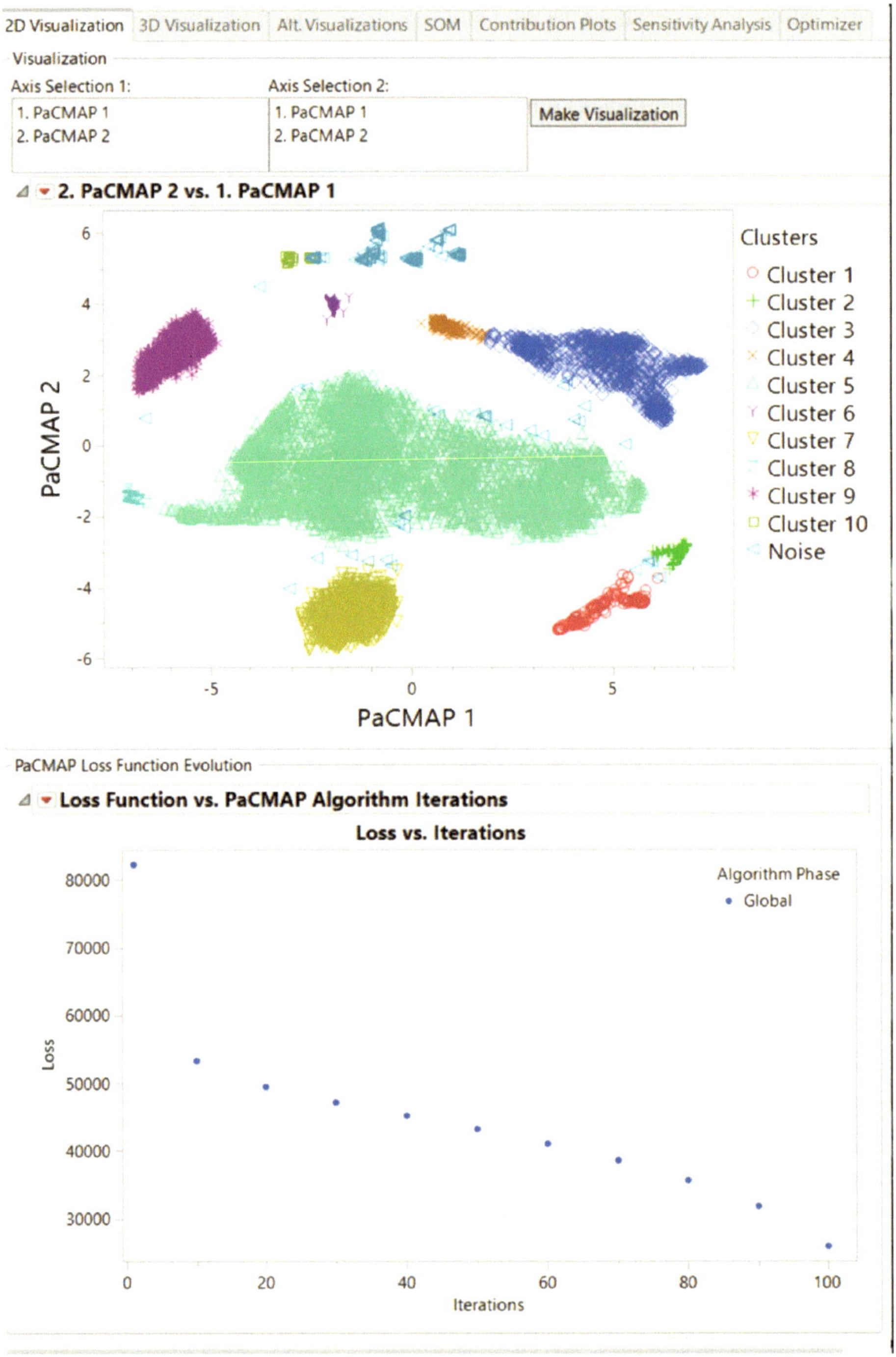

FIGURE A.4 FASTMAN-JMP main window with preliminary results and additional features for further analysis.

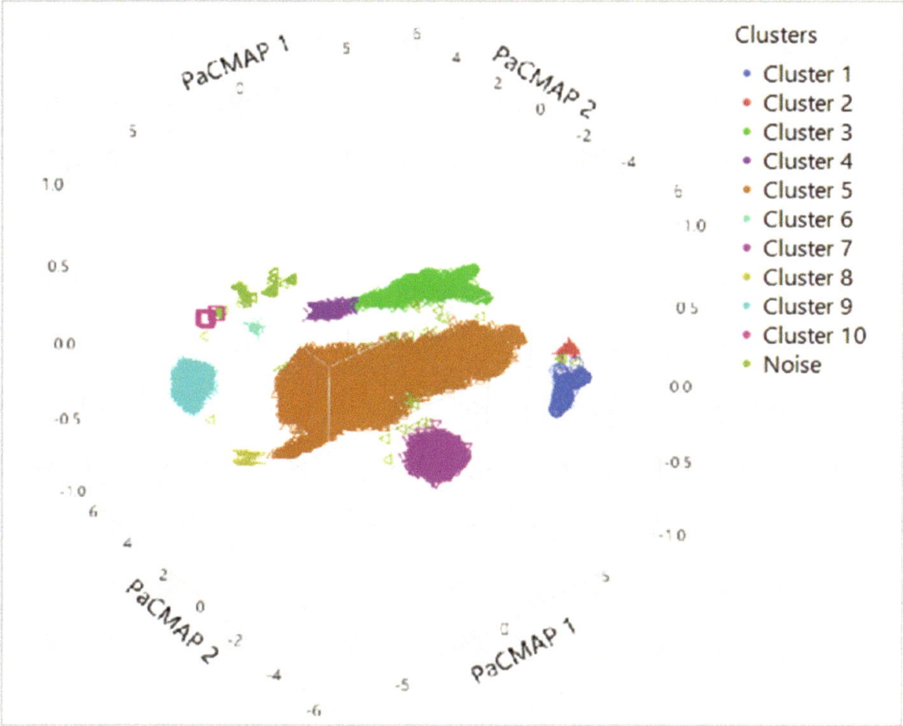

FIGURE A.5 FASTMAN-JMP main window 3D visualization and other tab page options.

Given the previously discussed properties of self-organizing map (SOM), its visualization plays a crucial role in validating clustering results for unlabeled data within this workflow. The SOM visualization tab on the main window offers users the capability to generate raw SOM plots for the entire dataset or individual components within it. Furthermore, users can project any of the final clusters onto the SOM plots. In Figure A.6, you can see the raw SOM plot generated for the entire dataset, with cluster 6 projected onto it, denoted by the white dots superimposed on the SOM.

A.2.1 Variable Contribution

As previously discussed, variable contribution plots serve the purpose of identifying the variables that contribute to the separation between classes. They assist users in pinpointing the process variables associated with the fault state. Two distinct calculation algorithms are available for generating these contribution plots, and users can choose between them through tab page selections: "PCA" and "SGS". Figure A.7 illustrates how both of these contribution calculation methods are accessible to users on the contribution plots tab page of the main window, enabling a comparison between any two given clusters.

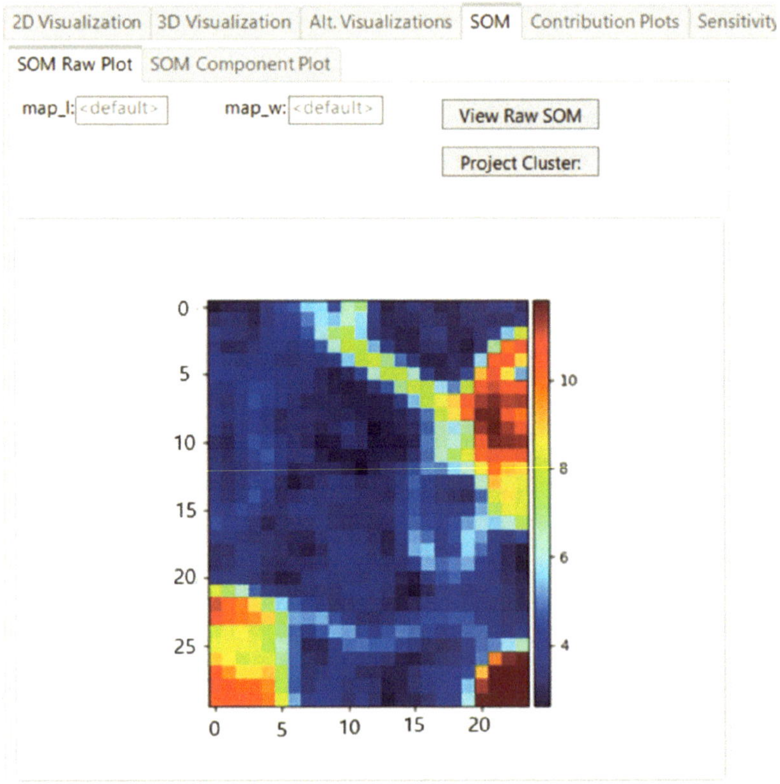

FIGURE A.6 FASTMAN-JMP main window SOM tab page option.

As demonstrated in Figure A.7, these contribution plots can highlight multiple variables as potential significant contributors, along with their respective percentage contributions. This information can then guide process operators or engineers in the right direction when attempting to comprehend the root cause of the fault state or any deviation from normal operation.

A.2.2 Sensitivity Analysis

After completing the entire process from cleaning to clustering, users may wish to investigate the sensitivity of the final result to specific parameter selections. For instance, if PaCMAP was selected as the dimensionality reduction method and the user is curious about how the number of iterations affects the final clustering outcome, they can conduct a sensitivity analysis by specifying a starting value, an ending value, and an interval value (e.g., starting at num_iters = 50 and ending at num_iters = 450 with intervals of 50). The program will then generate the final clustering results for all these values, as demonstrated in Figure A.8.

In addition to that, we calculate several pseudo-accuracy metrics for clustering in each of the sensitivity analysis runs, including the Silhouette score, the

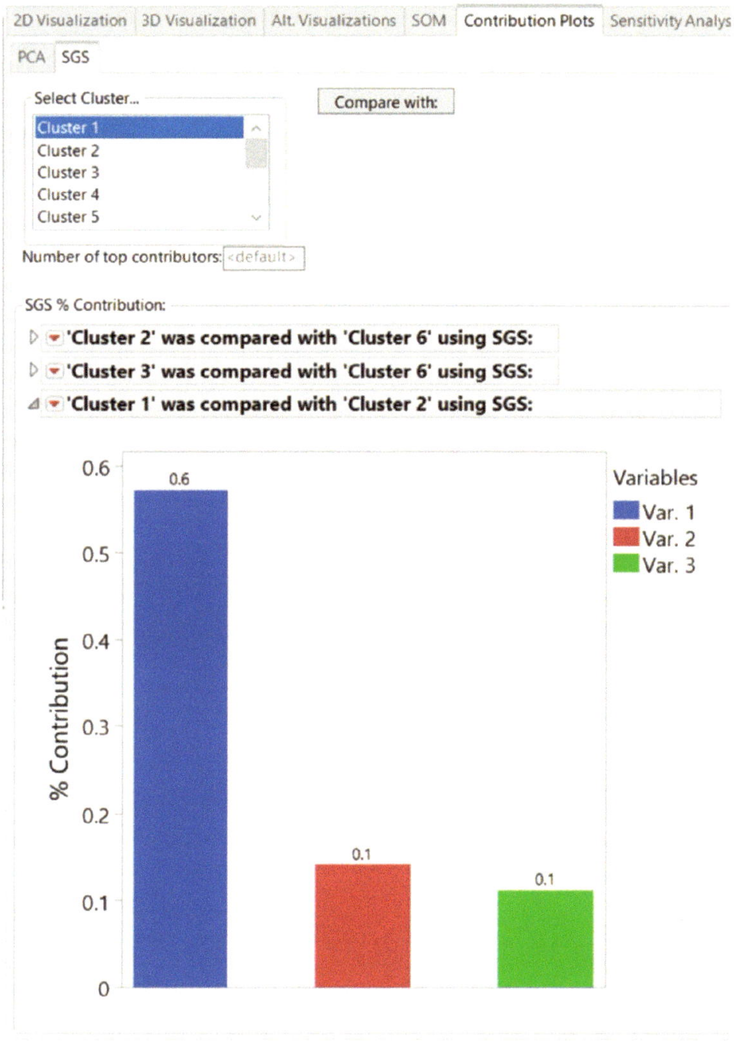

FIGURE A.7 FASTMAN-JMP main window variable contribution plot tab page option.

Calinski-Harabasz Index, the Davies-Bouldin Index, and the S_Dbw validity index. The Sensitivity Analysis section, found in Figure A.8, presents a comprehensive plot illustrating the evolution of these scores. The *x*-axis of this plot represents the parameter subjected to sensitivity analysis.

The Silhouette score and Calinski-Harabasz score are plotted together as they both indicate better clustering with higher scores. On the other hand, the Davies-Bouldin Index and S_Dbw validity index are plotted together, as lower scores for these metrics signify better clustering. It's worth noting that each metric evaluates clustering accuracy in distinct ways, and as a result, they may occasionally yield divergent conclusions.

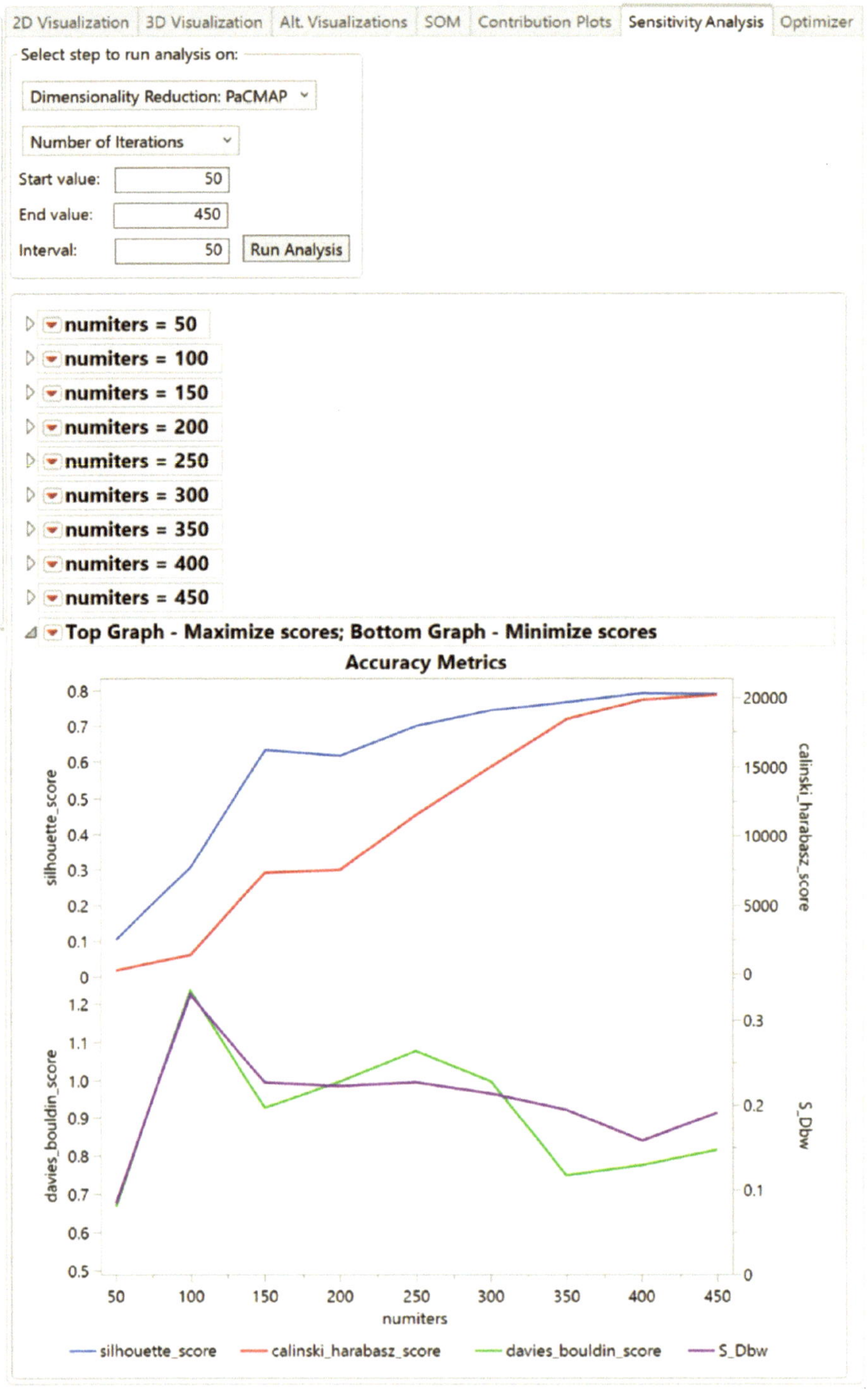

FIGURE A.8 FASTMAN-JMP main window sensitivity analysis tab page option.

A.2.3 MULTIMODAL CLASSIFICATION OPTIMIZER

Figure A.9 depicts an example Pareto plot generated by employing the multimodal classification optimizer tool. In this plot, the optimal solution space, situated in the lower right-hand corner, aims to maximize the Silhouette score while minimizing the Davies-Bouldin Index. Each ensemble in the Pareto plot output presents multiple solutions, indicated by a single color representing various points on the graph, with some solutions being superior to others. This variability arises from the algorithm's exploration of different hyperparameter values during optimization.

This automatic optimization process serves as an invaluable initial step for FASTMAN-JMP users, offering insights into potential combinations of dimensionality reduction and clustering methods, along with their respective hyperparameters. These suggested configurations have the potential to outperform the methods previously applied to the specific dataset under consideration.

In the case of Figure A.9, the optimization results suggest that employing UMAP dimensionality reduction in conjunction with K-means clustering is likely the most effective approach for the given datafile. Detailed hyperparameter information associated with these results can be accessed by hovering over the graph points or through a separate data table that opens within JMP upon algorithm completion.

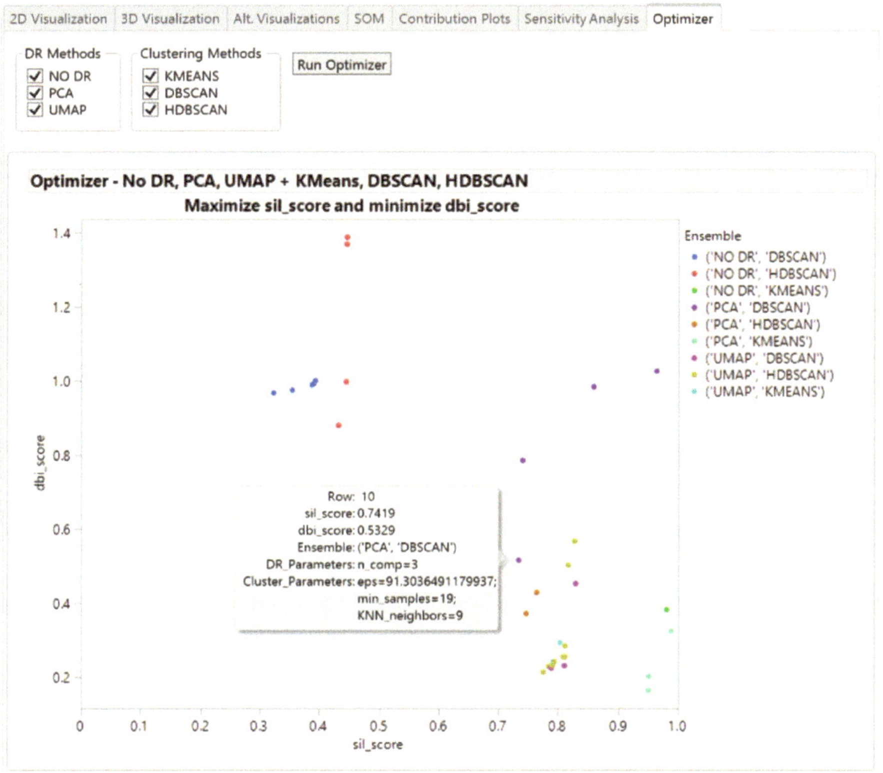

FIGURE A.9 FASTMAN-JMP main window multi-objective optimizer tab page option.

REFERENCES

1. Software for Advanced Knowledge Discovery and Fault Detection (FASTMAN). (2020), https://pse.che.lsu.edu/resources.php
2. Seghers E. (2023), A Data-Driven Multivariate Process Monitoring Platform for Knowledge Discovery and Model Building in Industrial Applications, MS. Thesis, Louisiana State University.
3. JMP® (1989–2021), Version 16.2. SAS Institute Inc.

Appendix B: Tennessee Eastman Process (TEP)

The Tennessee Eastman Process (TEP), first introduced as a process control challenge problem by Downs and Vogel [1], is a realistic simulation based on a real chemical process and used as an important benchmark in the evaluation of process control and process monitoring techniques.

The process uses four gaseous reactants (A, C, D, and E) to produce two liquid products (G and H) in competition with an unwanted byproduct (F) and in the presence of an inert (B). There are four competing reactions related to temperature through Arrhenius laws, two producing desired products and the others producing the byproduct. The entire simulation includes five unit operations: a reactor, condenser, separator, compressor, and a stripper and has 41 measured variables along with 12 manipulated variables for a total of 53 process variables. Downs and Vogel also defined 20 disturbances or faults for the challenge process (see Tables B.1–B.3).

TABLE B.1
TEP Measured Variables

Number	Variable Name	Units
1	A feed	kscmh
2	D feed	kg/h
3	E feed	kg/h
4	A and C	kscmh
5	Recycle flow	kscmh
6	Reactor feed rate	kscmh
7	Reactor Pressure	kPa gauge
8	Reactor level	%
9	Reactor temperature	°C
10	Purge rate	kscmh
11	Product separator temperature	°C
12	Product separator level	%
13	Product separator pressure	kPa gauge
14	Product separator underflow	m^3/s
15	Stripper level	%
16	Stripper pressure	kPa gauge
17	Stripper underflow	m^3/s
18	Stripper temperature	°C
19	Stripper steam flow	kg/h
20	Compressor work	kW
21	Reactor cooling water outlet temperature	°C
22	Separator cooling water outlet temperature	°C

TABLE B.2
TEP Manipulated Variables

Number	Variable Name	Units
23	D feed flow	%
24	E feed flow	%
25	A feed flow	%
26	A and C feed flow	%
27	Purge valve	%
28	Separator pot liquid flow	%
29	Stripper liquid product flow	%
30	Reactor cooling water flow	%
31	Agitator speed	%

The data was generated using the control scheme and Simulink code from Ricker, who applied a decentralized control strategy to the process which involved partitioning the plant into subunits and the creation of a controller for each. The process flow sheet is given in Figure B.1.

TABLE B.3
TEP Process Faults Description

Fault No.	Description	Type
1	A/C feed ration, B composition constant	Step
2	B composition, A/C ratio constant	Step
3	D feed temperature	Step
4	Reactor cooling water inlet temperature	Step
5	Condenser cooling water inlet temperature	Step
6	A feed loss	Step
7	C header pressure loss – reduced availability	Step
8	A, B, C feed composition	Random variation
9	D feed temperature	Random variation
10	C feed temperature	Random variation
11	Reactor cooling water inlet temperature	Random variation
12	Condenser cooling water inlet temperature	Random variation
13	Reaction kinetics	Slow drift
14	Reactor cooling water valve	Sticking
15	Condenser cooling water valve	Sticking
16	Unknown	
17	Unknown	
18	Unknown	
19	Unknown	
20	Unknown	

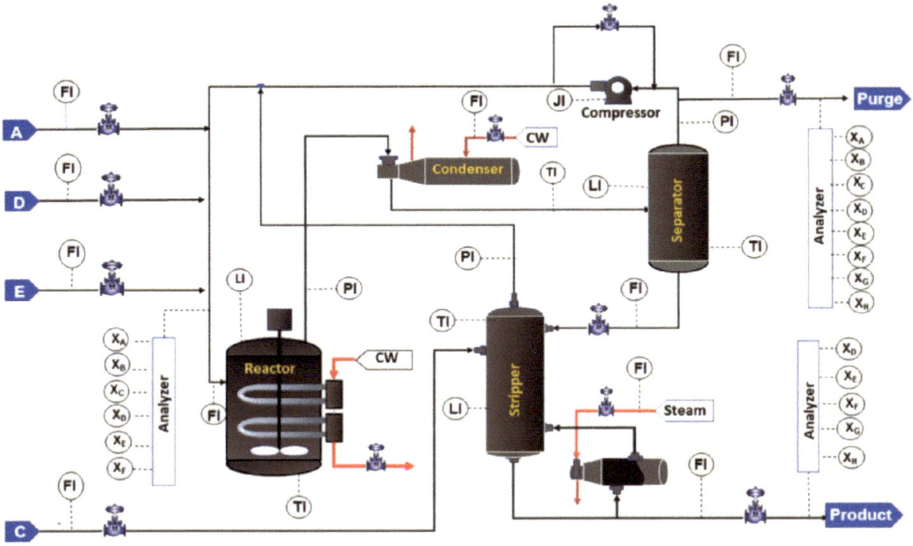

FIGURE B.1 Process schematic with control scheme.

REFERENCE

1. Downs J. J., Vogel E. F. (1993), A plant-wide industrial process control problem. Computers and Chemical Engineering, 17(3), pp. 245–255.

Appendix C: High-Temperature PEM Fuel Cell Modelling

C.1 PROCESS DESCRIPTION

A schematic representation of a high-temperature polymer electrolyte membrane fuel cell (HT-PEMFC) is given in Figure C.1. During operation of a HT-PEMFC, oxygen flows into the cathode (right-hand side in Figure C.1) while hydrogen flows into the anode (left-hand side). Platinum or platinum group metal (PGM) nanoparticles act as catalyst promoting the oxygen oxidation reaction (ORR) and the hydrogen reduction reaction (HRR) in the cathode and anode, respectively. These reactions result in the flow of positive charges, protons, through the polymer electrolyte membrane (PEM), while negative charges, electrons, flow through an external circuit from the anode to the cathode where species are recombined to form water. A so-called ionomer binder holds the PEM and the catalyst nanoparticles together. The PEM, the catalyst nanoparticles, and the ionomer binder conform the membrane-electrode assembly (MEA), which is largely responsible for the performance of the fuel cell.

The model incorporates the influence of 10 input variables (hydrogen stoichiometric ratio, S_{H_2}; oxygen stoichiometric ratio, S_{O_2}; temperature, T; pressure, P; membrane ion exchange capacity, IEC_{mem}; ionomer binder ion exchange capacity, IEC_{io};

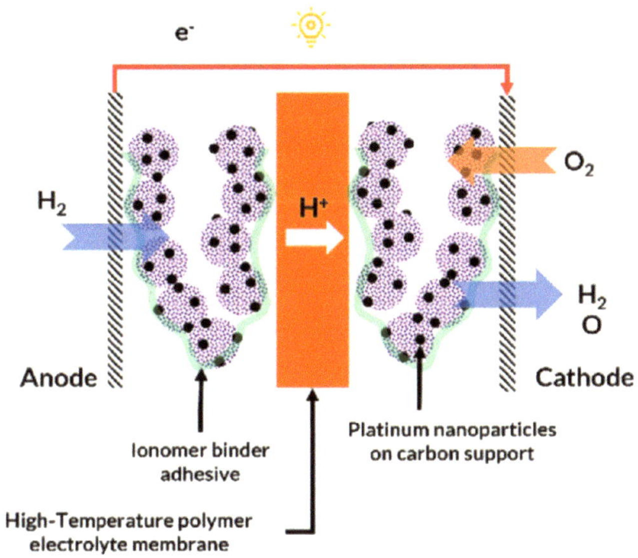

FIGURE C.1 Schematic representation of a high-temperature polymer electrolyte fuel cell.

membrane thickness, δ_{mem}; ionomer binder thickness, δ_{io}; carbon monoxide to hydrogen ration in feed stream, CO/H_2; and platinum loading, L_c) as well as current density (i) to predict the voltage of the fuel cell. Seven fitting parameters were included in the model: reference exchange current density for the anode and the cathode, $i_{0-ref}^{an,cat}$; an empiric parameter for the limiting current at the cathode and the anode side, $K_{an,cat}$; the charge transfer coefficient for the cathode and the anode, $\alpha_{an,cat}$; an empiric parameter for the effect of phosphoric acid (PA) in mass transfer, γ; and an empiric parameter for the concentration overpotential, B [10].

C.2 KNOWLEDGE-BASED MODEL AND ASSUMPTIONS

A 0D, lumped, knowledge-based model was built off known physics-derived relationships. The model also features empirical expressions to describe concentration polarization overpotential losses that hail from transport phenomena in complex porous structures. A lumped model was selected over a distributed-parameters one to reduce the number of parameters to be estimated while focusing on descriptors of interest that factor in how H_3PO_4 content in the membrane and CO in the feed gas stream impact HT-PEMFC polarization. The key assumptions for the model were continuum level, steady state operation, ideal gas behavior, no gas crossover, no water transport considerations, and uniform distribution of catalyst and ionomer binder and gas transport properties across the electrodes and gas diffusion layers. This Appendix presents the model equations that include mass balances and reaction rate expressions, parameters for these equations, and descriptions for individual overpotential terms that govern HT-PEMFC polarization.

C.3 REACTIONS AND OVERALL MASS BALANCES

The present model includes the chemical reactions for hydrogen oxidation in the anode (R.1) and oxygen reduction in the cathode (R.2).

$$H_2 \rightarrow 2H^+ + 2e^-, \tag{R.1}$$

$$\frac{1}{2}O_2 + 2e^- + 2H^+ \rightarrow + H_2O. \tag{R.2}$$

From the anode and cathode reactions, the following species balances can be written:

$$\frac{dN_{H_2}}{dt} = \left(S_{H_2} - 1\right)\frac{I}{z_{H_2}F}, \tag{C.1}$$

$$\frac{dN_{O_2}}{dt} = \left(S_{O_2} - 1\right)\frac{I}{z_{O_2}F}, \tag{C.2}$$

$$\frac{dN_{H_2O}^{cat}}{dt} = \frac{I}{z_{H_2O}F}. \tag{C.3}$$

C.4 POLARIZATION BEHAVIOR AND OVERPOTENTIAL LOSSES

The total cell voltage is calculated as follows:

$$E(i) = E_{oc} - \eta_{act} - \eta_{con} - \eta_{ohm}. \tag{C.4}$$

Here η represents the cell overpotentials. The theoretical open circuit voltage is calculated as a function of temperature T and the species partial pressures P_i as follows [1]:

$$E_{oc} = 1.23 - 0.9 \times 10^{-3}(T - 298.15) + \frac{RT}{2F}\ln\left(P_{H_2}\sqrt{P_{O_2}}\right). \tag{C.5}$$

From the Butler-Volmer equation [2], the activation overpotential, including the effect of CO, can be expressed as follows:

$$\eta_{act} = \frac{RT}{\alpha_{an}F}\ln\left(\frac{i}{i_0^{an}(1-\theta_{CO})^2}\right) + \frac{RT}{\alpha_{cat}F}\ln\left(\frac{i}{i_0^{cat}}\right), \tag{C.6}$$

where θ_{CO} is the CO coverage calculated as follows [3]:

$$\theta_{CO} = 19.9\exp\left(-7.69 \times 10^{-3}T_{an}\right) + 0.085\ln\left(\frac{y_{CO}}{y_{H2}}\right). \tag{C.7}$$

For the exchange currents, the flowing expressions are used, introducing a correction for the PA doping [4, 5]:

$$i_0^{an} = a_C L_C \left(\frac{P_{H_2}H_{H_2}^{CL}}{C_{H_2}^0}\right)^{0.5} \exp\left(-\frac{E_C^{H_2}}{R}\left(\frac{1}{T_{an}} - \frac{1}{T_{ref}^0}\right)\right) i_{0-ref}^{an} \exp\left(-\gamma(1-m_{io})\right), \tag{C.8}$$

$$i_0^{cat} = a_C L_C \left(\frac{P_{O_2}H_{O_2}^{CL}}{C_{O_2}^0}\right) \exp\left(-\frac{E_C^{O_2}}{R}\left(\frac{1}{T_{cat}} - \frac{1}{T_{ref}^0}\right)\right) i_{0-ref}^{cat} \exp\left(-\gamma(1-m_{io})\right). \tag{C.9}$$

m_{io}, the acid mass fraction, is calculated from the IEC as:

$$m_{io} = 0.0902 IEC_{io} + 0.0352. \tag{C.10}$$

In the above equations, γ is an empirical parameter that helps account how the PA content alters the exchange current density values.

Cheddie et al. proposed an empirical expression for the calculation of Henry's constant for the oxygen in PA [5]. In this model, we approximate the Henry's constant for oxygen as a function of temperature and PA mass fraction in the ionomer binder:

$$H_{O_2}^{CL} = \left(1.0 \times 10^{-6}\right)\exp\left(-\frac{900m_{io}^2 - 1500m_{io} + 650}{T}\right). \tag{C.11}$$

Hydrogen mass transport is not considered as a limiting factor. Therefore, for the hydrogen, the Henry's constant is simply assumed to be $H_{H_2}^{CL} = 4H_{O_2}^{CL}$ [6].

The concentration overpotential is determined from a semi-empirical equation as proposed by Corrêa et al. [7]:

$$\eta_{con} = -B\left(\ln\left(1 - \frac{i}{i_{lim}^{an}}\right) + \ln\left(1 - \frac{i}{i_{lim}^{cat}}\right)\right). \quad (C.12)$$

The limiting current, $i_{lim}^{an,cat}$, is affected by the diffusivity in the catalyst layer (CL) and the partial pressure and the CL thickness [8]. In this work, an empirical parameter, K_{cat}, is used as a constant for predicting the liming current in the cathode as a function of partial pressure, diffusion, temperature, and ionomer binder thickness:

$$i_{lim}^{cat} = \frac{K_{cat} D_{O_2}^{CL} P_{O_2}}{T\delta_{CL}}. \quad (C.13)$$

Considering the effect of carbon monoxide coverage for the anode, a similar expression is used but modified to include carbon monoxide coverage:

$$i_{lim}^{cat} = \frac{K_{an} D_{H_2}^{CL} P_{H_2}}{T\delta_{CL}} (1 - \theta_{CO})^2. \quad (C.14)$$

It is assumed that

$$K_{an} = 0.5 K_{cat}. \quad (C.15)$$

The oxygen diffusivity in PA is approximated from the PA mass fraction in the ionomer binder as follows [5]:

$$D_{O_2}^{CL} = \left(1.0 \times 10^{-6}\right) \exp\left(-\frac{4500\ m_{io}^2 - 10000 m_{io} + 4010}{T}\right). \quad (C.16)$$

For the hydrogen diffusivity, it is assumed to be $D_{H_2}^{CL} = 2D_{O_2}^{CL}$.

The ohmic overpotential accounts for the ionic resistance of both the membrane and the ionomer binder, while all other sources of ohmic losses (e.g., electrical conduction through contacts) are considered negligible [9]. The resistances are considered to be in series so that the global resistance is additive combination of the individual membrane and ionomer resistances:

$$\eta_{ohm} = i\left(\frac{\delta_{mem}}{\kappa_{mem}} + \frac{\delta_{io}}{\kappa_{io}}\right), \quad (C.17)$$

where κ_{mem} and κ_{io} are the ionic conductivities of the membrane and the ionomer binder, which are predicted from a machine learning model using experimental data.

REFERENCES

1. Rasheed R. K. A., Liao Q., Caizhi Z., Chan S. H. (2017), A review on modelling of high temperature proton exchange membrane fuel cells (HT-PEMFCs). International Journal of Hydrogen Energy, 41(5), pp. 3142–3165.
2. Mann R. F., Amphlett J. C., Peppley B. A., Thurgood C. P. (2006), Application of Butler-Volmer equations in the modelling of activation polarization for PEM fuel cells. Journal of Power Sources, 161(2), pp. 775–781.
3. Dhar H. P., Christner L. G., Kush A. K. (1987), Nature of CO adsorption during H2 oxidation on relation to modeling CO poisoning of a fuel cell anode. Journal of the Electrochemical Society, 134(12), pp. 3021–3026.
4. Barbir F. (2013), Fuel cell electrochemistry. In Barbir, F. (Ed), PEM Fuel Cells, pp. 33–72. Academic Press.
5. Cheddie D. F., Munroe N. D. H. (2007), A two-phase model of an intermediate temperature PEM fuel cell. International Journal of Hydrogen Energy, 32(7), pp. 832–841.
6. Bernardi, D. M., Vebrugge, M. W. (1992), A mathematical model of the solid-polymer-electrolyte fuel cell. Journal of The Electrochemical Society, 139(9), pp. 2477–2491.
7. Corrêa J. M., Farret F. A., Canha L. N., Simões M. G. (2004), An electrochemical-based fuel-cell model suitable for electrical engineering automation approach. IEEE Transactions on Industrial Electronics, 51(5), pp. 1103–1113.
8. He W., Wang B., Dickerson J. H. (2012), Overall concentration polarization and limiting current density of fuel cells with nanostructured electrodes. Nano Energy, 1, pp. 828–832.
9. Abdin Z., Webb C. J., Gray E. M. (2016), PEM fuel cell model and simulation in Matlab-Simulink based on physical parameters. Energy, 116, pp. 1131–1144.
10. Briceno-Mena L. A., Venugopalan G., Romagnoli J. A., Arges C. G. (2021), Machine learning for guiding high-temperature PEM fuel cells with greater power density, Patterns, 2(2), p. 100187, doi: 10.1016/j.patter.2020.100187

Appendix D: Distance Metrics for Clustering

D.1 INTRODUCTION

Clustering algorithms are for the most part based on the notion of similarity. In turn, for most practical applications similarity is equivalent to *distance*. We say that two points (a, b) in an n-dimensional space are similar if they are close to each other and are dissimilar if they are far apart. This intuition is clear when $n < 3$ and all features are real numbers since we can easily visualize the space and data points in it. However, for higher dimensions and types of data other than real numbers, different strategies for measuring distance might be required. This appendix introduces a few of the most used metrics. The reader is encouraged to read the detailed and thorough discussion of distance in the context of machine learning presented by Tan et al. [1].

D.2 MINKOWSKI METRICS

D.2.1 Manhattan

Manhattan distance, also known as city block distance, calculates the distance between two points by summing the absolute differences between their coordinates along each dimension.

$$d_M(a,b) = \sum_{i=1}^{n} |a_i - b_i|. \tag{D.1}$$

D.2.2 Euclidean

Euclidean distance is the straight-line distance between two points in Euclidean space, computed using the square root of the sum of squared differences in coordinates.

$$d_E(a,b) = \left(\sum_{i=1}^{n} (a_i - b_i)^2 \right)^{(1/2)}. \tag{D.2}$$

D.2.3 Minkowski

The idea of the Manhattan and the Euclidean distance can be generalized to what is called the Minkowski distance:

$$d_{Mi}(a,b) = \left(\sum_{i=1}^{n} |a_i - b_i|^p \right)^{(1/p)}. \tag{D.3}$$

D.3 SPATIAL METRICS

Spatial metrics are distance measures suitable for spatial data analysis, considering geographical or spatial attributes of the data.

D.3.1 Canberra

Canberra distance is a weighted version of Manhattan distance, emphasizing the importance of smaller differences between values and adjusting for situations where variables may have different scales.

$$d_M(a,b) = \sum_{i=1}^{n} \frac{|a_i - b_i|}{|a_i| + |b_i|}. \tag{D.4}$$

D.4 ANGULAR METRICS

Angular metrics focus on measuring the angle between vectors, particularly relevant in scenarios where directionality or orientation is significant.

D.4.1 Cosine

Cosine similarity measures the cosine of the angle between two vectors in a multi-dimensional space, indicating their similarity regardless of their magnitude.

$$\mathrm{s}_{\cos}(\mathrm{a},\mathrm{b}) = \cos\left(\theta_{(\mathrm{a},\mathrm{b})}\right) = \frac{a \cdot b}{\|a\|\ \|b\|}. \tag{D.5}$$

The cosine distance is defined as the complement of the cosine similarity as $\mathrm{d}_{\cos}(\mathrm{a},\mathrm{b}) = 1 - \mathrm{s}_{\cos}(\mathrm{a},\mathrm{b})$.

D.5 BINARY DATA METRICS

Binary data metrics are specialized distance measures designed for binary or categorical data.

D.5.1 Hamming

Hamming distance quantifies the dissimilarity between two binary strings of equal length by counting the number of positions at which the corresponding symbols are different. For example, the Hamming distance between 000 and 110 is 2.

Rather than a mathematical equation, the Hamming distance is more easily described by an algorithm:

```
Input: a, b objects in the n-dimensional space
Count ← 0
For i in n:
        If a[i] != b[i]:
                count += 1
output: count
```

D.5.2 JACCARD

Jaccard distance calculates the dissimilarity between two sets by dividing the size of their intersection by the size of their union, providing a measure of dissimilarity between binary vectors or sets.

$$d_J(a,b) = \frac{\sum_{i=1}^{n}(a_i \wedge b_i)}{\|a\|^2 + \|b\|^2 - \sum_{i=1}^{n}(a_i \wedge b_i)}. \tag{D.6}$$

REFERENCE

1. Tan P. N., Steinbach M., Kumar V. (2016), Introduction to Data Mining. Pearson Education India.

Index

Note: **Bold** page numbers refer to **tables** and *Italic* page numbers refer to *figures.*

D

E

F

T

U

V

W

Z